Encyclopedia of Environmental Issues

ENCYCLOPEDIA OF ENVIRONMENTAL ISSUES

Volume III

Powell, John Wesley—Zoos

Editor
Craig W. Allin
Cornell College

Project Editor
Robert McClenaghan

Salem Press, Inc.
Pasadena, California   Hackensack, New Jersey

Managing Editor: Christina J. Moose
Research Supervisor: Jeffry Jensen
Acquisitions Editor: Mark Rehn
Photograph Editor: Karrie Hyatt
Project Editor: Robert McClenaghan
Copy Editor: Doug Long
Production Editor: Janet Alice Long
Layout: William Zimmerman

∞ The paper used in these volumes conforms to the American National Standard for Permanence of Paper for Printed Library Materials, Z39.48-1992(R1997)

Library of Congress Cataloging-in-Publication Data

Encyclopedia of environmental issues / editor, Craig W. Allin; project editor, Robert McClenaghan.
p. cm.
Includes bibliographical references and index.
ISBN 0-89356-994-1 (set : alk. paper). — ISBN 0-89356-995-X (v. 1 : alk. paper) — ISBN 0-89356-996-8 (v. 2 : alk. paper) — ISBN 0-89356-997-6 (v. 3 : alk. paper)
1. Environmental sciences—Encyclopedias. 2. Pollution—Encyclopedias. I. Allin, Craig W. (Craig Willard) II. McClenaghan, Robert, 1961-
GE10 .E52 2000
363.7'003—dc21 99-046373
CIP

Second Printing

PRINTED IN THE UNITED STATES OF AMERICA

Contents

List of Articles by Category

List of Articles by Category

Rocky Flats, Colorado, nuclear plant releases
Rocky Mountain Arsenal
SANE
Silkwood, Karen
Soviet nuclear submarine sinking
Superphénix
Three Mile Island nuclear accident
Union of Concerned Scientists
Windscale radiation release
Yucca Mountain, Nevada, repository

PHILOSOPHY AND ETHICS
Antienvironmentalism
Coalition for Environmentally Responsible Economies
Deep ecology
Eaubonne, François Marie-Thérèse d'
Ecofeminism
Ecology, concept of
Ecotage
Ecoterrorism
Environmental education
Environmental ethics
Environmental justice and environmental racism
Gaia hypothesis
Green marketing
Green movement and Green parties
Intergenerational justice
Lovejoy, Thomas
Monkeywrenching
Naess, Arne
NIMBY
Privatization movements
Public trust doctrine
Schumacher, Ernst Friedrich
Social ecology
Speciesism
Valdez Principles
Vegetarianism
Wise-use movement

POLLUTANTS AND TOXINS
Agent Orange
Agricultural chemicals
Alar
Asbestos
Chloramphenicol
Dichloro-diphenyl-trichloroethane (DDT)
Dioxin
Diquat
Heavy metals and heavy metal poisoning
McToxics Campaign
Malathion
Medfly spraying
Mercury and mercury poisoning
Methyl parathion
Noise pollution
Odor control
Pesticides and herbicides
Plastics
Pollution permits and permit trading
Polychlorinated biphenyls
Pulp and paper mills
Radioactive pollution and fallout
Space debris
Thermal pollution

POPULATION ISSUES
Ehrlich, Paul
Osborn, Henry Fairfield
Population-control and one-child policies
Population-control movement
Population growth
United Nations Population Conference
World Fertility Survey
Zero Population Growth

PRESERVATION AND WILDERNESS ISSUES
Abbey, Edward
Adams, Ansel
Alaska National Interest Lands Conservation Act
Antarctic Treaty
Arctic National Wildlife Refuge
Aswan High Dam
Biosphere reserves
Boulder Dam
Brower, David
Burroughs, John
Carter, Jimmy
Convention on Wetlands of International Importance
Convention Relative to the Preservation of Fauna and Flora in Their Natural State
Cross-Florida Barge Canal
Dams and reservoirs

Experimental Lakes Area
Exxon Valdez oil spill
Flood control
Fluoridation
Groundwater and groundwater pollution
Irrigation
Lake Erie
Law of the Sea Treaty
Mediterranean Blue Plan
Mexico oil well leak
Monongahela River tank collapse
Ocean pollution
Oil spills
Prior appropriation doctrine
Rhine river toxic spill
Riparian rights
Runoff: agricultural
Runoff: urban
Sacramento River pesticide spill
Santa Barbara oil spill
Sea Empress oil spill
Sedimentation
Teton Dam collapse
Tobago oil spill
Torrey Canyon oil spill
Water conservation
Water pollution
Water pollution policy
Water quality
Water rights
Water-saving toilets
Water treatment
Water use
Watersheds and watershed management
Wells
Wolman, Abe

WEATHER AND CLIMATE
Climate change and global warming
Cloud seeding
El Niño
Greenhouse effect
Heat islands
Intergovernmental Panel on Climate Change
Kyoto Accords
Manabe, Syukuro
Molina, Mario
National Oceanic and Atmospheric Administration
Ozone layer and ozone depletion
Rowland, Frank Sherwood
Weather modification

APPENDICES
Time Line of Environmental Milestones
Directory of Environmental Organizations
Directory of U.S. National Parks
Glossary

Encyclopedia of Environmental Issues

Powell, John Wesley

BORN: March 24, 1834; Mount Morris, New York
DIED: September 23, 1902; Haven, Maine
CATEGORY: Preservation and wilderness issues

John Wesley Powell was the first white explorer of the Grand Canyon, a geologic and geographic surveyor of the American West, director of the U.S. Geological Survey (1881-1892), and a leader of the conservation movement in the United States.

Best known for his explorations of the western United States, John Wesley Powell contributed to both the romantic and scientific views of those lands. Many stories have been told of his exploits in leading surveying expeditions through the uncharted lands, especially his trip down the Grand Canyon. Newspapers across the United States reported on the progress of his travels. People preparing for recreational whitewater trips still read Powell's accounts of the challenges he faced on those rivers. The personal drama of the stories was heightened by the fact that Powell had lost his right arm in the Battle of Shiloh during the Civil War.

Powell's first contributions to scientific knowledge of the American West were the materials he returned to the Smithsonian Institute and the reports he made to Congress. Among the most enduring of his environmental science contributions was his initiative in generating topographic maps of the United States. He envisioned a set of maps that would cover the entire country, but the project was too big for his lifetime. Still, that initiative continues to have direct impacts on people who enter public lands, whether for recreational or commercial purposes.

Wallace Stegner's biography of Powell, *Beyond the Hundredth Meridian* (1954), brought Powell's views back to the attention of environmental management officials. Stegner presented Powell not only as a leading figure in the organization of science in government, but also as an environmental policy visionary who was ahead of his time. Powell's *Report on the Lands of the Arid Region of the United States* (1878), written a few years before he became director of the U.S. Geological Survey in 1881, advised against using the checkerboard grid—the standard land survey approach used in the eastern United States—to survey the West. Arguing that the lack of water in the West needed to be the dominant factor in dividing land for agricultural purposes, Powell preferred an approach in which 160 acres would not be standard. He saw 80 acres as a sufficient amount of irrigable land for a homestead and thought that 2,560 acres might be needed for pasturage. Though the National Academy of Sciences supported his recommendations, they were not accepted by Congress.

The value of Powell's influence on federal land policy remains in dispute, but in the late twentieth century the Bureau of Reclamation

John Wesley Powell, the first white man to travel through the Grand Canyon, directed the U.S. Geological Survey from 1881 to 1892. (Library of Congress)

called him "the father of irrigation development," and at least two secretaries of the Interior Department (Stewart Udall and Bruce Babbitt) claimed to be inspired by Powell.

Larry S. Luton

SEE ALSO: Conservation; Forest and range policy; Grand Canyon; Land-use policy; Range management; Watersheds and watershed management

Power plants

CATEGORY: Energy

A variety of fuel sources are used for electric power plants, including coal, oil, natural gas, flowing water, and radioactive elements. All methods of generating electricity have environmental consequences.

In the United States the major types of electric power plants are coal (50 percent), nuclear (20 percent), hydroelectric (10 percent), oil (10 percent), and natural gas (10 percent). Developing technologies such as solar, geothermal, wind, and tidal power contribute less than 1 percent to the total. In other countries, the relative utilization of fuels may be quite different depending on what local resources are available. France has little coal and oil, so it depends mostly on nuclear power (75 percent). China's effort toward greater industrialization is largely based on its coal resources.

The world's known reserves of oil and natural gas are declining in spite of extensive exploration to find new deposits; coal, however, is still abundant. Environmental problems arise from mining, transportation, and the combustion process. Surface mining of coal defaces the landscape and causes erosion. Deep mining is a hazardous occupation for miners because of long-term lung effects and the possibility of underground explosions. A typical coal-burning power plant uses about 7 million tons of fuel per year, causing a hazard along shipping routes. Oil is transported by pipelines and ocean tankers, both of which are vulnerable to accidents that can harm fish and wildlife.

When coal, oil, or natural gas are burned at an electric power plant, high-pressure steam is produced to turn a turbine that is connected to a generator. Spent steam is condensed and recycled, but the heat removal causes thermal pollution. Some coal contains as much as 6 percent sulfur impurity, which is converted to gaseous sulfur dioxide during burning. Nitrogen from the air also forms gaseous products. When these gases are released into the atmosphere, they combine with water to form sulfuric acid and nitric acid, which eventually return to earth as acid rain. The harmful effects of acid rain include loss of fish in acidified lakes, deterioration of forests, and surface erosion of marble buildings and monuments. Acid rain can be reduced by using cleaner fuel, such as natural gas or low-sulfur coal. Also, sulfur dioxide can be removed before it leaves the smokestack by means of chemical "scrubbers," but such technology adds expense and cannot be retrofitted to older plants.

When fossil fuels are burned, carbon combines with oxygen from the air to form gaseous carbon dioxide. When the carbon dioxide mixes with the atmosphere, it acts like the glass panels of a greenhouse, allowing solar radiation to enter but preventing infrared heat from escaping, causing global warming. If global warming takes place, the level of the oceans could rise, causing disastrous flooding of coastal communities. Also, a hotter and drier climate would create worldwide problems for agriculture. The only way to reduce the production of carbon dioxide is to burn less coal, oil, and natural gas.

The United States has about one hundred nuclear power plants in operation. The main objections to nuclear power are the fear of an accident that may release radioactivity into the environment and the problem of long-term nuclear waste storage. Nuclear plants also cause thermal pollution in a manner similar to fossil fuel facilities. The major benefit of nuclear power is that no chemical burning takes place, so no fly ash, acid rain, or greenhouse gases are produced.

Water power is a renewable, nonpolluting resource, but it does have environmental problems. When a new dam is built, the stored water

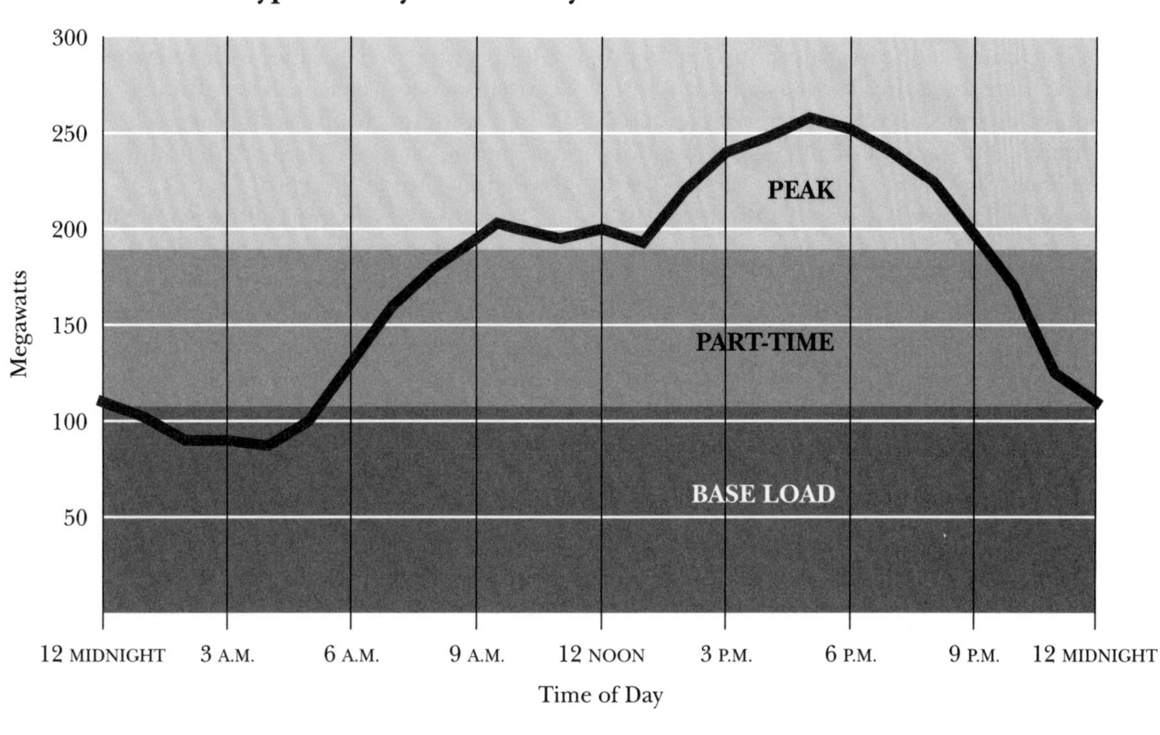

often fills a scenic canyon or floods farmland near the river. Farmers strongly object to the potential hazard of high-voltage lines going across their land. During extended droughts, as happened in the Pacific Northwest during the 1980's, consumers are faced with electric power shortages. The greatest hazard of a dam is the large quantity of water stored behind it. If a dam fails, as the Teton Dam in Idaho did in 1976, the devastation downstream can be very costly.

Geothermal energy is obtained from underground heat sources, but it is available near the earth's surface at only a few locations, such as in Iceland. Tidal power is limited geographically to places on the coastline where there is a large difference between high and low tides. Wind power, collected by windmills, requires an average wind speed of at least 10 miles per hour to be practical. The main problems encountered are the noise created by the rotating blades, maintaining a sixty-cycle frequency as the wind speed varies, and preventing damage to the blades from high-velocity wind gusts and ice in the winter.

Solar energy from sunlight produces no air pollution or waste products and is replenished daily, so it seems to be an ideal fuel. One major problem is that solar radiation is not available at night, so an energy storage system is needed. One type of collector uses hundreds of mirrors that reflect sunlight up to a boiler on top of a tower, which then generates steam to run a conventional turbine and generator. A second type of collector uses photovoltaic cells that convert sunlight directly into electricity. The energy conversion efficiency is fairly low, and mass-producing photocells that can operate in an outdoor environment with long-term reliability is still a challenging problem. Some environmentalists have proposed that the future of solar energy may lie not in building traditional, centralized power plants, but rather in setting up small collector systems for individual communities.

Hans G. Graetzer

SUGGESTED READINGS: Volker Mohnen, "The Challenge of Acid Rain," *Scientific American* (August, 1988), and Richard Balzhiser and Kurt Yeager, "Coal-Fired Power Plants of the Future," *Scientific American* (September, 1987), are both highly recommended. Also, the entire September, 1990, issue of *Scientific American* deals with the topic of energy. For a balanced discussion of the benefits and hazards of nuclear power, see *Nuclear Power: Both Sides* (1982), by Michio Kaku and Jennifer Trainer. George J. Mitchell, U.S. senator from 1980 to 1994, gives an alarming analysis of global warming in *The World on Fire: Saving an Endangered Earth* (1991).

SEE ALSO: Acid deposition and acid rain; Fossil fuels; Geothermal energy; Hydroelectricity; Nuclear power; Solar energy; Thermal pollution; Tidal energy; Wind energy.

Predator management

CATEGORY: Animals and endangered species

Predator management involves efforts to control, maintain, or reintroduce a wild population of predatory species without endangering other wildlife or domestic livestock. Weighing the needs of humans and their domesticated animals against the intrinsic value of predators that are assuming their natural place in the ecosystem has become a great source of controversy among conservationists, hunters, ranchers, and animal rights activists.

Humans have always competed with animals for food and survival. This competition intensified with the advent of ranching, where large herds of domesticated animals pose a ready food source for wild predators, particularly wolves and coyotes. For much of human history, the concept of predator management has been limited to shooting on sight. Even well into the twentieth century, predators were seen as "nuisance" animals with no intrinsic value to the environment and, as such, were trapped, snared, shot, and poisoned.

Large-scale predator kills continued in the United States into the 1990's. Wolves and other large predators are still killed in Alaska and the Pacific Northwest to protect big game animals such as moose, elk, and caribou. In 1994 Alaskan governor Tony Knowles halted the state's wolf-kill program after a nationwide television broadcast showed trapped wolves suffering in snares. While wildlife groups enthusiastically endorsed the move, hunters claimed that the state had bowed to outside pressure.

Some people in remote areas of the United States, including Native Americans, rely on big game animals for food and economic survival, but the primary reason for protecting these nonendangered species from predators is so people have something to shoot during hunting season. Many hunters liken themselves to large predators with regard to their place in the food chain. The difference is that natural predators bring down the old and the injured, while hunters seek only prime specimens, which weakens the gene pool of the game animals instead of supporting natural selection.

Congress began providing predator control assistance to ranchers in 1915, though it did not become a federal obligation until the Animal Damage Control Act of 1931. Still in effect, this act authorizes government agents to kill an estimated 1.5 million predators per year, including coyotes, black bears, mountain lions, bobcats, and badgers. Most of the slaughter of these so-called nuisance animals occurs in the western states under the guise of livestock protection. According to figures from the U.S. Humane Society, in 1994 the government spent more than $38 million, nearly as much as it spent to protect endangered species, to kill predators in the United States. In an attempt to combat negative publicity, the department charged with controlling these predators was renamed Wildlife Services in 1997.

In 1990 predators throughout the United States killed 500,000 sheep, with coyotes being the primary culprit. The adaptive creatures, once described by writer Mark Twain as "a long, slim, sick and sorry-looking skeleton, with a gray wolf-skin stretched over it," now inhabit every state except Hawaii. A 1991 estimate by the United States Department of Agriculture placed their numbers at 1.4 million. Wolves are also a problem for ranchers, and their 1995 reintroduction into Yellowstone National Park in Wyoming remains a controversial issue.

The long-term effects of predator kills are not yet known. Studies indicate that slaughtering 40 percent or more of the wolves in a given area over a four-year period may result in a short-term increase in moose and caribou. However, not all solutions to predator problems are lethal. Live capture and relocation is another option—provided that there is a suitable habitat available and the animal has not become a nuisance killer. The problem with relocation is that some of these animals have become accustomed to humans and the easy meals that come from living in their company. This is especially a problem with black bears, which are true omnivores that are just as willing to forage through backyards and garbage cans as through their natural habitats.

Sterilization, by either surgical or chemical means, is another viable option. Surgical sterilization is not as ridiculous as it seems, especially in the case of wolves. As a result of the hierarchical nature of their society, only the dominant (or alpha) pair of wolves in each pack mate and bear pups, so only two animals from each pack need to be sterilized to prevent the group from breeding. Another effective, if unusual, method of protecting livestock is the use of guard animals such as donkeys or llamas. This method is particularly effective against coyotes, which instinctively fear anything new in their environment.

Cooperation seems to be the most effective option. A century of wide-scale hunting left the Mexican Wolf on the brink of extinction, and the remaining animals were trapped in the mid-1970's. In 1996 hunters, ranchers, and conservationists formed an unlikely alliance to support the reintroduction of the endangered predator to parts of the southwestern United States. Sharing the range with wolves will encourage ranchers to closely manage their livestock, which should improve grazing and make their operations more profitable. The Mexican Wolf Recovery Plan, approved in 1982, includes a compensation fund to reimburse ranchers for losses and incentives that reward landowners when pups are born on private property.

P. S. Ramsey

Suggested readings: For an overview of public policy toward predators and their place in the ecosystem, see *Preserving Nature in the National Parks* (1998), by National Park Service historian Richard West Sellars. Another good source of information on predator management is David Petersen, "The Killing Fields: Animal Damage Control Activities," *Wilderness* 57 (summer, 1994). *Of Wolves and Men* (1978), by Barry Holstun Lopez, documents the long, often tumultuous, relationship between wolves and humans, and their competition for resources.

See also: Endangered species and animal protection policy; Extinctions and species loss; Forest and range policy; Grazing; Hunting; Wildlife management; Wolf reintroduction.

Preservation

Category: Preservation and wilderness issues

The preservation ethic calls for maintaining large tracts of wilderness in an undisturbed state. Preservation is typically contrasted with conservation, which allows managed exploitation of resources for economic purposes.

Preservation emerged as a response to the large-scale disposal of public lands and to such economic activities as mining and logging, which destroyed much of the landscape of the western United States during the nineteenth century. John Muir, who is usually cited as the first American preservationist, condemned the common perception of wilderness as an economic resource. Muir and other preservationists argued that the American wilderness possessed an inherent value that must be protected from commercial exploitation. Most historians point to the battle over damming Hetch Hetchy Valley at the beginning of the twentieth century as the first major conflict between conservationists and preservationists.

Although preservationists successfully excluded most commercial development from the national park system, conservation remained the dominant ethic in land-use management in the

United States for the first half of the twentieth century. However, preservation gained a powerful new constituency in the years after World War II. A rising standard of living allowed many Americans the opportunity to pursue leisure activities. This group, which typically lived in urban or suburban areas, supported preservation because it allowed them to enjoy outdoor activities such as camping in an unspoiled setting.

The growing demand for wilderness areas influenced government officials charged with setting land-use policy, culminating in 1964 with the passage of the Wilderness Act. Throughout the 1960's and 1970's, politicians responded to the American public's continued support for preservation. However, during the 1980's and into the 1990's, government officials increasingly favored multiple-use management over preservation.

The history of preservation showed that it served as a response to the success of capitalism as an economic system. Ironically, resource exploitation allowed Americans the income and free time to enjoy the outdoors while at the same time significantly altering the environment. To many Americans the negative consequences of exploitation of natural resources were beginning to outweigh the benefits that development brought to society. Environmentalists articulated this concern, which even appeared in the wilderness legislation of the 1960's. Politicians, including President Lyndon Johnson, began discussing the importance of beauty and the value of nature.

Critics contended that the preservation ethic was an expression of middle- and upper-class Americans who had already enjoyed the benefits of economic exploitation of resources; the needs of lower-class citizens who would profit from the continued development of natural resources were disregarded in favor of leisure activities for the affluent. However, preservation constituted more than an approach to land-use management that called for the establishment of natural playgrounds for middle-class Americans. It also served as more than a muddled critique of capitalism. It expressed a value system, and as such it could be counted as a philosophy that had important political and social implications.

Preservation is usually contrasted with conservation as a means of detailing its philosophic import. While conservation understands nature in terms of resources that have value to human economies, preservation regards nature as possessing additional value to humans, bringing aesthetic and even spiritual qualities to human life. Conservationists perceive forests, minerals, and wildlife as separate categories, while preservationists argue that such categorization is artificial. They claim that nature must be understood as a whole, with the apparently individual parts intimately interconnected in ways that humans, who are a part of the natural system, cannot fully understand. Thus, the failure to preserve wilderness has consequences that ripple through the entire ecosystem, including human society.

Some environmentalists argue that the distinction between conservation and preservation, while useful for defining advocacy groups and management plans, is simplistic and misleading. Rather than perceive the two as mutually exclusive, adversarial positions, they maintain that preservation and conservation are two tools for understanding the human relationship to the environment. Some radical environmentalists dismiss preservation on the grounds that, like conservation, it assesses nature according to human values, albeit a wider range than conservation does. They maintain that preservation ultimately fails as a philosophy because it does not understand that nature exists wholly apart from any value system. Thus, preservation is faulty because it does not lead to the complete reform of human perceptions necessary to end the destruction of the environment.

Despite such criticisms, preservation remained an important environmental ethic at the close of the twentieth century. In the United States and elsewhere, preservationists could claim that their efforts had saved wilderness regions that otherwise would have been devastated by mining, logging, or other activities. However, these victories required preservationists to address difficult issues regarding the management of wilderness areas. Issues including scientific study within wilderness preserves, fire suppression, and the reintroduction of wildlife such as wolves all sparked lively debate. Moreover, pre-

A ranger sits with high school students and discusses management of national parks, which is based on the preservation ethic. (Jim West)

servationists had to press for air- and water-quality standards outside the preserved areas in order to protect the integrity of the wilderness.

Thomas Clarkin

SUGGESTED READINGS: Charles Davis, editor, *Western Public Lands and Environmental Politics* (1997), includes an article on wilderness preservation policy at the federal level. R. McGreggor Cawley, *Federal Land, Western Anger: The Sagebrush Rebellion and Environmental Politics* (1993), discusses the importance of preservation during the 1960's. Bryan G. Norton's *Toward Unity Among Environmentalists* (1991) argues that conservation and preservation should not be considered mutually exclusive concepts. Roderick Nash's *Wilderness and the American Mind* (1967) is a comprehensive history of preservation. Max Oelschlaeger's *The Idea of Wilderness from Prehistory to the Age of Ecology* (1991) offers a precise definition of preservation.

SEE ALSO: Deep ecology; Muir, John; National parks; Nature preservation policy; Wilderness Act; Wilderness areas.

Prior appropriation doctrine

CATEGORY: Water and water pollution

The prior appropriation doctrine is a set of principles that governs the distribution of water in the United States. This doctrine is mainly followed in arid and semiarid regions in the western United States.

Water rights in the United States are governed by two vastly different sets of principles: riparian and prior appropriation. Riparian rights are based on the principle of equality of access to water. Riparian proprietors expect access to the water adjacent to their land without respect to when the land was acquired. As a general rule, riparian principles apply in areas that get more than 50 centimeters (20 inches) of rainfall per year.

Prior appropriation is in effect in the arid and semiarid western states—that is, areas of fewer than 50 centimeters of rainfall per year. The doctrine is generally concerned with the priority of water use rather than equality of distribution. Therefore, individuals who first settled in prior appropriation areas have first rights to water for their needs. All subsequent arrivals have a right to water resources in accordance with the time of their arrival and the needs of earlier claimants. Prior appropriation appears to have developed around concepts that were refined during early mining activities in the western gold fields. Water was often used during the mining operation, and the first user of water was protected against later claims. These ideas came to be applied to irrigation agriculture in areas of limited water.

As a result of this type of water appropriation, subsequent users of water resources may find that their permitted allotment falls short of their needs during dry periods. For instance, in the spring, as the mountain snowpack melts, a particular stream may provide sufficient water for all users. However, once the snowpack has melted and the summer dry season begins, stream flow diminishes. In order to guarantee sufficient flow for the earliest or prior appropriator, water users who settled later will have their available water reduced. Resolution of conflicts and determination of appropriation are carried out by commissioners chosen through election or appointment to administer the water allocation process.

Since water availability is not equal throughout the arid and semiarid West, the degree to which prior appropriation principles are applied varies. It is the dominant method of determining water rights in the Rocky Mountain states. The West Coast states of California, Oregon, and Washington have versions of the doctrine that have been modified to meet their specific needs. The Great Plains states from North Dakota through Texas have also developed a somewhat mixed doctrine because of their transitional location between regions of riparian doctrine in the East and prior appropriation in the West.

Jerry E. Green

SEE ALSO: Riparian rights; Water rights.

Privatization movements

CATEGORY: Philosophy and ethics

Privatization efforts in the United States are designed to protect individual property rights by weakening governmental programs that protect the environment.

Efforts to privatize the environment stem from two related approaches. One group of advocates for privatization argues that privatization of the environment will provide better protection than present governmental regulatory programs. The second group, like the first, emphasizes the primacy of individual property rights in the American political tradition, contending that Americans have the right to do what they want with their property. This latter group often displays scant concern for the environment.

Seventeenth century English philosopher John Locke's views concerning the primacy of individual property rights infuse all efforts to privatize the environment. One group, however, maintains that protecting the environment is also an important goal. They argue that common pool resources such as water or wilderness are subject to the "tragedy of the commons" because there is no incentive not to exhaust (or pollute) a resource that belongs to everyone. One response to the "tragedy of the commons" is extensive governmental regulation.

Advocates of privatization decry this approach on two grounds. First, they argue that it flies in the face of American political traditions regarding the importance of private property that go back to the founding of the nation. Second, they contend that governmental regulation will only exacerbate tendencies to use up or pollute a resource because if a resource belongs to everyone, it belongs to no one, and no one will take responsibility for protection. Privatizing a resource will guarantee protection by the property owner. For example, an owner of a forest will be disinclined to see it harmed by pollution and will take appropriate action to protect it. Such privatization advocates often consider themselves ardent environmentalists and contend that privatization will better protect the environment than governmental regulation. Although they contend that humans have primacy over nature, they maintain that protecting the environment is good for humankind and should be done.

A second group of advocates of privatizing the environment also starts from the Lockean perspective of the primacy of individual property rights. More so than the first group, they are often suspicious of government, especially the national government. This group, however, views the environment as something to be exploited for the benefit of individual property owners. They do indicate that a wise owner might wish to leave something for future generations, but this is an individual decision. Some advocates of this form of privatization view natural resources as inexhaustible or easily substitutable through the operation of market forces. Governmental regulation of the environment is wrong in their view because it interferes with individual property rights, not because it may be an ineffective means of protecting the environment.

Both groups of advocates of privatization are opposed to the "command and control" policies of environmental protection used in the United States since the 1970's. They argue that governmental regulations such as the Clean Air Act or efforts at preserving species interfere with Americans' rights to do whatever they wish with their property. During the late 1970's and early 1980's, a movement in the western United States known as the Sagebrush Rebellion often resorted to court action (and sometimes violence) to try to weaken various environmental laws and regulations, such as the Endangered Species Act, that limited the exercise of property rights. Others described a policy known as "wise use" that maintained that natural resources should be used wisely, which usually meant providing a profit for their owners. This movement achieved some success during the presidency of Ronald Reagan as agencies such as the Environmental Protection Agency scaled back enforcement of environmental regulations. Some advocates of privatization called for massive sales of federal lands in the western states, saying that the national government should not own natural resources.

These efforts at privatization have generally had a negative impact on the environment. What land sales occurred benefited large businesses, not small property owners. Meanwhile, the attack on environmental regulation slowed efforts to achieve a cleaner environment. One aspect of the privatization efforts of the 1980's that has had a more positive impact is the attempt to improve environmental quality through market-based solutions that take property rights into account. These programs are based on the concept that individual property owners will be more effective and efficient in protecting the environment, as they enhance their own property, than a program based on governmental regulation of industry. While programs such as emissions trading have not always been completely successful, they represent an innovative means of dealing with environmental issues such as air pollution.

The privatization movement has often been a part of a larger movement that is suspicious of government and wants to decrease its impact. As such, it flies in the face of the early twentieth century Progressives, such as Theodore Roosevelt, who maintained that the protection of the environment was the responsibility of everyone, including the government. Advocates of privatization rely on the workings of the market to protect the environment but do not readily accept that the market does not always work in favor of the environment. Their narrow interpretation of the views of the Founders on property rights sometimes verges on distortion. The Founders were concerned with protecting individual property, but they were also concerned with achieving the common good, even if this required governmental action.

The privatization movement is both an intellectual movement and an advocacy movement concerned with changing the relationship of government and society. It has had some successes and has raised some important questions. It has not always exerted a positive influence on the environment in spite of the efforts of some of its advocates.

John M. Theilmann

SUGGESTED READINGS: Terry L. Anderson and Donald R. Leal, *Free Market Environmentalism* (1991) advocates privatization as a mean of protecting the environment. The Sagebrush Rebellion is described in R. McGreggor Cawley, *Federal Land, Western Anger: The Sagebrush Rebellion and Environmental Politics* (1991), but see also Charles Davis, editor, *Western Public Lands and Environmental Politics* (1997).

SEE ALSO: Land-use policy; Sagebrush Rebellion; Wise-use movement.

Public trust doctrine

CATEGORY: Philosophy and ethics

The legal theory that government is the custodian of common properties for the use of the public traces its origins to Roman law and its development to English common-law decisions. The public trust, in common with the better-known private trust, involves an asset managed by a trustee in accordance with the wishes of the trustor for named beneficiaries. Unlike the private trust, however, it is not based on a written document but on the legal fiction that the creator gave trust assets to government to hold for the people.

In *Shively v. Bowlby* (1894), the U.S. Supreme Court concluded that the original states held the tidewaters and their lands in trust and that states subsequently admitted to the union held the same title. In the hands of state and federal judges, the public trust doctrine has become a continually growing and sometimes conflicting body of state common law that has defined the meaning of trust assets and terms, the extent of state government's trustee duties, and the identity of trust beneficiaries.

Judicial decisions have expanded the scope of state trust assets beyond tidal waters and their use for navigation. Now the public trust assets include all tidal waters, not just navigable ones, and their use extends to recreation, fishing, and wildlife habitat. The state trust assets also include rivers, wetlands, lakes, parks, trees, wildlife, and may even include an entire ecosystem. The public trust also lies latent within private property rights and provides the public with access to beaches and streams on private land. The doctrine also limits the use of privately held wetlands. In *Marks v. Whitney* (1971), the California Supreme Court held that the state's public trust easement for navigation, commerce, and fisheries forbids an owner to fill and develop them.

State governments, as trustees, have an obligation to regulate, manage, develop, and preserve trust property for the benefit of the public. State legislatures and agencies may not make decisions that merely reflect current political, economic, and social needs because they have a fiduciary duty to a present and future public constituency. This duty does not, however, restrict trust assets to their present use but requires state governments to identify the impacts on trust assets in their planning processes and balance competing trust interests in their decision making. In *National Audubon Society v. Superior Court* (1983), the California Supreme Court held that the Los Angeles Department of Water and Power did not have the unrestricted right to appropriate the waters of streams that flowed into Mono Lake and to divert them to the Los Angeles Aqueduct. The state Water Resources Board had a duty to take public trust interests into account in the planning and allocation of water resources in a manner that protected the lake's trust uses.

State governments may violate their trust responsibilities in three ways. First, they may alienate trust assets by selling them to a private party. In *Illinois Central Railroad v. Illinois* (1892), the U.S. Supreme Court held that the state legislature had violated its public trust duty when it made a grant to the railroad of 1,000 acres of submerged Chicago waterfront property. When the legislature subsequently withdrew the grant, it did not take private property without just compensation because the railroad had held only a revocable title.

State governments may also divert ownership of public trust assets from one public use to another by transferring them from one government agency holding public trust lands to another agency with a mission to develop public property. In *Robbins v. Department of Public Works* (1968), the Massachusetts Supreme Judicial Court held that the state park agency's transfer of the Fowl Meadows wetlands to the state department of public works for highway use violated the agency's responsibility to protect the parklands for the public.

Finally, state agencies may derogate trust assets by destroying them. The Los Angeles Department of Water and Power could not destroy Mono Lake by its continued appropriation of lake waters, nor may private people or corporations pollute trust assets. When a privately owned oil barge dumped thousands of gallons of crude oil into coastal marshes, killing wildlife and destroying their environment, the court in *In re Steuart Transportation Company* (1980), held that Virginia had the sovereign right, derived from the public trust doctrine, to protect the public interest in the preservation of wildlife resources.

Public trust doctrine and its principles have been incorporated into state constitutional provisions and environmental statutes. The Pennsylvania constitution, along with those of ten other states, recognizes that its government is the trustee of the state's public resources with the power to maintain and preserve the natural environment. The Michigan Environmental Protection Act, which has served as a model for similar state statutes, provides that the government hold in trust the air, water, and public resources of the state and bestows upon the courts the authority to decide whether those resources have been or are likely to be polluted, impaired, or destroyed. The public trust doctrine may even apply to the federal government because it holds in trust the public lands and waters of the nation, and, as *In re Steuart Transportation Company* has suggested, it has a duty to protect and preserve the public's interest in those natural resources.

William C. Green

SUGGESTED READINGS: Joseph L. Sax, "The Public Trust Doctrine in Natural Resource Law: Effective Judicial Intervention," *Michigan Law Review* 68 (1970), is the seminal work. The public trust doctrine has its critics, including Lloyd R. Cohen, "The Public Trust Doctrine: An Economic Perspective," *California Western Law Review* 29 (1992). Jack Archer et al., *The Public Trust Doctrine and the Management of America's Coasts* (1994), studies the doctrine's vitality in coastal areas, while John Hart, *Storm over Mono: The Mono Lake Battle and the California Water Future* (1996), explores its usefulness in the conflict over an inland lake.

SEE ALSO: Intergenerational justice; Land-use policy; Nature preservation policy; Privatization movements; Sustainable forestry; Water use.

Pulp and paper mills

CATEGORY: Pollutants and toxins

The pulp and paper industries form one of the most important segments of the economy. However, the chemicals used in the primary industrial processes in pulp and paper mills present multiple sets of environmental problems, including noxious odors, potentially toxic solid waste, and polluted effluents.

Paper is manufactured by reducing vegetable matter to a liquid slurry, removing excess liquid from this pulp, and forming the remaining fibers into a mat. Materials used to produce paper include wood, cotton, hemp, and flax, depending on the desired product, but all require similar processing involving large amounts of water. Most papers contain a least a portion of wood, and if wood is used, the cellulose fibers that can be formed into paper must be separated from the lignin that surrounds them. Once the fibers are successfully separated, the resulting pulp mat may require bleaching to produce a paper that is sufficiently white for commercial use.

All the processes involved in manufacturing pulp and paper present potential environmental hazards, but the kraft process used to reduce wood fibers to pulp may present the greatest dangers. Prior to the invention of the kraft process, pulp mills used mechanical methods to crush plant matter and used rags to separate the unwanted material from the fibers necessary to make paper. The runoff from the mill was often dirty and thus posed some hazards to the environment, but because it was a mechanical process, no new chemical compounds were formed.

In contrast with mechanical methods, the kraft process uses chemicals to break down the lignins in wood. Industry experts report that over 70 percent of the world's pulp is produced using the kraft process. Well into the late twentieth century many pulp and paper mills discharged the untreated wastewaters from the pulping and bleaching processes directly into rivers and lakes. While the kraft process has allowed production of paper from woods once considerable unsuitable for paper, the resulting pulp is generally more difficult to bleach than pulps produced using other methods. The preferred bleaching agent has been chlorine dioxide, an acid that reacted easily with the residual lignin compounds in the pulp to form chlorinated organic compounds. Scientists refer to such compounds as adsorbable organic halogens (AOX), while the general public more commonly uses terms such as organochlorines or dioxins.

Researchers have shown that many of these compounds have mutagenic and carcinogenic properties. Excessive exposures may lead to birth defects, reproductive difficulties, and some forms of cancer. AOX compounds pose special threats because many accumulate in the fatty tissues of animals rather than passing through the body and therefore tend to become concentrated as they move up the food chain. Residents of the Great Lakes region of the United States, particularly children and women of childbearing age, are advised to limit their consumption of

A paper mill in Rumford, Maine. The United States has enacted legislation designed to curb the emissions of dioxin from such factories. (AP/Wide World Photos)

lake trout, a fish that is known to have high concentrations of AOX in its fat.

Because of the environmental dangers associated with organochlorines, many countries, including the United States, have implemented AOX emissions standards. Some experts believe that the only acceptable standard would be zero discharge of AOX emissions. While some manufacturers have argued that achieving zero emissions would be prohibitively expensive, particularly if they were forced to modify existing mills, others are exploring options such as a closed loop mill. In a closed loop facility, no liquid waste leaves the factory. Effluents are treated on site and cycled back through the system. Another option is to convert to oxygen bleaching, or totally chlorine free (TCF) bleaching. Mills in European countries that converted to TCF bleaching discovered that operating costs were actually lower than with previously used chlorine processes. In addition to posing less of a threat to the environment, oxygen-based bleaching compounds are less corrosive. This means equipment within a mill lasts significantly longer before requiring replacement.

In addition to the well-publicized hazards of AOX, pulp and paper mills present a number of other additional environmental problems. People living near pulp mills often complain about noxious odors. Strong odors are often a by-product of the chemical digesting process used in pulping wood, but to date researchers have been unable to find any harmful effects other than psychological discomfort. Millworkers and their families, however, believe additional research is necessary.

The solid waste produced by a mill may also negatively impact the environment. The exact composition of the waste may be difficult to determine, particularly if a mill uses recycled materials. Most pulp and paper mills routinely use the waste sludge and liquids from the digester as boiler fuel for steam turbines to generate electricity. This reduces the volume of potential solid waste but cannot completely eliminate it. The industry is attempting to recycle as much of the solid waste from mills as possible, but it remains a long way from achieving optimum efficiency in this area.

The ever-growing appetite for paper means that hundreds of thousands of acres of forest are harvested annually to be processed into pulp. Besides posing obvious questions regarding deforestation, possible erosion, and destruction of unique ecosystems and wildlife habitat, harvesting exposes workers and the general public to dust, debris, and exhaust fumes from logging equipment.

Nancy Farm Männikkö

SUGGESTED READINGS: Patrima Bajpai and Pramod K. Bajpai, *Organochlorine Compounds in Bleach Plant Effluents: Genesis and Control* (1996), provides a comprehensive discussion of important issues in paper and pulp mills, although some readers may have trouble with the technical jargon. *Advances in Bleaching Technology: Low AOX Technologies for Chemical and Recycled Fibers* (1997), edited by Ken L. Patrick, discusses a variety of advances in eliminating chlorine bleaches in pulp mills. Harry W. Edwards, *Pollution Prevention Assessment for a Manufacturer of Folding Paperboard Cartons* (1995), provides a brief overview of pollution prevention in general.

SEE ALSO: Birth defects, environmental; Dioxin; Odor control.

R

Radioactive pollution and fallout

CATEGORY: Pollutants and toxins

Radioactive pollution occurs when human activity results in the release of radioactive materials into the environment. Radioactive particles that fall to the earth following the detonation of a nuclear weapon are known as fallout. Both types of pollution may adversely affect the health of biological systems with which they come in contact.

People are well aware of the use of radioactive elements in nuclear weapons and reactors. However, radioactive elements are sometimes used for their physical or chemical properties rather than their radioactivity. Their presence may, nevertheless, cause small amounts of radioactive pollution. For example, uranium and thorium compounds have been used for centuries to give ceramic glazes brilliant orange and yellow hues. (Although pieces produced in the United States after about 1950 are probably safe to use, other ceramics with uranium or thorium glazes are probably best left for occasional show unless they are tested and known to be safe.)

In addition, trace amounts of uranium have been used to color porcelain teeth, and tiny amounts of uranium or thorium are present as impurities in some tinted contact lenses and eyeglass lenses. Gas lantern mantles have long used thorium to produce a bright, white glow. Although the hazard to the public is believed to be negligible, nonthorium mantles are now being produced. Tungsten electrodes for arc-welding may contain 2 percent thorium for easier starting and greater weld stability. The resulting radiation dose to both the welder and the public is very small. Smoke detectors use tiny amounts of radioactive americium to ionize air in a small chamber. This allows a current to flow. Smoke particles reduce this current and trigger the detector. Americium has a 458-year half-life, so it will remain radioactive long after the smoke detector is discarded. However, the small amounts of radioactivity involved make it safe to discard smoke detectors with normal trash.

INDUSTRIAL, MEDICAL, AND SCIENTIFIC USE

Since small amounts of radioactivity are easily detected, both industry and the medical field use radioactive elements as tracers. For example, radioactive technetium may be injected into a patient's vein so its progress can be followed with radiation detectors. Imaging systems can then reveal constrictions in the patient's heart or arteries. The technetium used has a six-hour half-life. The nondecayed portion of technetium is eventually eliminated from the body and passes into the sewage system. No special precautions are taken because the radioactivity quickly decays and is greatly diluted with normal waste. Worldwide, more than twenty million diagnostic procedures using various radioactive elements are done each year.

Radioisotope thermoelectric generators (RTGs) were developed to supplement the power from solar cells or replace them on space missions where sunlight is too weak. The Jupiter mission's *Galileo* spacecraft had two RTGs, while the Saturn mission's *Cassini* spacecraft had three. A standard RTG uses the heat from the radioactive decay of 11 kilograms (24 pounds) of plutonium dioxide to produce electricity. Since the process involves no moving mechanical parts, it is very reliable. RTGs are built to withstand the explosion of the spacecraft during launch as well as the heat of reentry.

Although used many times, the United States has had only three accidents involving RTGs. In 1964 a navigational satellite failed to reach orbit and burned up over the Indian Ocean. Its RTG was of an earlier design that also burned up, as was then intended. The resulting plutonium oxide dust settled out of the stratosphere over the next several years. Greatly diluted across the globe, it barely increased background radiation. In 1968 a spacecraft was destroyed after launch by the range safety officer, and its RTGs were recovered intact. In 1970 the lunar module of the damaged Apollo 13 spacecraft reentered the atmosphere over the South Pacific. Its RTG plunged into the 6-kilometer-deep (3.7 miles) Tonga Trench. Although it was never recovered, surveys have shown no release of radioactivity.

Depleted Uranium

Uranium is a widely distributed trace element. Its estimated concentration in the earth's crust of 20 parts per million makes it thirteen times more abundant than tungsten and four thousand times more abundant than gold. Pure uranium is a lustrous, silver-white metal, and although it is radioactive, the activity consists of alpha particles and low-energy gamma rays that can be readily shielded to safe levels. The ease with which it can be safely handled under favorable conditions is leading to its increasing use. Many view this as dangerous.

Natural uranium consists of three different forms, or isotopes: about 0.7 percent is uranium 235, 99.3 percent is uranium 238, and only a trace amount is uranium 234. Weapons-grade uranium must be enriched to 93 percent uranium 235, while reactor fuel is generally enriched to a least 3 percent uranium 235. Extraction processes leave uranium that has only 0.2 percent uranium 235. This is called depleted uranium (DU). Depleted uranium is only 60 percent as radioactive as natural uranium, but there is a lot of it. The United States has an estimated 611,000 tons of depleted uranium.

Rather than provide storage for it as low-level radioactive waste, the Department of Energy (DOE) has been actively seeking uses for depleted uranium. Almost twice as dense as lead, it makes a good radiation shield, and it is used to shield radioactive isotopes shipped to hospitals. The Lockheed Martin Corporation mixes depleted uranium in a special concrete to form Ducrete to be used as radiation shielding in shipping casks for spent reactor fuel.

Depleted uranium packs a great deal of weight into a small volume; therefore, it is used to make counterweights in commercial aircraft and the tips of Tomahawk cruise missiles. Its density and hardness led to its use in the armor of the M1A1 Abrams tank and armor-piercing ammunition. Powdered uranium is pyrophoric

Children in Warsaw, Poland, drink an antiradiation iodine solution in April of 1986 to counteract the effects of fallout from the Chernobyl nuclear accident in the neighboring Soviet Union. (Reuters/Stringer/Archive Photos)

(burns spontaneously in air). Some depleted uranium in the armor-piercing rounds burns upon impacting a hard surface. This makes the round more effective against tanks because it helps it penetrate armor and often ignites a secondary explosion. While most of the uranium is expected to remain within several meters of the target, a significant amount may not. Some uranium oxide particles formed from burning uranium are smaller than 5 microns and can be carried by the wind for many miles. Such small radioactive particles can also lodge in the lungs of organisms and are potentially hazardous.

An estimated 330 tons of depleted uranium was used during the 1991 Persian Gulf War. Depleted uranium rounds were so effective that many nations want to add them to their arsenals. The U.S. Navy uses a Phalanx Gatling gun to defend its ships against planes and missiles. Concerned about possible radiation exposure of the ship's crew, the Navy switched from depleted uranium to tungsten rounds for the Phalanx gun in the mid-1980's. Pressure is mounting for the army to make a similar switch.

FALLOUT FROM NUCLEAR WAR

Fallout is the name given to radioactive particles that rain down from the debris cloud of a nuclear explosion. Whether the fallout is local or global depends chiefly on the yield of the weapon. Nuclear weapon yields are measured in terms of how many tons of the high explosive trinitrotoluene (TNT) would be required to release the same energy. With the improved accuracy of missiles, nuclear powers have reduced weapon yields to a few hundred kilotons or fewer. One exception is the People's Republic of China, which still maintains some warheads of 3 to 5 megatons.

Yields of fewer than 100 kilotons produce local fallout, which consists of particles that fall to the ground within twenty-four hours of the explosion. Local fallout is generally most intense near ground zero; however, winds may carry local fallout hundreds of kilometers or more. Larger weapons loft debris higher into the air so that small particles may drift for several days before falling to the ground. Particles lifted into the stratosphere may remain there for months or longer and be carried around the world. The radioactivity of fallout decreases with time so that the longer it remains aloft, the less dangerous it is. Therefore, local fallout is far more hazardous than global fallout.

When a nuclear weapon explodes, it instantly becomes an expanding fireball of radioactive vapor. The radioactivity chiefly comes from the debris of the nuclear fission of uranium. The explosion also produces a torrent of neutrons that can transform some normal elements into radioactive elements. Wherever the fireball touches the ground, dirt and debris are sucked into the air. As the fireball expands and cools, radioactive vapor condenses into radioactive particles and debris that are pulled into the fireball. Since hot air rises, the fireball rises and forms the hallmark mushroom-shaped cloud. Fallout begins within minutes as the heaviest radioactive pebbles rain down near the stem of the mushroom cloud. Fine particles are carried downwind from ground zero and continue to fall to the earth over the next several hours.

If the wind is steady, the radioactivity of the fallout accumulated on the ground could be described with a series of concentric, elongated ovals. The near ends of all of the ovals would touch at ground zero. The innermost oval would have the highest radioactivity. The next oval outward would mark lower activity, and its far end would extend farther from ground zero. For a 1-megaton bomb, the oval that contains a lethal dose of fallout during the first forty-eight hours would be 1,000 square kilometers (386 square miles) in area. However, fallout radioactivity decays relatively quickly, so that after one year the lethal oval would cover only 1 square kilometer (0.38 square miles). At first, 20,000 square kilometers (7,722 square miles) would be contaminated badly enough that an unprotected person might show signs of radiation sickness within two weeks. After one year, that area would decrease to about 20 square kilometers (7.72 square miles).

Depending upon such factors as targeting strategies, timing, and weather, fallout from a large-scale nuclear attack could kill tens of millions of people and endanger hundreds of millions more. Since fallout is radioactive dust, it accumulates most readily on horizontal surfaces such as the ground and the roofs of buildings. If caught in a fallout zone, the best strategy would be to get as far away from the fallout as possible and place as much mass between the person and the fallout as possible. For example, the shelter of a simple basement can reduce the radiation dose to 5 or 10 percent of that of a person in the open.

Charles W. Rogers

SUGGESTED READINGS: *Waging Nuclear Peace: The Technology and Politics of Nuclear Weapons* (1985), by Robert Ehrlich, is one of the best popular-level books on all aspects of nuclear war, including fallout. Other good treatments of fallout and its effects can be found in *Last Aid: The Medical Dimensions of Nuclear War* (1982), edited by Eric Chivian et al.; *The Medical Implications of Nuclear War* (1986), edited by Fredric Solomon and Robert Marston; and *Arsenal: Understanding Weapons in the Nuclear Age* (1983), by Kosta Tsipis. *Environmental Radioactivity from Natural, Industrial, and Military Sources* (1997), by Merril Eisenbud and Thomas Gesell, is an excellent summary of the many ways people encounter radioactivity.

SEE ALSO: Limited Test Ban Treaty; Nuclear accidents; Nuclear and radioactive waste; Nuclear testing; Nuclear winter; Yucca Mountain, Nevada, repository.

Radon

CATEGORY: Atmosphere and air pollution

Radon is a radioactive gas that naturally occurs in rocks. Although it accounts for approximately 50 percent of the normal background radioactivity in the environment, radon gas can pose a health hazard if it accumulates in houses and buildings.

Unsafe levels of radon gas have been detected in structures built over soils and rock formations containing uranium. One of the radioactive products of uranium is radium, which decays directly to radon. Every 3 square kilometers (1.2 square miles) of soil to a depth of 15 centimeters (6 inches) contain about 1 gram (0.035 ounces)

A billboard on the outskirts of Pruntytown, West Virginia, warns of the dangers of radon in the area. (Jim West)

of radon-emitting radium. Three forms of radon are generated in the decay of uranium in rocks and soils. The potential health risks are posed by the radon isotope with an atomic mass of 222 (radon 222), which has a 3.8-day half-life. Radon 220 and radon 219 also form in rocks and soils, but these isotopes have half-lives of fifty-six seconds and four seconds, respectively. The shorter half-lives of these isotopes compared to radon 222 give them a much greater chance to decay within the rocks and soils before becoming airborne. Thus they are of lesser radiological significance.

Certain regions, such as the Reading Prong stretching from southeastern Pennsylvania to northern New Jersey and portions of New York, contain higher concentrations of radium in their rocks and soils. Similar regions occur across the nation and around the world.

Radon gas is chemically inert, and within its 3.8-day half-life, the gas can become airborne and enter a building through small fissures in its foundation. Indoor radon levels are typically four or five times more concentrated than outdoor levels where air dilution occurs. Contributions to indoor radon levels also come from building materials, well water, and natural gas.

Airborne radon itself poses little hazard to health. As an inert gas, inhaled radon is not retained in significant quantities by the body. The potential health risk arises when radon in the air decays, producing nongaseous radioactive products. These products can attach themselves to dust particles or aerosols. When inhaled, these particles can be trapped in the respiratory system, causing irradiation of sensitive lung tissue. Sustained exposure may result in lung cancer. The U.S. Environmental Protection Agency (EPA) has estimated that between 5,000 and 20,000 lung-cancer deaths each year are attributable to radon products. The EPA recommends remediation measures if radon levels in a building exceed four picocuries per liter of air. Remediation techniques to relieve indoor radon pollution usually involve ventilating basements and foundation spaces to outside air.

Anthony J. Nicastro

SEE ALSO: Environmental illnesses; Indoor air quality; Radioactive pollution and fallout.

Rain forests and rain forest destruction

CATEGORY: Forests and plants

Rain forests are ecosystems containing diverse plant and animal species that occur in regions of the world that receive an average annual rainfall of 50 to 800 centimeters (20 to 315 inches). These forests may be found in the tropics around the equator (between the tropics of Cancer and Capricorn) or in temperate regions such as the Pacific Northwest of North America and the southeastern United States.

Rain forests play significant roles in the preservation and maintenance of global climates. The largest rain forest, located in Brazil, covers most of the Amazon River and its basin. Other countries with large rain forests include Indonesia, Cameroon, and Malaysia. Countries with smaller rain forests include Australia, China, the Ivory Coast, and Nigeria and countries of the Caribbean Islands, northern South America, and the Pacific Islands. Biodiversity in these forests is very high for both plants and animals. In the tropical rain forests, more than a hundred species of plants may be found within a hectare of land. Many scientists believe that the tropical rain forests, which cover only 7 percent of the earth's surface, may house from 60 to 70 percent of the earth's plant and animal species. Because of this, tropical rain forests constitute important laboratories for scientists worldwide.

Trees in rain forests may grow to more than three hundred feet in height. These tall trees, along with their middle-layered or middle-storied plants, have leaves that spread like a sea to form what is generally called the "forest canopy." Large buttress roots such as those of the kapok trees of Brazil and the fig trees in Sabah, Malaysia, anchor the large trees while at the same time functioning to absorb essential nutrients and moisture from the soil. Nutrient recycling is critical to the luxuriant growth of plants in the tropical rain forests. It is also essential for the numerous epiphytes that grow in the lower to middle stories of the rain forests.

Although temperate forests have not received

the same level of media attention as tropical rain forests, they are just as fascinating and are similarly threatened with regard to loss of biodiversity. These forests, which occur in places such as the Pacific Northwest and southeastern United States, are characterized by deciduous tree species that shed their leaves during the winter months, or by evergreen pine tree species. Among the many important species of trees are the large redwoods (Sequoias) of Northern California. These giant trees can grow to heights of more than three hundred feet and are among the oldest trees in the world. Other temperate rain forests may be found in Central and Northern Europe, temperate South America, and New Zealand.

Importance and Destruction

The luxuriant growth of the plants in tropical rain forests gives the impression that rich and fertile soils underlie them. However, as slash-and-burn agriculture and other production techniques have demonstrated, the soils are generally infertile. The luxuriant growth is the result of the tremendous amount of nutrient recycling that takes place among species in a humid environment. Thus, the removal of native plants to clear space for agriculture depletes necessary biomass for recycling and eventually exposes the soils to rapid erosional forces. Hence, after a few years of high crop productivity, yields decline dramatically, leading to further clearing of more forests.

Despite this productivity problem, the tropical rain forests play significant roles in the lives of the people that inhabit them, and they are a source of important materials for economic development. Indigenous peoples have developed the means to cohabit with other living organisms in these unique environments. However, civilization, industrialization, and population growth are gradually threatening the homelands and the very existence of these native peoples.

Perhaps most important, the abundant plant life in rain forests enhances the global environment through photosynthesis, an ongoing biological process that removes carbon dioxide from the atmosphere and produces oxygen. This process is an important factor in the prevention of global warming.

Rain forest destruction has been occurring for many centuries. The process, which first started in the temperate rain forests of Europe and North America, greatly intensified in tropical areas during the last half of the twentieth century. Rain forests are said to have covered more than 25 million square kilometers (9.65 million square miles) of the earth's surface at one time, but they are now vanishing at such an alarming rate that countries such as Thailand, Nigeria, Indonesia, Ivory Coast, and Malaysia have already lost more than 50 to 60 percent of their original forestland. Brazil is losing an average of 50,000 square kilometers (19,000 square miles) of forests every year. In the Congo Basin, rain forest losses threaten the very existence of several indigenous peoples. In Indonesia, which is second only to Brazil in the amount of tropical rain forest land it contains, deforestation is estimated at 12,000 square kilometers (4,600 square miles) per year.

Factors Contributing to Destruction

Many factors are responsible for the loss of tropical rain forests. These include population growth, farming systems (especially slash-and-burn agriculture), energy requirements (in terms of firewood or fuel wood), industrialization, wood consumption, and mineral consumption (such as the intense gold mining occurring in the Amazon Basin).

Population growth in many countries containing tropical rain forests increased by an average of 3 to 5 percent per year during the last fifteen years of the twentieth century. Some of this growth was the result of high birth rates, while some was caused by migration. For example, following the Sahelian drought of the early 1980's, Ivory Coast received an estimated 1.5 million refugees, thereby putting significant pressures on its resources. The situation in Nigeria, with a population of more than 110 million people, is no different. It is estimated that at the current rate of destruction, Nigeria may be without rain forests by the early to mid-twenty-first century. Meanwhile, as cities expand and new megalopolises are created in industrialized countries, temperate rain forests are losing ground.

Related to population growth is the growing need for energy. Many countries that still have abundant rain forests use wood as their main source of energy. Therefore, along with intensive logging for export, cutting wood as biomass for fuel is helping to accelerate the destruction of the rain forests. This situation is best exemplified in Thailand and Indonesia, as well as Madagascar, Malaysia, and many Central American countries. In Madagascar, the loss of woody vegetation has led to severe erosional problems.

Another important factor contributing to rain forest destruction is slash-and-burn agriculture, in which land is cleared, burned, and subjected to crop and livestock production for a few years, followed by new slashing and burning of undisturbed forest lands. While crop yields are initially high, they rapidly decline after the second and third years. Subsistence agriculture, upon which slash-and-burn agriculture is based, was originally associated with small populations. Land could be allowed to go fallow for many years, leading to full regeneration of secondary forests under natural conditions. Increased population, along with other factors, has greatly reduced the length of time land can be allowed to go fallow. Hence, more forests are destroyed to keep up with demand. More than 30 percent of the more than 12,000 square kilometers (4,600 square miles) of rain forest lost annually in Indonesia is attributable to slash-and-burn agriculture.

HALTING RAIN FOREST DESTRUCTION

Reforestation efforts have been under way for many years in Europe and North America to restore lost temperate rain forests. However, such reforestation activities are usually approached from an industrial perspective. Plants of only one or two species are grown, giving rise to monocultures. This is different from a natural regeneration approach that eventually restores the original biodiversity. Such efforts are yet to be fully implemented in tropical rain forests. Perhaps this may be beneficial in an environment where diseases and insects may wipe out an entire monocultured forest within a short time.

Approaches such as using agroforestry systems to compliment and reduce slash-and-burn agriculture are one way by which tropical rain forest destruction is being halted or slowed. Other approaches include the establishment of nature preserves and large botanical gardens, as has been successfully done in Costa Rica, or the establishment of national parks such as the Korup National Park in Cameroon. Such nature preserves have also been put in place for a number of temperate rain forests in the United States and elsewhere. Kew Gardens in Great Britain is a unique place for germplasm preservation of a large number of plant species.

Another strategy is debt forgiveness in exchange for tropical rain forest preservation. The United States has forgiven a number of Central and South American countries loans amounting to almost five billion dollars, provided these countries set aside large tracts of forestland as nature preserves. The International Board of Plant Genetic Resources (IPBGR), an agency of the Food and Agricultural Organization (FAO), has begun preserving the germplasm of forest genetic materials in addition to its preservation of traditional crop genetic resources.

Oghenekome U. Onokpise

SUGGESTED READINGS: Useful overviews of rain forests and related issues may be found in *The Rain Forests: A Celebration* (1992), edited by L. Silcock, and compiled by the Living Earth Foundation, and J. Terborgh's *Diversity and the Tropical Rain Forests* (1992). William J. Peters and Leon F. Neuenschwander's *Slash and Burn: Farming in the Third World* (1988) summarizes scientific information on slash-and-burn agriculture and the social, cultural, economic, and political effects of this ancient system of farming.

SEE ALSO: Biodiversity; Deforestation; Mendes, Chico; Old-growth forests; Rainforest Action Network; Slash-and-burn agriculture.

Rainforest Action Network

DATE: founded 1985
CATEGORY: Forests and plants

The Rainforest Action Network (RAN) is a nonprofit volunteer organization working to protect tropical rain forests and the human rights of people living in tropical rain forests. Founded in

1985 by Randall Hayes, the network publicizes dangers associated with rain forest destruction and focuses public attention on companies involved in their degradation.

Rainforest Action Network activists erect a miniature oil derrick and pour oil into the Arco Plaza fountain in downtown Los Angeles to protest Arco's plans to drill for oil in Ecuador. (Reuters/Rainforest Action Network/Archive Photos)

The RAN's conservation mission is achieved through education, the support of activists working in countries with rain forests, and through direct grassroots activities. Campaigns encourage the public to influence corporate executives through product boycotts and policymakers through letter writing, petition drives, and nonviolent demonstrations. Other activities include the organization of coalitions among scientific, environmental, and grassroots organizations worldwide; holding conferences and seminars; and providing technical or financial assistance to native communities and nongovernment organizations (NGOs) in rain forest countries.

At the international level, members participate in letter-writing campaigns targeted at the leaders of countries that permit the destruction of rain forests. Information sharing and coordination of activities is facilitated through a cooperative alliance with other environmental and human rights groups in more than sixty countries. At the local level, RAN members are organized within grassroots organizations known as Rainforest Action Groups. Members of these local organizations are encouraged to write policymakers, coordinate nonviolent demonstrations, organize product boycotts, and participate in educational or direct-action campaigns. Through a program known as Protect-an-Acre, the network supports local organizations within rain forest countries that initiate projects to protect the ecological or cultural integrity of forest communities.

In 1987 the RAN led a nationwide boycott of Burger King fast-food restaurants to raise public awareness concerning the purchase of beef from companies involved in expanding pastureland for cattle at the expense of rain forests. This campaign led to a 12 percent drop in Burger King's sales, prompting company officials to cancel $35 million in contracts for beef raised in Central America and discontinue the company's purchase of beef fed on former rain forest lands. The network also led a successful boycott of products produced by the Mitsubishi Corporation, encouraging corporate executives to discontinue operations that degrade forests and native cultures. Another program of letter writing and legal advocacy work prompted the government of Nicaragua to instruct the Solcarsa Company to halt the cutting of trees within old-growth rain forests located inside an area known as the North Atlantic Autonomous Region.

Publicity generated by the RAN has also resulted in large companies changing their policies concerning the purchase and resale of old-growth wood products from U.S. forests. For example, Home Depot stores no longer sell old-growth redwood. Other companies from which the network has won concessions include Scott Paper, Sony, Conoco, and Coca-Cola Foods. The RAN has also encouraged organizations such as the World Bank to deny funding for companies involved in ecologically destructive activities within rain forests. Since timber harvesting is the leading agent in rain forest destruction, the network has become a strong advocate for sustainable alternatives to pulp, paper, and tropical wood used in furniture and construction.

Thomas A. Wikle

See also: Deforestation; Logging and clear-cutting; Old-growth forests; Rain forests and rain forest destruction; Slash-and-burn agriculture; Sustainable forestry.

Range management

Category: Forests and plants

Rangelands are regions used as a source of forage by free-ranging domesticated and wild animals. They are not suitable for cultivation. Rangelands provide tangible products such as wood, water, and minerals and intangibles such as natural beauty, open space, and wilderness; accordingly, range management requires consideration of these multiple uses.

Rangelands comprise about one-half of the earth's surface and include a variety of areas, such as temperate grasslands, tropical savannas, arctic and alpine tundras, desert shrublands, shrub woodlands, and forests. While most are semiarid, they often feature riparian zones and can include wetlands. Rangeland plants coevolved with the herbivores that depended on them for food, which were in turn eaten by carnivorous predators. A balance of nature resulted, subject to annual variation and unpredictable natural catastrophes, that led to relative stability of plant and animal species and populations. Prior to the development of agriculture ten thousand years ago, this balance of nature permeated the entire earth. Since then, one-half of the land has been used for the agriculture, industry, and habitation of humans.

Rangelands remain unsuitable for crop cultivation because of physical limitations, such as inadequate precipitation, rough terrain, poor drainage, or cold temperatures. They have been home to numerous nomadic peoples and their animals around the world. While not subject to intensive use, they are affected by human activities. Because rangelands support an estimated 80 percent of livestock production worldwide, provide habitat for many wild animals, and yield numerous tangible and intangible products, their viability and sustainability are important.

Mismanagement of rangelands is commonly caused by overgrazing by domesticated or wild herbivores. Continued heavy grazing leads to denuding of the land, erosion by the elements, and starvation of the animal species. Because decreased plant cover changes the reflectance of the land, climatic changes can follow that prevent the regeneration of plant life, leading to desertification. Semiarid regions are particularly prone to overgrazing because of low and often unpredictable rainfall; however, these are the areas of the world where most livestock have been relegated. Overgrazing has contributed to environmental devastation worldwide. Largely uncontrolled grazing by livestock is partly responsible for the formation of the desert of the Middle East, degradation of rangeland in the American West in the late nineteenth and early twentieth centuries, and the devastation of parts of Africa and Asia. Likewise, feral horses are damaging environments in the western United States and the Australian outback. Overgrazing by wildlife can be deleterious as well. Removal of predators can lead to overpopulation, excessive grazing, starvation, and large die-offs, such as the Kaibab Plateau deer disaster in Arizona during the 1920's and 1930's.

Proper management can prevent the deterioration of rangeland. Managing rangelands involves controlling the number of animals and enhancing their habitat. Carrying capacity, which is the number of animals that can be supported indefinitely on a given unit of land, must not be exceeded. Optimizing, instead of maximizing, the number of animals will sustain a healthy plant community, referred to as good range condition. For private land, optimizing livestock numbers is in the long-term self-interest of the landowner, although it is not always seen as such. For land that is publicly owned or owned in common, or that has unclear or disputed ownership, restricting animals to the optimum level is particularly difficult to achieve. As Garrett Hardin describes in his 1968 essay "The Tragedy of the Commons," personal, short-term benefits often lead to long-term disaster.

Restricting livestock is physically easy through herding and fencing, although it can be politically difficult and expensive. Controlling charismatic feral animals, such as horses, or wildlife when natural predators have been eliminated and hunting is severely restricted is much more problematic. As for habitat improvement, various approaches can increase carrying capacity for either domesticated or wild herbivores. Removing woody vegetation by controlled burning or mechanical means will increase grass cover, fertilizing can stimulate forage growth, and reseeding with desirable species, often native plants, can enhance the habitat. Controlling plant pests (noxious weeds) and animal pests (such as grasshoppers and rabbits) may be necessary as well. Effective rangeland management requires matching animals with the forage on which they feed.

Riparian areas—linear strips of land on either side of a river or stream—are particularly susceptible to overgrazing. Animals naturally congregate in areas with water, lush vegetation, and shade, and they can seriously damage them by preventing grasses from regrowing and young trees from taking root, as well as by compacting the soil and fouling the watercourse. The ecosystem can be devastated, subjecting the land to erosion and threatening the survival of plant and animal species. While herding and fencing can be used to control animals in these areas, a less expensive method to encourage movement away from rivers or streams is to distribute sources of water and salt, which range animals crave and will seek.

Effective range management also accounts for the multiple demands placed on these areas. The balance between livestock and wildlife is an acute source of controversy between ranchers and environmentalists in the American West, especially because much of the land used for grazing livestock is publicly held. Issues include the low cost of grazing permits on public versus private land; the Emergency Feed Program, which reimburses ranchers for one-half the cost of feed during droughts and other disasters; the lack of control for wild animal populations; the livelihood of ranchers; and the contribution of range livestock to the country's food supply. Other uses of rangeland include extraction of lumber, minerals, and energy, as well as meeting the recreational and aesthetic needs of people.

James L. Robinson

Suggested readings: Balanced treatments of range management are found in chapter 13 of *Natural Resource Conservation* (1998), by Oliver S. Owen, Daniel P. Chiras and John P. Reganold, and chapter 7 of *Contemporary Issues in Animal Agriculture* (1999), by Peter R. Cheeke. General texts include *Range Management: Principles and Practices* (1997), by Jerry L. Holechek, Rex D. Pieper, and Carlton H. Herbel, and *Rangeland Ecology and Management* (1994), by Harold F. Heady and R. Dennis Child.

See also: Erosion and erosion control; Grazing; Hardin, Garrett; Kaibab Plateau deer disaster; Soil conservation.

Reclamation

Category: Land and land use

Reclamation is the process of returning disturbed land areas to stable and productive uses after a mineral such as coal has been removed by mining.

Failure to reclaim mined land may result in substantial loss of biological productivity of the land surface, significant degradation of water bodies, and hazards to human health and safety. Degradation of land and water is attributable to erosion and sedimentation of soils, acid mine drainage, and damage to groundwater aquifers. Hazards to human health and safety include open mine shafts, mine fires, subsidence of the ground above underground tunnels, clifflike surfaces called "highwalls," and landslides on steep slopes. Land areas disturbed by mining that have not been reclaimed are called abandoned mine lands.

The amount of land to be reclaimed at any mine is determined by the amount of mineral removed and the type of mining operation. Underground mines require little reclamation except near the tunnel entrance. Surface or strip mines disturb larger areas and volumes of soil and rock than underground mines, therefore requiring more reclamation. In all forms of surface mining, rock and soil located above and between seams of the mineral are removed to expose the mineral for extraction.

Before soil and rock can be removed, all vegetation covering the land surface to be mined must be removed. Next, topsoil and subsoils are excavated and used in an adjacent area that is being reclaimed, or they are separated from other overburden rock and stockpiled for later use. Segregation and reuse of fertile topsoils during mining is critical to the later success of reclamation efforts because a suitable growing medium is essential to reestablish viable plant communities. Reclamation of abandoned mine lands, where topsoil was not separated from other overburden, is generally more difficult and expensive than reclamation at operating mines.

Reclamation encompasses three activities: backfilling and grading, replacement of topsoil, and revegetation. Backfilling and grading occur after the mineral has been removed. Overburden is replaced in the mined area to reestablish a stable land surface that is consistent with the surrounding area and compatible with an intended postmining land use. Front-end loaders,

Topsoil has been placed over the site of a mining operation. The mound awaits revegetation. (Ben Klaffke)

heavy trucks, bulldozers, and graders are used to fill and grade the contour of highwalls, overburden piles, and depressions to approximate original slopes. The resulting surfaces must be stable—that is, not prone to landslides or erosion—and should blend into the surrounding natural topography.

Mining operations sometimes unearth natural materials that are toxic or acid forming. When exposed to the atmosphere, these materials may alter the acidity of surrounding soil or water bodies, destroying the organisms that live there. To prevent environmental contamination, they must be isolated from surface water and groundwater, soils, and vegetation. This generally means placing them below the root zone of plants during backfilling and grading. During backfilling and grading, heavy equipment repeatedly crosses the work area, causing compaction of the ground surface. Prior to redistribution of topsoil, it may be necessary to rip up this surface to relieve compaction. This helps prevent slippage of topsoil by creating a roughened surface and aids root penetration by vegetation, thus improving surface stability.

After backfilling and grading are completed, a layer of topsoil is spread over the graded overburden to a depth determined by the intended postmining land use and the amount of topsoil available, often 1.2 meters (4 feet) or more. Topsoil stockpiled for more than two or three years begins to lose nutrients, beneficial bacteria, and fungi that aid in plant establishment, so soil tests are used to determine what soil amendments may be needed. Where nutrients are lacking, they are replenished using fertilizers similar to those used on home lawns and gardens.

Revegetation must occur soon after placement of topsoil to control the effects of wind and water erosion. A fast-growing annual grass or cereal grain cover crop, as well as mulch, may be used to stabilize the soil until the first normal planting season. Shrubs and small trees may also be planted. The goal of revegetation is the establishment of a diverse, permanent vegetative cover of a seasonal variety native to the area, or of a variety that supports the intended postmining land use. In the eastern United States, where water is plentiful, it may take five years to determine if reclamation is successful; in the semiarid western United States, this may require ten years.

Common uses of reclaimed mined land include cropland agriculture, commercial forestry, recreation, tourism, public works (such as airfields, roads, housing developments, and industrial sites), and fish and wildlife conservation.

In the United States, reclamation of land disturbed while mining coal has been required since 1977 under national legislation known as the Surface Mining Control and Reclamation Act (SMCRA). Principal responsibility for applying this law rests with the U.S. Department of the Interior's Office of Surface Mining Regulation and Enforcement and state regulatory authorities, with programs approved under the statute. A small fund is available to reclaim abandoned mine lands, financed by fees on each ton of coal produced by active mining operations.

The requirements of SMCRA apply only to coal mines operating since 1977. Lands disturbed by the mining of gold, silver, nickel, copper, bauxite, limestone, and other minerals and industrial materials are not subject to uniform national standards but may be subject to requirements for reclamation imposed by some states. However, such state regulations vary greatly. The knowledge and technology necessary for successful reclamation of land disturbed by mining are available for almost all ecological systems, except desert and alpine climate conditions that do not favor the rapid plant growth necessary to stabilize reclaimed soil.

Michael S. Hamilton

Suggested readings: For a discussion of reclamation in hostile environments, see C. Wayne Cook, R. M. Hyde, and P. L. Sims, *Guidelines for Revegetation and Stabilization of Surface Mined Areas in the Western States* (1974). *Reclamation of Surface Mined Lands* (1988), edited by Lloyd R. Hossner, uses plain language to describe the environmental consequences of failing to reclaim mined lands.

See also: Acid mine drainage; Groundwater and groundwater pollution; Erosion and erosion control; Restoration ecology; Strip mining; Surface mining; Water pollution.

A worker sorts recyclable materials before they are taken to the processing center. There are more than nine thousand curbside recycling programs in the United States. (Jim West)

Recycling

CATEGORY: Waste and waste management

Recycling is the process of collecting solid waste materials in order to reuse them as raw material or finished products. Recycling can reduce the amount of waste entering landfills, conserve natural resources and energy, reduce pollution, and create jobs.

Recycling involves several steps. First, a recyclable waste is collected and delivered to a processing facility, which often entails the participation of waste generators. The waste is then sorted and prepared for delivery to a manufacturing facility, where it is used to manufacture a new recycled-content product, which is then purchased by consumers. When it once again becomes a waste, the product may be recycled again. Environmental benefits and costs are realized at each point in this cycle.

When a waste material is recycled, it acts as a substitute for a virgin material in the manufacturing process. For example, trees—a virgin material—are pulped to make paper. When producing recycled-content paper, waste paper is used instead of trees. Almost any waste is theoretically recyclable, meaning that a manufacturing process can be devised to use it to make a new product. However, in a practical sense, something is recyclable only if all the recycling steps function in accord. If any one part of the cycle is too small or big, problems are created. For example, if community recycling programs collect used phone books but few manufacturing facilities use them, much of the collected material cannot be recycled.

The traditional methods for dealing with municipal solid waste—landfills and incinerators—entail significant cost. Likewise, it is not necessary for recycling programs to realize a profit; however, if their costs are significantly higher than other methods, they should be used only if they offer significant nonmonetary benefits.

When a community recycles, the volume of waste landfilled or incinerated is reduced. This can extend landfill life and reduce the cost and environmental impact of landfills and incinerators. Recycling saves natural resources, as the recycled waste replaces virgin materials that are extracted from the earth. The use of recycled materials in manufacturing processes often, though not always, creates less pollution and requires less energy than the use of virgin materials. Recyclable materials are often sold to manufacturers, which creates revenue. Finally, recycling industries are often labor intensive and therefore tend to create entry-level jobs. However, recycling also uses energy and can generate pollution, primarily while collecting and transporting recyclables. Each community is faced with a different mix of opportunities and challenges. Determining when, what, and how to recycle can be a complex process.

Collection

Collection involves gathering recyclables and hauling them to a location where they can be processed. The most common recyclable collection methods are drop-off centers and curbside collection programs. Drop-off centers are centralized locations that receive and store recyclables brought by residents of the surrounding area. Over twelve thousand such centers operate in the United States.

Drop-off centers are cost-effective for the collection agency. Many of the collection costs are

Percentages of Recovered Municipal Solid Waste, 1970 and 1994

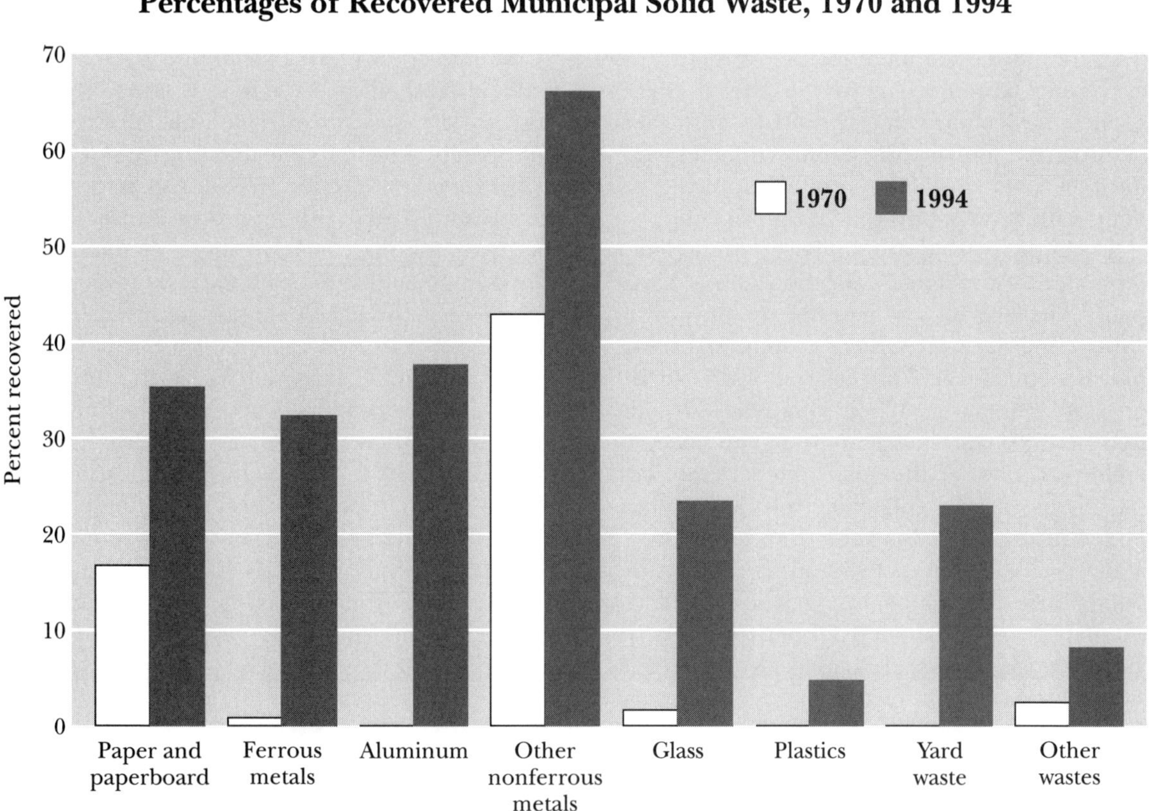

Source: U.S. Department of Commerce, *Statistical Abstract of the United States, 1996*, 1996. Primary source: Franklin Associates, for Environmental Protection Agency.

Note: Total U.S. municipal solid waste generated in 1994 was about 209 million tons; amount recovered was about 49 million tons, or 23.6 percent. (Not included in the waste generation figure are mining, agriculture, industrial, and construction wastes, junked automobiles and equipment, or sewage.)

borne directly by residents, as each must personally transport recyclables to the central location. If users make special recycling trips, drop-off center recycling increases automobile driving, with all the related pollution and fuel consumption. Thus, drop-off centers are best located at sites that are already frequented by residents, such as shopping centers, allowing combined trips. In 1995 the Environmental Protection Agency (EPA) reported costs between $35 and $95 per ton for the majority of eighteen drop-off centers studied.

In curbside collection programs, a collection agency—such as a public works department or private waste hauler—gathers recyclables from each residence, generally at the curb. There are over nine thousand such programs in the United States. These programs tend to have higher participation rates than drop-off centers because they are more convenient for the resident. However, they are more costly to the collection agency: Collection vehicles must be purchased, maintained, and operated, the collection crew must be paid. Even the cost of supplying residents with storage bins can be significant.

Curbside collection can be commingled or separate. Commingled collection allows participants to place mixed materials in one or two containers. Separate collection requires participants to place each material at the curb in a separate container. When commingled collection is used, materials may be separated by the collection crew at the curb into separate compartments on the collection vehicle or placed mixed in the collection vehicle and separated at a processing center. A 1993 study by the National Solid Waste Management Association (NSWMA) estimated costs of about $120 per ton for a typical route with materials loaded onto the collection vehicles commingled.

Processing and Transportation

Recyclable processing facilities are often called material recovery facilities (MRFs). These facilities accept recyclables or mixed wastes. Processes performed at MRFs include removing contaminants, sorting, baling, compaction, and transfer to transport vehicles or containers. It is important to remember that MRFs produce commodities for manufacturing facilities. Manufacturing equipment is designed to input materials with specific properties. If an MRF supplies recyclables with unacceptable properties (for example, too many contaminants), an entire load may be rejected.

There are two types of MRFs: dirty and clean. Dirty MRFs accept the entire waste stream, from which recyclables or compostables are separated. Though this is convenient for residents, it can be expensive and may result in inferior quality. Clean MRFs only accept recyclable materials that have already been separated from waste by drop-off or curbside collection programs. Though less convenient for the waste generators, processing at clean MRFs is generally less expensive and produces a higher quality material. A 1992 NSWMA study estimated processing costs at clean MRFs at about $50 per ton.

At both types of MRFs a combination of manual and mechanical separation is often used. Materials are moved through the facility using mobile equipment, plus mechanical or pneumatic conveyors. Few environmental impacts are associated with MRFs, although dust, noise, and repetitive motions can have negative impacts on workers. In addition, truck traffic may negatively affect residents in the area surrounding the MRF.

MRFs load clean, sorted recyclable materials onto trucks and trains. These materials are then transported to manufacturing facilities. Environmental impacts include vehicle exhaust, leaking fluids, and natural resource and energy depletion. The environmental and economic cost of long transportation distances may outweigh the benefit of recycling. Thus, even if a manufacturing facility that can use a recyclable waste exists, it may be best to not recycle if it is too far away.

Manufacture and Sale

A majority of the benefits of recycling occur during the manufacturing process. Commonly recycled municipal wastes include paper, glass, metals, plastics, tires, oil, and yard waste. On a weight basis, paper constitutes the largest component of municipal solid waste (about 40 percent). Among the many types of paper that are recycled are newspaper, corrugated, computer, high grade office, and mixed. About 40 percent of municipal waste paper is currently recycled. It is used to make many products, including newsprint, writing paper, paperboard, corrugated containers, tissues, toilet paper, and roofing paper. Making new paper from waste can save 30 to 40 percent of the energy that would be used with virgin materials. Recycling 1 ton of newspaper can save seventeen trees, a renewable but threatened resource. However, paper fibers cannot be recycled forever. Each time a paper fiber is repulped, it breaks into shorter fragments, resulting in a lower-quality product. For this reason, few 100 percent recycled-content paper products are made.

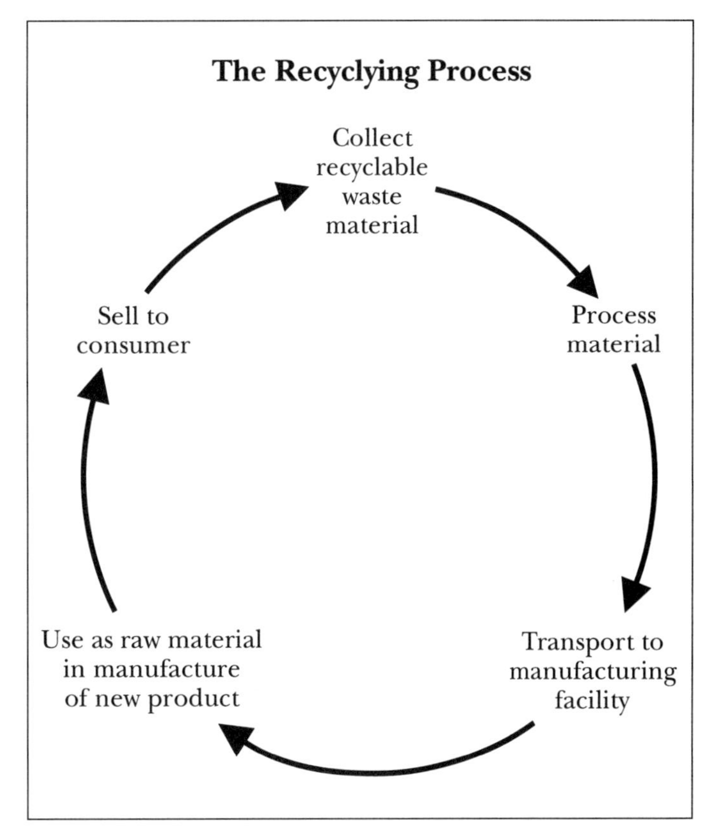

Glass represents about 6 percent of municipal solid waste. It is generally separated for recycling into three categories: green, clear, and brown. About 36 percent of glass beverage containers are recycled, primarily into new glass containers and bottles. Using waste glass instead of virgin materials can prolong furnace life and conserve energy, about 0.25 percent for each percent of waste glass used. However, the prime ingredient in glass is sand, a plentiful resource. Since glass materials are not changed by the recycling process and contaminants are not a large problem, glass can be recycled indefinitely.

Metals represent about 8 percent of municipal solid waste. Almost all metals can be recycled if free of foreign materials such as plastics, fabrics, and rubber. In 1996 about 58 percent of the steel food cans and 52 percent of aluminum packaging were recycled. Making new aluminum from waste can use 95 percent less energy compared to making aluminum from bauxite. Most metal products are made of alloys, mixtures of metals, and other elements, such as carbon. To ensure specific properties are obtained, few 100 percent recycled-content metal products are made.

Plastics account for about 9 percent of the waste stream by weight but may account for about twice that in volume. Most manufacturers now identify the resin type with a number—ranging from one to seven—surrounded by three arrows, located on the bottom of the container. When recycling plastics, the different resins are not mixed (except in products such as plastic lumber) because they have different properties. The most commonly recycled plastics are number one (polyethylene terephthalate, used to make soft drink bottles) and number two (high-density polyethylene, used to make plastic milk jugs). Plastic materials make excellent containers because of their high strength and low density; however, their lightness also makes them expensive to collect on a per-ton basis. Recycling plastic saves petroleum, a finite and important natural resource.

A given waste is not recycled until it is made into a new product that is purchased by a consumer. Old newspaper is not recycled until it is purchased as a cereal box, for example. The environmental impact of the distribution and sale of a recycled-content product is similar to that of a virgin material product. For example, both may be transported long distances to market, while junk mail and fliers may be used to advertise each product.

Jess W. Everett

SUGGESTED READINGS: For a complete guide to recycling, see *The McGraw-Hill Recycling Handbook* (1993), edited by Herbert Lund. For a detailed engineering treatment of recycling programs, see *Integrated Solid Waste Management* (1993), by George Tchobanoglous, Hillary Theisen, and Samuel Vigil. For the political side of recycling, see Mathew Gandy's *Recycling and the Politics of Urban Waste* (1994). Finally, many recycling books exist for children; *Recycling: A Handbook for Kids* (1996), edited by Gail Gibbons, is recommended.

SEE ALSO: Composting; Landfills; Resource recovery; Solid waste management policy; Waste management.

Refuse-derived fuels

CATEGORY: Energy

Refuse-derived fuel (RDF) is solid fuel material originating from municipal solid waste (MSW) that has been processed to remove the inert components and concentrate the combustible portion.

Raw MSW is a notoriously poor fuel because of its high moisture and low heat content. In addition, mass-burn incineration of MSW has a reputation for the production of a broad range of atmospheric pollutants. Finally, the ash produced may become concentrated in potentially toxic elements, such as cadmium or arsenic. One goal of RDF production is to improve the combustion of MSW by producing a fuel of lower moisture content, more uniform size, greater density, and lower ash content than raw MSW. Another goal is to avoid land disposal of a valuable resource.

A typical process for the production of RDF involves passage of raw wastes through a screen to separate small, inert materials (such as stones, soil, or glass); pulverization of larger particles in a shredding device; separation of ferrous metals by magnetic extraction; and segregation of the lightweight, mostly organic fraction in an upward air stream ("air classification"). The shredded organic waste fraction can be used directly as a fuel ("fluff RDF"), or fluff RDF can be compressed into high-density pellets or cubettes ("densified RDF"). The latter material is popular because it is easy to transport and store and because of its adaptability in handling and combustion.

RDF can be utilized as a cofuel with coal or fired separately. A major advantage of RDF over coal is that its sulfur content is markedly lower (0.1 to 0.2 percent compared to 5 percent or more for some coal samples). Additionally, nitrogen content is much lower in RDF. Both sulfur and nitrogen are among the more notorious precursors to acid rain. Also, as a result of processing, RDF contains smaller amounts of potentially toxic metals than MSW.

RDF possesses only 50 to 60 percent of the calorific value and 65 to 75 percent of the density of typical bituminous coals. As a consequence, considerably larger weights of RDF must be burned to obtain performance similar to that of coal. The use of RDF in a boiler may therefore adversely impact the performance of the air pollution control system and ash removal system. The chlorine content of RDF is higher than typical coal. There are occasional concerns with respect to odors and dust production in storage, as well as with particulate, carbon monoxide, and hydrogen chloride discharges during combustion.

John Pichtel

SEE ALSO: Alternative fuels; Landfills; Resource recovery; Waste management.

Renewable resources

CATEGORY: Resources and resource management

Renewable resources are natural resources that are continuously produced or regenerated faster than they are depleted. Such resources are mainly derived from solar radiation, such as forest resources or renewable energy resources; however, water, soil, wildlife, plants, and wetlands can also be considered renewable resources.

Renewable energy sources, such as wind energy and hydroelectricity, are mainly derived from solar energy in one form or another. Direct solar radiation is usually converted into heat, which can be used for such purposes as heating homes or water. Solar water heating has been used in the southern United States since at least the early twentieth century. In mild climates such as southern Florida, it can easily furnish all the hot water requirements of a typical home. It has also been widely used in tropical countries throughout the world.

Passive solar heating is the heating of a building by solar radiation that enters the building through south-facing windows (north-facing in the Southern Hemisphere). A properly designed passive solar home must have enough interior heat capacity, usually in the form of concrete floors or walls, to be able to keep the house from overheating on a sunny day and to store excess heat for release at night. In many parts of North America and Europe, passive solar homes have proved to be economical, since the passive system is part of the house itself (its windows, walls, and floors) and thus adds little or no extra cost.

Active solar heating, in contrast to passive solar heating, uses air or liquid solar collectors that convert solar radiation into thermal energy, which is stored and distributed using a mechanical system (fans or pumps). Active systems have fallen into disfavor because they are technically and economically inferior to passive solar systems.

Direct solar radiation can also be used to produce electricity by several different methods. Solar power towers use a field of movable mirrors to reflect sunlight onto a tower where the solar heat is used to generate electricity in place of a fossil or nuclear fuel. This concept has been tested in California and is technically, but not yet economically, feasible. Ocean thermal conversion plants generate electricity using the difference in temperature between the warm, solar-heated upper portions of the ocean and the colder water farther down. This concept has been partially tested, but no complete plant has yet been built.

Photovoltaic or solar cells are semiconductor devices that generate electricity directly from solar (or other electromagnetic) radiation. Originally developed for artificial satellites after World War II, photovoltaic cells work well and are already in use for many purposes, such as solar calculators and radios, patio lights, electric fences, traffic signal controls, and corrosion prevention on metal bridges. However, they are not yet cheap enough to be economical for generating electricity in homes.

A less direct type of renewable solar energy is hydroelectricity—electricity generated by water turbines that are turned by water flowing down a river or dropping from a dammed water reservoir. The kinetic energy of the moving water is derived from gravitational potential energy of water at greater heights, and that energy is ultimately derived from solar radiation that evaporated the water from the oceans, allowing it to rain down in the mountains. An interesting type of hydroelectric plant is one that derives its energy from rising and falling tides. The only major example of such a tidal energy power plant is on the Rance River in northern France. An older form of water power was the waterwheel used by millers until well into the nineteenth century.

Wind energy is also a renewable solar energy resource because it is the uneven heating of the earth's land and water areas by the sun that causes winds. Wind energy has long been used as an energy source: It powered the sailing ships that explored the globe, and it powered the windmills used in Asia and Europe since the Middle Ages to grind grain and pump water. Since the 1920's wind turbines have been used to generate electricity in rural areas of the United States, and they are now being used by some electric utilities.

Biomass is often defined as the total mass of living organisms in an ecosystem, including both plants and animals, but the term is also used for nonliving biological materials such as wood from dead trees. The energy stored in biomass is solar energy that has been stored by photosynthesis.

Biomass is an important resource not only for human life, since all human food is biomass in one form or another, but also for society in general since biomass materials can be used as a source of energy and organic chemical compounds, including therapeutic drugs. Biomass energy resources include solid, liquid, and gaseous fuels. The solid fuels include wood (the major energy resource used in the United States until about 1880) and agricultural wastes such as corn stover and sugar cane bagasse. These are currently being used on a small scale for home heating and electric power generation. Liquid biomass fuels include methanol and ethanol, both of which can be used in motor vehicle engines. The major gaseous biomass fuel is methane, the main constituent of the fossil fuel called natural gas; methane is generated by the anaerobic (oxygen-starved) decomposition of manure and other organic materials.

Laurent Hodges

SUGGESTED READINGS: A great deal of information about renewable resources can be found in the *Dictionary of Renewable Resources* (1997), edited by Hans Zoebelein. A readable popular account of renewable energy resources is Jennifer Carless, *Renewable Energy: A Concise Guide to Green Alternatives* (1993). For biological uses of renewable resources, see *Introduction to Forest and Renewable Resources* (1995), by Grant W. Sharpe et al. A good Internet resource is the National Renewable Energy Lab Web site (www.nrel.gov).

SEE ALSO: Alternative energy sources; Biomass conversion; Energy conservation; Hydroelectricity; Solar energy; Wind energy.

Resource recovery

CATEGORY: Waste and waste management

The recovery of various resources through recycling programs or certain aspects of production processes is a means of slowing the use of both nonrenewable and renewable resources. Although there are substantial environmental and economic rewards for resource recovery, there has not been a great deal of impetus in this direction in either industrialized or nonindustrialized nations.

Spurred by wartime necessity, the United States practiced various forms of recycling during World War II. In the postwar years a "throwaway" culture developed in which it was considered cheaper and more convenient to discard used or unwanted materials. Since the 1970's, this approach has come under fire from environmental groups and government. In particular, the problems associated with municipal solid waste such as pollution and finding landfill space have led many local governments to start recycling programs such as bottle deposit laws (by the late 1990's ten states had bottle deposit laws) and sorting and curbside pickup of recyclables such as paper, glass, and certain types of plastics. Some landfills are also selling the methane gas generated as a result of decomposition of organic wastes. In some cases, waste incinerators are being used to generate electric power, another form of resource recovery.

Recycling operates at two levels: primary and secondary. Primary recycling occurs when the original waste material is remade into the same material (for example, newspapers recycled to make newsprint or glass that is melted down to make new glass products). In secondary recycling, waste material is made into something else. Tires, for example, can be shredded and incorporated into asphalt, and some plastics can be reused in some types of outdoor clothing or carpet fibers.

Recycling programs have met with varying degrees of success. In some cases the market price for material has declined and made recycling programs costly to operate. By the late 1990's more than 50 percent of waste paper was being recycled in the United States, either by industry itself or by consumers. In other areas the record was not as good, with a 20 percent recovery rate for glass, a 12 percent recovery rate for old tires, and nearly 40 percent recovery for metals. Scrap aluminum, for example, takes only 10 percent of the energy required to make aluminum from virgin ore, so the environmental and economic savings generated by recycling are substantial.

Internationally, some automobile makers, notably Saab and Volvo, encourage recycling of used cars.

Many consumers are reluctant to recycle, saying that it is more convenient to throw material out than to sort trash. Some cities and counties, however, charge their inhabitants by the volume of trash generated, providing an economic incentive to recycle. Industry, too, often opposes recycling. Although glass bottles are easier to recycle, the soft drink industry uses plastic bottles, which are cheaper to transport, and generally opposes bottle laws. Congress passed the Resource Conservation and Recovery Act (RCRA) in 1976 to deal with municipal solid waste. The bulk of this legislation dealt with landfills, but it indicated that using fewer materials and reusing materials were preferred alternatives to landfilling. The RCRA provided incentive to local governments and industry to utilize alternatives to landfilling waste, such as resource recovery.

Some firms have found that there are substantial economic rewards to resource recovery. According to the concept of industrial ecology, waste should not be seen as a disposable product but rather as potential raw material for future use. Such an approach diminishes the strain put on renewable and nonrenewable resources. The primary incentive is increased profitability, but, if followed, this approach would have a positive environmental impact by promoting sustainability. Advocates even contend that some wastes that are not presently useable should be stored with an eye to future use. The chemical industry, for example, is already making extensive use of by-products and is recovering and recycling materials. However, a market will have to be created for industrial ecology to be successful. A hard-goods manufacturer has no incentive to become a maker of polymer feedstock unless a ready market exists.

The positive potential for resource recovery to reduce pollution and slow the use of natural resources is great. It is difficult to achieve resource recovery because many governments, firms, and individuals operate in the present, in which it is more economical or convenient to throw away materials than to reuse them. Solutions are slowly coming, and they often involve a variety of approaches. In some cases, "command and control" governmental regulations, such as the RCRA, provide an incentive for action. At the other end of the scale, voluntarism, such as community-based recycling programs that emphasize the achievement of improved environmental quality, can be helpful in achieving resource recovery as well as raising public consciousness. Market-based incentives, whether in the form of profits or the avoidance of the costs of acquiring new resources or disposing of waste, also play an important role in achieving resource recovery.

The concept of industrial ecology emphasizes the economic incentives of resource recovery for firms and governments in both industrialized and developing nations. Individual recycling of postconsumer waste is a good starting point for resource recovery, but it will take industrial action directed toward waste minimization to achieve a full-scale program of resource recovery. Ultimately, there are limits to how many times a resource can be recycled, but without a program of resource recovery, resources will be used up more quickly, and more pollution will be generated.

John M. Theilmann

SUGGESTED READINGS: On recycling, see John E. Young, "Discarding the Throwaway Society," *Worldwatch Paper 101* (1991), and the entire July-August, 1992, issue of *EPA Journal*. For information on industrial ecology, see *Industrial Ecology and Global Change* (1994), edited by R. Socolow et al., and R. A. Frosch, "Industrial Ecology: Adapting Technology for a Sustainable World," *Environment* 10 (1995).

SEE ALSO: Landfills; Recycling; Solid waste management policy; Sustainable development; Waste management.

Restoration ecology

CATEGORY: Ecology and ecosystems

Restoration ecology is concerned with converting ecosystems that have been modified or degraded by human activity to a state approximating their original condition.

Federal laws often dictate ecological restoration following strip mining, construction, and other activities that alter the landscape. As a part of the management of natural areas that have been disturbed to some degree, several options are available. One is to do nothing but protect the property, allowing nature to take its course. In the absence of further disturbances, one would expect the area to undergo the process of ecological succession. Theoretically, an ecosystem similar to that typical of the region, and including an array of organisms, would be expected to return.

One might, therefore, ask why ecological restoration is mandated. For one thing, succession is often a process requiring long periods of time. As an example, the return of a forest following the destruction of the trees and the removal of the soil would require more than one century. Also, ecosystems resulting from succession may be lacking in species typical of the region. This is true when succession is initiated in an area where many exotic (alien) species are present or where certain native species have been eliminated. Succession can produce a new ecosystem with a biodiversity comparable to the original one only if there is a local source of colonizing animals and seeds of native plants. Also, satisfactory recovery by succession is unlikely if the soil has been heavily polluted by heavy metals or other substances caused by industrial land use.

Once it has been decided that a given ecosystem is to be restored, success requires that a plan be designed and followed. Although the specifics may vary greatly, all restoration projects should follow five basic steps: Envision the end result, consult relevant literature and solicit the advice of specialists, remove or mitigate any current disturbances to the site, rehabilitate the physical habitat, and restore indigenous plants and animals.

Much can be learned from restoration projects that have been conducted in various parts of the world involving a wide variety of ecosystems. A classic ecological restoration of a prairie was conducted in Wisconsin beginning in the 1930's. Because most North American prairies have been converted to agricultural uses, many opportunities exist for prairie restoration. In such projects it is often necessary to eliminate exotic plants by mechanical means or by application of herbicides. Native prairie grasses and forbs can be established by transplantation or from seed. It may also be necessary to introduce native fauna from nearby areas. Periodic pre-

The Crosswinds Marsh in Michigan was created to replace wetlands destroyed during the construction of a new runway at Detroit Metro Airport. (Jim West)

scribed burning is often necessary to simulate natural fires common in prairies.

After decades of loss of wetlands in the United States, governmental policy is now "no net loss." Thus, when a wetland is destroyed by development, it is required that a new wetland be created as compensation. Before introducing native biota, it is necessary to alter the hydrology of the new site.

Thomas E. Hemmerly

See also: Bioremediation; Ecology; Ecosystems; Reclamation; Wetlands.

Rhine river toxic spill

Date: November 1, 1986
Category: Water and water pollution

In 1986 a chemical leak near Basel, Switzerland, caused a massive loss of river life.

The Rhine River, Europe's major commercial inland waterway, originates in the east-central Swiss Alps and flows northwest for 1,320 kilometers (820 miles) through or along the borders of Liechtenstein, Austria, Germany, and France before entering the North Sea at the Netherland coast.

On November 1, 1986, firefighters responded to a large fire at the Sandoz chemical factory in Schweizerhalle, Switzerland, near Basel. Drums filled with various chemical substances stored in the warehouse ruptured and spilled their contents during the blaze. Water that was sprayed onto the fire carried these chemicals into catch basins. However, a catchment wall in the riverside warehouse collapsed, releasing approximately 1,200 tons of agricultural chemicals such as pesticides, fungicides, herbicides, fertilizers, and dyes. The contaminated water moved downstream at 3.2 kilometers per hour (2 miles per hour), forming a 56-kilometer-long (35-mile-long) slick and turning the Rhine red.

A water-use crisis ensued in heavily populated downstream areas as an estimated 30 tons of highly toxic waste entered the Rhine in approximately two hours. An estimated 500,000 fish were instantly killed, predominately from 1,800 kilograms (4,000 pounds) of the mercury-based fungicide Tillex, and German farmers had to move livestock away from the river. Fishing in the Rhine and its tributaries was banned, and alternate supplies of drinking water were required all the way to Amsterdam, Holland.

Anger against the Swiss government intensified with knowledge of their failure to warn neighboring countries for twenty-four hours. Subsequent river analysis revealed high concentrations of Atrazine, a pesticide not stored in the Sandoz site. Another chemical firm, Ciba-Geigy, later admitted that an accident had occurred at their factory one day prior to the Sandoz fire that had released approximately 397 liters (105 gallons) of Atrazine into the Rhine.

A meeting held in Basel to discuss the incident ended with protestors pelting officials with dead eels. Considerable public and scientific interest in the Rhine's restoration ensued, and twelve major pollution incidents were reported within thirty days. More reports followed as persistent chemicals permeated the riverbed and moved into the groundwater. Although the Sandoz plant had not broken chemical storage laws, managers had badly underestimated the risks by choosing not to act upon recommendations made five years earlier by an insurance company.

Threats of lawsuits and government penalties caused Sandoz's stock to drop 15 percent in the Zurich stock exchange in the two weeks following the spill. The International Commission for Protection of the Rhine began a three-phase, fifteen-year pollution abatement program called the Rhine Action Plan (later changed to the Salmon 2000 project) and instituted a strict international alarm system for similar incidents. Following the Sandoz incident, police loudspeakers warned citizens to shut windows as clouds of sulfurous fumes left the Rhine, but warnings were only broadcast in German, leaving the Italians and Turks guessing. The mass return of salmon and other species such as sea trout and flounder during the 1990's, along with a larger fishing industry than before, made the Rhine's recovery one of the great environmental success stories of the twentieth century.

Daniel G. Graetzer

SEE ALSO: Agricultural chemicals; Environmental health; Pesticides and herbicides; Water pollution.

Right-to-know legislation

CATEGORY: Human health and the environment

During the 1970's and 1980's the U.S. government enacted legislation to address such environmental problems as air pollution, water pollution, and health threats from toxic substances. Much of this legislation includes provisions for informing the general public about hazards posed by pollution and toxic substances.

The Comprehensive Environmental Response, Compensation, and Liability Act (CERCLA), commonly known as the Superfund, was originally enacted in December, 1980. Its primary purpose was to address the problem of inactive hazardous waste sites. The mandates of CERCLA were controversial, largely because industry had to perform extensive and costly cleanups no matter how long ago the waste was discarded or how well a company had exercised responsibility in handling hazardous wastes.

Sections 102 and 103 of CERCLA provide for reporting procedures. Section 102 outlines reportable quantities set by the Environmental Protection Agency (EPA): Any hazardous substances that have not been assigned a reportable quantity under the Clean Water Act must be reported if they are found in quantities of 1 pound or more. Section 103 stipulates that the National Response Center must be notified as soon as a person in charge of a vessel or facility learns of any release of a hazardous substance that is equal to or greater than the reportable quantity for that substance. Failure to report a release carries a maximum criminal penalty of three years in prison for a first offense and a five-year sentence for repeat offenses; a maximum civil penalty equal to $25,000 per day may be assessed.

The original Superfund act expired in September, 1985. On October 17, 1986, the U.S. Congress reauthorized $9 billion and made major revisions in CERCLA with the Superfund Amendments and Reauthorization Act (SARA). Title III of SARA is the Emergency Planning and Community Right to Know Act (EPCRA). EPCRA has four major components, three of which address reporting requirements. Section 302 requires any facility that produces, uses, or stores any substances on the EPA's List of Extremely Hazardous Substances in quantities that equal or exceed the specified limit to notify the State Emergency Response Commission (SERC). Section 304 provides for emergency release notification to the community emergency coordinators for the Local Emergency Planning Committee (LEPC) of areas likely to be affected and to the SERC of any state likely to be affected.

Sections 311 and 312 form the major right-to-know components of EPCRA and are closely intertwined with the Hazard Communication Standard of the Occupational Safety and Health Administration (OSHA). Using the basic framework of the OSHA standard, these sections require that the general public and employees in facilities that produce or use hazardous chemicals be given information regarding their presence. Certain items are exempt, such as food, additives, drugs, cosmetics, gasoline used in a family car, petroleum products used for household purposes, and substances used for medical and research purposes.

Further provisions are made in clauses designated Tier I and Tier II. As of March 1, 1988, and annually thereafter, industries are required to submit chemical inventory forms for the material safety data sheet (MSDS) substances that are regulated by OSHA. Tier I information must be submitted to the LEPC, the SERC, and the fire department with jurisdiction over a given facility. Tier II information is more detailed: It must include the location of each chemical and a description of how the hazardous chemical is stored, unless a facility is excused on the basis of a trade secret plea. Any facility submitting Tier I or Tier II statements must, on request, allow onsite inspections by local fire departments. While the regulations are silent as to availability of Tier I information, any person may obtain Tier II information by submitting a written request. Any citizen may also request an MSDS on any hazardous chemical, even if the facility does not use it in reportable amounts.

The purpose of section 313 is to provide communities with information about chemical releases from local facilities. Owners and operators of certain manufacturing facilities must submit annual reports to the EPA and designated state officials on environmental releases of listed toxic chemicals. This report applies to routine and permitted releases as well as accidental releases. Furthermore, certain suppliers must notify their customers if products they distribute contain chemicals that are subject to section 313 mandates. Failure to submit MSDSs or lists of MSDS chemicals are punishable by civil penalties of up to $10,000 per day for each violation. Noncompliance with the annual inventory requirements may result in a penalty of up to $25,000 for each violation.

In addition to EPCRA's right-to-know title, some other legislation includes reporting provisions. Examples are the Occupational Safety and Health Act (OSHA) of 1970, which mandates that employees be apprised of hazards to which they are exposed, as well as information regarding symptoms of contamination and appropriate emergency treatment, and the Toxic Substances Control Act (TSCA) of 1976, which includes notice provisions for products for domestic use.

Victoria Price

Suggested readings: A concise account of environmental legislation is Edward E. Shea's *Introduction to U.S. Environmental Laws* (1995). *SARA Title III Regulations and Keyword Index* (1993), compiled by McCoy and Associates, provides helpful details of the SARA revision. A comprehensive discussion of environmental law appears in *Environmental Law and Enforcement* (1994), by Gregor I. McGregor. Cynthia A. Lewis and James M. Thunder, *Federal Chemical Regulation: TSCA, EPCRA, and the Pollution Prevention Act* (1997), takes an exhaustive look at the provisions of these acts.

See also: Clean Air Act and amendments; Environmental policy and lobbying; Hazardous and toxic substance regulation; Hazardous waste; Superfund.

Activists rally in support of right-to-know legislation that would give workers and the general public access to information on the health effects of pollutants and toxic materials. (Jim West)

Riparian rights

CATEGORY: Water and water pollution

Riparian rights are the benefits associated with the use of water for owners of land bordering on or adjacent to bodies of water. Such rights typically include the right to hunt, fish, and irrigate.

The U.S. system of riparian rights for water usage traces its roots back to English common law. Riparian rights and other water laws attempt to reconcile the various demands of users and potential users with the supply of clean water. Increasing demands for water use, coupled with the relative scarcity of adequate water resources and the growing recognition of the importance of habitat protection, have resulted in changing and evolving systems of water law and regulation in the United States.

Two basic water law systems exist for dealing with water-use conflicts in the United States. The eastern states, historically viewed as having adequate water resources, predominantly follow the doctrine of riparian rights. Under the riparian system, water-use rights are based on land ownership and require equal sharing among users in times of shortage. Western states, historically seen as lacking abundant water resources, adopted a system based on the prior appropriation doctrine. The prior appropriation establishes priority among competitive users: Earlier settlers have more right to water use. The western water law systems developed as a means to encourage economic development by allowing water rights to be separated from land ownership. The right to consume water accessible from one's land is the primary difference between riparian rights and appropriative rights.

Riparian rights treat water resources as a common property, which permits individual users to freely use the resource while spreading the cost of additional resource use to all owners of the resource. As Garrett Hardin warned, overuse of common property resources can result in a "tragedy of the commons." The increasing demand for water, as well as the demand to use it more efficiently and responsibly, has created tensions within the existing schemes of water rights and management. Erratic precipitation and rising per capita water use have led to a number of water emergencies. Likewise, many states now recognize the importance of leaving water instream to protect fish and wildlife, and to preserve aesthetic and recreational values.

Environmental concerns and water scarcity issues have led to significant modifications of water law systems. The only restriction on water use under the traditional riparian system is a prohibition against "unreasonable harm" to another riparian user. The absence of an efficient, system-wide mechanism to determine unreasonable uses and manage water resources successfully has led eastern states to reconsider how to allocate water among competing uses. Today, most eastern states have administrative permit systems that regulate riparian rights. Under these systems, a single agency manages water quality and allocation issues. The agency is also charged with defining and maintaining some minimum water quality and flow. Such modifications to riparian right systems attempt to protect private values and further public interests. As the evolution of riparian rights suggests, successful methods of allocating water must balance a watershed's consumptive water resources needs with its ecological needs.

Michael D. Kaplowitz

SEE ALSO: Drinking water; Hardin, Garrett; Irrigation; Prior appropriation doctrine; Water rights.

Road systems and freeways

CATEGORY: Land and land use

The construction and heavy use of highways can have severe impacts on the environment, including air and water pollution, land degradation, and loss of open space. Since motorized transportation is a common means of travel, compromise is often necessary to balance the need for additional roads and the need to protect the environment.

The era of the freeway began during the 1950's when lobby groups in the United States encour-

Increased traffic congestion leads to increased air pollution problems; each vehicle is a mobile source of hazardous emissions. (Jim West)

aged a political vision of a nationwide high-standard, high-speed road network. General Lucius Clay led a committee that studied highway needs in the United States and advised President Dwight Eisenhower that the National System of Interstate and Defense Highways was needed. Costs of the more than 69,000 kilometers (43,000 miles) of highways built after 1956 have been shared by federal and state governments on a 90-10 (federal-state) matching basis. The federal share comes from the Highway Trust Fund, which receives revenues from federal taxes on fuels, lubricants, vehicles, and parts.

Although the interstate system accounts for only 1 percent of the total road mileage in the United States, it carries 20 percent of the traffic. However, building the interstate system also included the construction of urban freeways, the use of substantial amounts of land, and the disruption of neighborhoods. New urban highways induce more traffic, causing increased congestion on local streets and a need for more land for parking and other vehicular uses.

Studies of the highway system as a whole—including local roads and services—have found that motor vehicle user fees only cover two-thirds of public expenditures, not including the substantial nonmonetary external costs. Because many of the costs are hidden, driving is perceived as less expensive than transit alternatives. U.S. highway statistics for 1995 indicated the existence of more than 201 million registered motor vehicles (including 136 million automobiles) and 3.9 million miles of public roads. In addition, annual travel added up to 2.4 trillion vehicle miles, and road fatalities for the year totaled 41,798.

Traffic congestion is most severe in areas experiencing rapid growth in both total population and number of vehicles in use. In fact, rapid population growth tends to offset the beneficial impacts of remedies adopted to reduce traffic congestion. Low-density settlement generates more total automotive vehicle trip miles per day, which consume more energy and cause greater emission of pollutants. One important development dating from the 1970's was a greater emphasis on the people-moving capacity of urban and suburban freeways by giving priority to buses and high-occupancy vehicles. Such measures have had a beneficial effect on freeway effectiveness by reducing congestion.

Air pollution constitutes the most serious environmental impact caused by highway transportation, which is a mobile source of pollution. Developments that may generate traffic, such as parking lots for shopping centers, may be classified as indirect sources. Internal combustion and diesel engines are the principal source of carbon monoxides and hydrocarbons, account for nearly one-half of the nitrogen oxides, and are the chief source of particulate lead in the atmosphere. Highway emissions are directly related to traffic volume and density, vehicle type, speed, and mode (idle, acceleration, cruise, or deceleration). Increased speed produces a demand for increased power, which leads to more fuel consumed and greater emissions, but the vehicle departs an area more quickly. Long commuting trips and traffic congestion both increase the emissions discharged into the atmosphere.

Highways also consume open space, affecting plant and animal life, as well as climate and water runoff. Highways facilitate the spread of urban areas and often lead to low-density developments, which are difficult to provide with services. Highways that connect developed areas usually follow valleys and other flat terrain because construction in such areas is less expensive. Consequently, highways are often in close proximity to streams, lakes, and wetlands. Until recently, hydrologic features blocking a proposed road were primarily seen as obstacles to be bridged, filed, or moved at lowest cost. However, concern over protected animal and plant species has changed this, and more expensive routes or construction procedures must be used. Another concern is runoff into waterways, including salt (from winter plowing) and petroleum products. Among the more subtle and probably more serious impacts of road construction are changes in local hydrologic patterns, such as changes in the water table that affect vegetation. Erosion and sedimentation are also associated with construction-related activities.

Noise—excessive or unwanted sound—annoys people, interferes with conversation, disturbs sleep, and, in the extreme, is a danger to public health. Highway noise is troublesome to control. Attenuation is achieved through providing buffer zones, modifying highway alignment, shielding noise, and implementing traffic management measures. The Federal Highway Administration specifies a methodology to evaluate traffic noise and abutting land use and provides guidance on when noise abatement should be considered.

In balancing transportation goals and environmental goals, engineers must distinguish between mobility and access. Mobility simply means movement, while access means the ability to reach destinations and desired services. Increasing mobility through more roads, more vehicles, and more traffic may actually reduce access over the long term. For example, older neighborhoods tend to have stores, schools, and transit services within walking distance, while newer auto-dependent neighborhoods tend to be lower density with few local services and often no pedestrian facilities, so more trips require driving. Hence mobility increases but access declines.

Stephen B. Dobrow

SUGGESTED READINGS: For a good overview of the history of roads and the vehicles on them, see M. G. Lay, *Ways of the World* (1992). To understand environmental effects of highways from an engineering point of view, *Highway Engineering* (1995), by Paul Wright et al., is useful. *Divided Highways* (1997), by Tom Lewis, tells the story of building the interstate highway system and how it transformed American life. Stephen Goddard's *Getting There* (1994) looks at the struggle between rail and road during the twentieth century. Anthony Downs, *Stuck in Traffic* (1992), looks at ways to cope with peak-hour traffic congestion.

SEE ALSO: Automobile emissions; Environmental impact statements and assessments; Land-use policy; Noise pollution; Urbanization and Urban Sprawl.

Rocky Flats, Colorado, nuclear plant releases

CATEGORY: Nuclear power and radiation

A series of accidents at the Rocky Flats nuclear plant caused low-level radioactive contamination of off-site reservoirs and land areas.

The four hundred buildings of the Rocky Flats site stand 26 kilometers (16 miles) northwest of downtown Denver. From 1952 to 1989 the primary mission at the site was to produce trigger assemblies (pits) for nuclear weapons. This required the machining of plutonium, uranium, beryllium, and other metals. Plutonium was also recovered from obsolete weapons and recycled. Radioactive americium 241, a decay product of plutonium, was separated and recovered in this process.

During the four decades of the plant's operation, numerous accidents threatened the surrounding population and environment. For example, a fire on May 11, 1969, may have been started by spontaneous combustion of plutonium scrap. It caused $50 million in damage and raised the concern that plutonium aerosols may have contaminated the countryside. Traces of plutonium were found off-site, but the contaminated area was not consistent with the wind pattern at the time of the fire. This strongly suggested that there had been one or more previous episodes of off-site contamination.

Workers remove radioactive waste from a room in the Rocky Flats nuclear plant in 1996. The facility was permanently closed by Secretary of Energy James Watkins in 1992 because of repeated safety violations. (AP/Wide World Photos)

Investigation showed that five thousand drums of oil contaminated with plutonium during machining operations had been stored on-site for many years. Some of these drums had corroded and leaked oil into the soil. It seems likely that the off-site plutonium contamination came from blowing dust from the drum storage area. When the leaks were discovered in 1959, rust retardant was added. Eventually, the drums were removed, the worst of the contaminated soil was hauled away, and the remaining contamination was asphalted over.

Some plant workers went public with their concerns that proper safety guidelines were not being followed. One infamous example involved a substance known as "pondcrete." Hazardous chemical waste mixed with low-level radioactive waste was placed in solar ponds to reduce its volume by evaporation. Sludge dredged from these ponds was mixed with Portland cement and placed in large, plastic-lined cardboard boxes. This was supposed to form a solid block of pondcrete that could be shipped elsewhere for burial, but much of the pondcrete remained mushy. Stacked in a parking lot and exposed to the weather, leaky material was washed into the soil by rainwater.

Requests by the public for specific information on hazards at the plant were frequently denied by officials, who claimed that such records were classified. A series of investigations finally resulted in a 1989 raid by seventy-five Federal Bureau of Investigation (FBI) agents and Environmental Protection Agency (EPA) staff members, who seized plant records. Finding evidence of mishandling of hazardous materials, Rockwell International, which managed the plant for the Department of Energy (DOE), was fined $18.5 million and stripped of its contract. The new management company, EG&G, was also accused of safety violations, despite the fact that it spent $50 million on repairs. In 1992 Secretary of Energy James Watkins ordered the permanent closure of the Rocky Flats nuclear plant.

According studies of tissue and urine samples, the doses of radiation received by people living near the plant were much less than those normally received by people from natural background radiation. As part of the clean-up and remediation process, a new water supply for the nearby town of Broomfield was constructed, along with catch basins and diversion ditches to control contaminated runoff. Further plans called for on-site radioactive and hazardous chemical waste to be either shipped off-site or processed and placed in two on-site storage cells.

Charles W. Rogers

SEE ALSO: Nuclear accidents; Nuclear and radioactive waste; Radioactive pollution and fallout.

Rocky Mountain Arsenal

CATEGORY: Nuclear power and radiation

Between 1942 and the 1980's, the Rocky Mountain Arsenal (RMA) served as a United States government facility where chemical weapons, combustible munitions, and rocket fuels were manufactured, stored, or destroyed. Releases of hazardous substances resulted in the designation of the RMA as a Superfund hazardous waste site in 1987.

Amid the growing involvement of the United States in World War II, approximately 43 square kilometers (27 square miles) of prairie and farmland northeast of Denver, Colorado, were selected in 1942 as the site of the Rocky Mountain Arsenal (RMA) for the production of chemical weapons. Subsequently, the site was used to manufacture, assemble, test, and demilitarize chemical and volatile munitions, including nerve gas, mustard agents, incendiary bombs, and rocket fuels. By the end of World War II, the RMA had produced over 100,000 tons of combustible munitions. In the late 1950's, superfuels for U.S. bomber planes and nuclear fuels to power rockets into space were being produced at the RMA. The production of rocket fuels was escalated during the 1960's to support the Titan missile program.

Since its beginning, the RMA has used more than six hundred different chemicals, including volatile organic compounds and heavy metals. Some of these chemicals, including rocket fuels, hydrazine, phosgene, halogenated volatiles, and isotopes of some of the heavy metals, contained radioactive materials that produced nuclear releases into the air, soil, and groundwater at the RMA site. Contamination occurred from burying toxic waste, using evaporative basins for disposal of hazardous liquid wastes, leaks in sewer lines, wind dispersion, and accidental spills.

In 1962 a waste-disposal well was drilled to a depth of more than 3,600 meters (12,000 feet) at the RMA. More than 662 million liters (175 million gallons) of treated waste material were eventually disposed of in the well. However, the pumping facilities were shut down in 1966 when it was determined that pressurized fluid injection was generating earthquakes in the Denver area.

Between 1970 and the early 1980's, facility activities focused on the destruction of chemical warfare materials. In the early to mid-1970's, the Colorado Department of Health detected a variety of toxic compounds, including low-level nuclear releases, in water and soil on the arsenal property. As a result, residential development, agricultural activities, use of on-site water for drinking, and consumption of fish and game taken at the RMA were all prohibited. All manufacturing activities at the RMA ceased in the 1980's, and it was placed on the National Priorities List of Superfund sites in July, 1987. As a Superfund site, the arsenal's only mission became environmental cleanup.

In 1996 the final RMA cleanup plan, called the Record of Decision, was signed. Citizens' concerns that potentially hazardous levels of toxic chemicals might be released into the air from contaminated soils during remediation operations led to the incorporation of a medical monitoring program for communities surrounding the RMA to provide periodic health checkups for detection and treatment of any illnesses related to chemical or nuclear releases. After cleanup, the RMA is scheduled to become an urban national wildlife refuge.

Alvin K. Benson

See also: Hazardous waste; Nuclear and radioactive waste; Radioactive pollution and fallout; Soil contamination; Superfund; Water pollution.

Roosevelt, Theodore

Born: October 27, 1858; New York, New York
Died: January 6, 1919; Oyster Bay, New York
Category: Preservation and wilderness issues

Theodore Roosevelt, president of the United States from 1901 to 1909, did more to boost conservation efforts in the United States than any other president.

Theodore Roosevelt was a sickly child of an aristocratic family in New York City. In his youth he

U.S. president Theodore Roosevelt (left) stands at Glacier Point above Yosemite Valley with friend John Muir. Roosevelt's presidency was characterized by consistent effort to conserve American wilderness. (Library of Congress)

collected animals both dead and alive, and he considered becoming a biologist while a student at Harvard. During the 1880's, after his first wife's death, Roosevelt sought consolation in the Dakota Badlands as a hunter and a rancher. Hunting was a passion with Roosevelt, but it was combined with a love of nature and considerable scientific knowledge about birds, animals, and plants. In 1887 he was a founding member and the first president of the Boone and Crockett Club, which, while devoted to hunting, also became one of America's earliest conservation organizations.

If the outdoors was Roosevelt's avocation, politics was his vocation. He held a number of political positions and became a national figure while leading the famous Rough Riders during the Spanish-American War of 1898. After serving as governor of New York, he was elected as vice president of the United States in 1900; when President William McKinley was assassinated in September, 1901, Roosevelt became the youngest president in U.S. history up to that time.

As president, Roosevelt's accomplishments in the field of conservation were momentous. In fact, the word "conservation" in its present sense only came into use during his presidency. It was a crucial time, for many of the nation's natural resources that had seemed inexhaustible to Thomas Jefferson one century earlier had been greatly depleted. Roosevelt had many friends who were equally devoted to nature, and while president he camped with John Burroughs in Yellowstone and John Muir in Yosemite. Unlike Muir, however, Roosevelt was not an environmental preservationist. Roosevelt agreed that much of the wild should be preserved in its natural state, but he and Gifford Pinchot, his chief forester and major conservation adviser, also believed that conservation meant wisely using nature's resources to benefit later generations.

Promoting water reclamation in the West by building dams and irrigation projects was among Roosevelt's successful presidential acts, as was the establishment of millions of acres of forest reserves (over the opposition of many in Congress) and the creation of fifty-one wildlife refuges. Five new national parks were added, including Colorado's Mesa Verde, Oregon's Crater Lake, and South Dakota's Wind Cave Park. Through the Antiquities (or National Monuments) Act of 1906, Roosevelt created eighteen national monuments, including Devils Tower in Wyoming, Petrified Forest and Grand Canyon in Arizona, and Muir Woods in California.

In 1908, toward the end of his second term, Roosevelt organized a three-day White House conservation conference for the nation's governors. There had been nothing like it before. It not only gave considerable publicity to the conservation cause but also resulted in thirty-six states establishing conservation commissions. In February, 1909, Roosevelt hosted a North American Conservation Conference and proposed an International Conservation Conference, but that dream died when he left the presidency.

Eugene Larson

See also: Burroughs, John; Conservation; Muir, John; National parks; Pinchot, Gifford; Wildlife refuges.

Rowland, Frank Sherwood

Born: June 28, 1927; Delaware, Ohio
Category: Weather and climate

Environmental chemist and activist Frank Sherwood Rowland was the first person to discover that chlorofluorocarbons (CFCs) released into the atmosphere were destroying the protective ozone layer.

In 1952 Rowland received his Ph.D. from the University of Chicago, where he studied under Willard F. Libby, recipient of the 1960 Nobel Prize in Chemistry for developing carbon-14 dating. Rowland became the first chairperson of the Chemistry Department at the University of California at Irvine in 1964. During the 1970's, he began to think about expanding his research into new areas. The use of the stable compounds chlorofluorocarbons (CFCs) as refrigerants and spray can propellants had become widespread during the 1960's. Rowland reasoned that because of their stability, CFCs should persist in the atmosphere for a long time. That led to the question of what happens to them as they accumulate in the atmosphere and enter the protective ozone layer at an altitude of about 12,000 meters (40,000 feet).

About this time, Mario Molina joined Rowland's research group as a postdoctoral fellow and undertook the CFC study. Rowland and Molina quickly discovered that the ultraviolet (UV) light streaming through the atmosphere has enough energy to break a bond in the CFCs, releasing the reactive chlorine atom. Chlorine is known to destroy ozone. Rowland and Molina published a paper in 1974 in the journal *Nature* warning that the increasing use of CFCs could result in the destruction of the ozone layer, allowing damaging UV radiation to reach earth.

Rowland and Molina had no direct proof that the ozone layer was being destroyed, but they called for a ban on CFC production. The CFC industry had grown to $8 billion per year, so Rowland and Molina knew that their proposal would be controversial. Du Pont was the first and largest producer of CFCs, and it led the defense of the industry. They argued that such a large industry should not be dismantled by an unproven hypothesis. As warnings about the possible environmental harm increased, however, public concern led many aerosol packagers to cease using CFCs.

The first evidence of ozone layer destruction came from measurements of Joseph Farman, an English scientist. Farman's data led to the discovery of the seasonal ozone hole over Antarctica. National Aeronautic and Space Administration (NASA) satellite data confirmed what Farman had reported. The CFC industry continued to resist a ban until 1988, when measurements showed a 6 percent overall decrease in the ozone layer. By this time Du Pont had developed alternatives to CFCs, and they agreed to cease production. An international agreement reached in Montreal led to a worldwide ban on CFC production effective January 1, 1996. It had been fourteen years since Rowland's initial warnings about ozone destruction. Rowland had not been content to report his scientific results but had actively advocated for the CFC ban before numerous commissions and government committees. For his work, Rowland received the Nobel Prize in Chemistry in 1995 jointly with Molina and Paul Crutzen, who had shown that nitrogen oxides also destroy ozone.

Francis P. Mac Kay

SEE ALSO: Chlorofluorocarbons; Greenhouse effect; Montreal Protocol.

Runoff: agricultural

CATEGORY: Water and water pollution

Agricultural runoff is water that flows into rivers and lakes from agricultural land and operations. It is one of the major sources of water pollution around the world.

Water from rain or melted snow flows from farmland into streams carrying herbicides, fungicides, insecticides, and fertilizers that farmers have used on the land. Much of the wastes of cattle, hogs, sheep, and poultry that are raised on feedlots flow into nearby streams. Water that returns to nearby rivers and lakes after irrigation may be polluted by salt, agricultural pesticides, and toxic chemicals. These organic materials and chemicals that are carried with soil eroding from agricultural land and transported by water runoff degrade the quality of streams, rivers, and lakes.

Agricultural runoff enters rivers and lakes from farmlands spread over large areas. Since it is a nonpoint source of water pollution, it is more difficult to control the pollution from these sources than pollution discharged from factories and sewage-treatment plants. Agricultural chemicals in the runoff have contaminated surface water in many areas of the United States. Rivers pick up sediment and dissolved salts from agricultural runoff as they flow down to the ocean. Salt concentration in the Colorado River increases from about 40 to 800 parts per million as it flows down from its headwaters to Mexico.

Nutrients such as potassium, phosphates, and nitrogen compounds from organic wastes and fertilizers are carried by agricultural runoff into rivers and lakes. Excessive nutrients stimulate the growth of plants such as pond weeds and

Runoff from agricultural land is often laden with toxic fertilizers and pesticides, which then pollute nearby waterways. (McCrea Adams)

duck weeds, plantlike organisms called algae, fish and other animals, and bacteria. As more grow, more also die and decay. The decay process uses up the oxygen in the water, depriving the fish and other aquatic organisms of their natural supply of oxygen. Some types of game fish such as salmon, trout, and whitefish cannot live in water with reduced oxygen. Fish that need less oxygen, such as carp and catfish, will replace them. If all the oxygen in a body of water were to be used up, most forms of life in the water would die. This condition is known as cultural eutrophication. In the late 1950's Lake Erie, 10,000 square miles in area, was reported to be eutrophic. Through stringent pollution control measures, however, the lake has been steadily improving.

The 1987 amendments to the Clean Water Act were the first comprehensive attempt by the U.S. government to control pollution from agricultural activities. In 1991 the U.S. Geological Survey began to implement the full-scale National Water-Quality Assessment (NAWQA) program. The long-term goals of the NAWQA program were to describe the status and trends in the quality of nation's water resources and provide a scientific understanding of the factors affecting the quality of these resources. In October, 1997, an initiative intended to build on the environmental successes of the Clean Water Act was announced. It focused on runoff from farms and ranches, city streets, and other diffuse sources. The plan called for state and federal environmental agencies to conduct watershed assessments every two years.

Irrigation is necessary for survival in many developing countries. Governments struggle to build advanced agricultural systems, and developments in agriculture have improved food production in the world. However, the growing reliance on fertilizers and other agricultural chemicals has contributed to the pollution of rivers and lakes. Therefore, interest has shifted to farming with reduced use of chemicals. Scientists are developing organic ways to grow food that require less fertilizer and fewer pesticides, while many farmers rotate their crops from year to year to reduce the need for chemical fertilizers. Instead of spraying their crops with harmful pesticides, some farmers combat damaging insects by releasing other insects or bacteria that prey upon the pests. Scientists are also developing genetically engineered plants that are resistant to certain pests.

Other strategies for minimizing pollution caused by agricultural runoff include maintaining buffer zones between irrigated cropland and sites where wastes are disposed, restricting application of manure to areas away from waterways, avoiding application of manure on land subject to erosion, reusing water used to flush manure from paved surfaces for irrigation, constructing ditches and waterways above and around open feedlots to divert runoff, constructing lined water-retention facilities to contain rainfall and runoff, applying solid manure at a rate that optimizes the use of the nitrogen it contains for a given crop, and allowing excess wastewater to evaporate by applying it evenly to land.

Watersheds are areas of land that drain to streams or other bodies of water. Most nonpoint pollution control projects focus their activities around watersheds because watersheds integrate the effects that land use, climate, hydrology, drainage, and vegetation have on water quality. In the United States, the National Monitoring Program was initiated in 1991 to evaluate the effect of improved land management in reducing water pollution in selected watersheds over a six- to ten-year period. Federal agencies involved in this program include the Environmental Protection Agency (EPA), the Department of Agriculture, the U.S. Geological Survey, and the U.S. Army Corps of Engineers. A composite index was constructed and published to show which watersheds had the greatest potential for possible water quality problems from combinations of pesticides, nitrogen, and sediment runoff. Watersheds with the highest composite score have a greater risk of water quality impairment from agricultural sources than watersheds with low scores. A 1997 initiative called for the EPA to develop and implement water quality criteria for nitrogen and phosphorous by the year 2000. These criteria would help states set site-specific standards to control nutrient pollution and thus reduce nutrient loading to rivers and lakes.

G. Padmanabhan

SUGGESTED READINGS: *Clean Water* (1992), by Karen Barss, explains the impact of agriculture on water pollution in simple language. *The Clean Water Initiative* (1998), by Claudia Copeland, addresses the concern over the nonpoint-source pollution. *Water Pollution* (1990), by Kathlyn Gay, discusses eutrophication and examines new strategies to combat water pollution.

SEE ALSO: Agricultural chemicals; Erosion and erosion control; Irrigation; Pesticides and herbicides; Water pollution; Watersheds and watershed management.

Runoff: urban

CATEGORY: Water and water pollution

Runoff, often referred to as stormwater, is rain or melted snow that flows over the ground and into rivers and streams. In urban areas, runoff can contribute to environmental problems such as flooding and water pollution. Environmental laws, including the Clean Water Act and the Endangered Species Act, require that cities and other urban governments reduce the quantity and increase the quality of the runoff that enters rivers and streams.

When rain falls on natural landscapes, much of it is caught by vegetation or soaks into the ground. A coniferous forest, for example, can intercept as much as 50 percent of annual rainfall. The rain that reaches the ground percolates into the soil and makes its way into the groundwater or travels slowly through the soil to reach the nearest stream hours, days, or even months later.

In developed areas, rainwater falls on impervious surfaces, roofs, roads, and other nonporous materials, where it is prevented from soaking into the ground. It then flows across those surfaces in large quantities, collecting and transporting pollutants, until it reaches a storm sewer, stream, or natural area. This large amount of water flowing into streams almost immediately after a storm can cause local flooding. Research has shown that the number of small floods increases up to ten times when a watershed reaches 20 percent urbanization. Conversely, since less water is allowed to enter the soil and groundwater, stream flows are greatly reduced in the dry season. This rapid fluctuation of water levels causes stream erosion and siltation and destroys habitat for fish and other aquatic life.

The pollution carried by urban runoff is usually called nonpoint-source pollution—pollution that is not easily traced to a particular point, such as the end of a pipe, but instead comes from all over the environment. Automobiles, combined with impervious surfaces, are responsible for a large part of the pollution in runoff. In addition to motor oil that is improperly dumped down storm drains, oil and other automotive fluids leak onto parking lots and roads and are picked up by runoff. Heavy metals, such as zinc and copper, accumulate from the dust of tire and brake wear and can be a major pollutant in urban streams. Lawns can be relatively impervious as well, and heavy storms or excessive watering can wash pesticides, herbicides, and fertilizers into streams, disrupting the already damaged aquatic environment. Airborne pollutants that fall on impervious surfaces, substances dumped on roads or in storm drains, bird and pet wastes, and street litter also contribute to the toxic mix of stormwater runoff.

An additional runoff problem is erosion of disturbed soil in areas of active building and development. Sediment can be washed into storm drains or directly into streams. This sediment covers gravel critical to aquatic insects and fish and fills the stream channel, contributing to downstream flooding.

The first attempts at urban stormwater management consisted of gutters and ditches to convey the excess water quickly to the nearest natural waterway. In 1869 famed landscape architect Frederick Law Olmsted, Sr., first used underground pipes to convey the muddy, horse-manure-laden runoff from the streets of the new community of Riverside, Illinois, to the nearby Des Plains River. Many cities then developed systems that combined storm sewers with sanitary sewers, but as communities grew and the acreage of impervious surfaces increased, these systems often became overloaded and then overflowed, sending stormwater and raw sewage into rivers and streams. These cities are now working to separate the storm and sanitary sewer systems, with engineering emphasis placed on storing or detaining runoff during storms and then releasing it to the streams later to reduce flooding.

This large-scale stormwater management relies on swales, ponds, and wetlands. Runoff from a town or development can be routed through a swale, a depression in the landscape that directs the runoff to another place or holds it long enough for it to evaporate or soak into the soil. Vegetation in the swale slows the water and also filters out some of the pollutants. A pond or wetland also detains the water and allows some of the pollutants to settle or be filtered by the plants. Some wetland organisms can actually break down oils and other pollutants into harmless elements.

In 1988 the reauthorization of the Clean Water Act required municipalities to do more to control non-point-source pollution, most of which reaches waterways via stormwater. Listing of various species of fish under the Endangered Species Act has also forced some cities to improve fish habitat by limiting pollution and sedimentation in local streams. This requires additional emphasis on pollution reduction, and one new approach is to involve citizens in the effort.

Since non-point-source pollution, and the runoff itself, accumulates drop by drop from many small sources, the small actions of many people throughout the watershed help to solve the problem. Residents are being encouraged to disconnect their roof drains, where appropriate, and divert the water to their own landscapes. In addition, rain barrels and cisterns can detain stormwater to be used for landscape watering and other purposes. Naturescaping, or gardening with native plants that are naturally adapted to the climate and resistant to native pests, requires fewer chemical applications and less supplementary watering. The careful use and disposal of landscape and household chemicals and automotive fluids can reduce pollutants in stormwater. In these ways, residents of a watershed can become part of the solution to the problem of urban flooding and local stream pollution.

Joseph W. Hinton

SUGGESTED READINGS: General-audience books on urban runoff are difficult to find, but one technical source, *Urban Runoff Quality Management* (1998), published by the American Society of Civil Engineers, contains excellent introductory material for the non-scientist. *Stormwater: Best Management Practices and Detention for Water Quality, Drainage, and CSO Management* (1993), by Urbonas and Stahre, is highly technical but again has good general information in the introductory sections. Further information can also be obtained from local city and county sewage or surface water management agencies.

SEE ALSO: Automobile emissions; Clean Water Act and amendments; Flood control; Water pollution; Watersheds and watershed management.

S

Sacramento River pesticide spill

DATE: July 14, 1991
CATEGORY: Water and water pollution

A train derailment near Dunsmuir, California, in 1991 spilled thousands of gallons of the pesticide metam sodium into the Sacramento River. Although the chemical contamination was short-lived, the ecological consequences have been long lasting.

On the night of July 14, 1991, a Southern Pacific train derailed at the Cantara Loop north of Dunsmuir, California, causing a chemical tank car to rupture and spill 19,000 gallons of the pesticide metam sodium into the upper Sacramento River. The Sacramento River pesticide spill, also called the Cantara spill, was California's largest inland ecological disaster. As the metam sodium rapidly mixed with the water, it released highly toxic compounds, virtually sterilizing one of the premier trout streams in California. The green contaminant plume eventually flowed 58 kilometers (36 miles) downstream into California's largest reservoir, Lake Shasta, where a string of air pipes placed at the bottom of the lake aerated the chemical. The aeration project accelerated the breakdown of the metam sodium, reducing toxic components to undetectable levels by July 29, 1991.

Studies were launched to identify and quantify the damage to natural resources. All the aquatic life in the Sacramento River between the Cantara Loop and Lake Shasta was killed. More than one million fish died, including more than 300,000 trout. Millions of insects, snails, and clams perished, along with thousands of crayfish and salamanders. Hundreds of thousands of trees, particularly willows, alders, and cottonwoods, eventually died from the spill, and many more were severely injured. The vegetative damage caused a corresponding dramatic loss of many wildlife species that depend on the river's vegetation for food and shelter. Birds, bats, otters, and mink either starved or were forced to move because their food sources were no longer available.

In addition to the devastating effects on wildlife and plant life, the spill produced many reported human health effects, the loss of recreational opportunities, and substantial economic loss for the residents of the Dunsmuir area. Although there was virtually no trace of the metam sodium about one month after the spill, it became clear that full recovery remained years away.

In 1992 the Department of Fish and Game planted more than 3,400 trees to accelerate the recovery of severely injured vegetation along the river. Some plants, such as elephant ears and torrent sedge, recovered after two growing seasons. During 1994 the trout population reached about one-half of what it was prior to the spill, and trout angling was again allowed on the river. However, strict regulations were established that would allow protection for the recovering wild trout fishery.

By late 1995 ospreys, dippers, sandpipers, and mergansers were all making good progress toward recovery. By 1996 many aquatic and insect populations were nearing numbers that existed prior to the spill; some species, however, particularly clams, snails, crayfish, and salamanders, were struggling to make a comeback. To accelerate recovery, state and federal trustee agencies awarded several million dollars for recovery projects, including research, recovery monitoring, habitat acquisition and restoration, resource protection, and public education.

Alvin K. Benson

SEE ALSO: Pesticides and herbicides; Water pollution.

Sagebrush Rebellion

Category: Land and land use

The Sagebrush Rebellion was a movement that originated in the western United States during the late 1970's and early 1980's that called for state, rather than federal, management of public lands. The rebellion evidenced the growing conflict between environmentalism and the economic exploitation of public lands.

The Sagebrush Rebellion was part of a larger historical trend that began during the nineteenth century in which settlers in the western United States struggled to gain control over public lands. Movements opposing federal management periodically arose, as they would again during the 1990's in the form of the wise-use movement. The Sagebrush Rebellion received national attention during the 1980's in part because of President Ronald Reagan's support and because it was part of a larger movement that demanded the reduction of the federal government's involvement in American life. It represented a response to the environmental movement that gained ground during the 1970's.

The federal government owns nearly 50 percent of the land in the western United States. In response to increased environmental activism, government regulations concerning the use of these public lands significantly increased during the 1970's. Several of these regulations threatened lumbering, cattle grazing, and other commercial activities that were permitted on public lands. Many westerners regarded the increased regulation as unfair, claiming that it stifled development and slowed economic growth in the region. Some maintained that the federal government was limiting opportunities in the West in order to protect established industries in the eastern United States.

In response, these westerners advocated the transfer of federal management to state governments, which were more amenable to development. They formed organizations such as the League for the Advancement of States Equal Rights and Sagebrush Rebellion to advance their cause. During the late 1970's two members of Congress from western states offered legislation at the federal level to transfer land management to the states; however, these initiatives gained little support.

The movement garnered national attention in 1979 when the Nevada legislature passed Assembly Bill 413, commonly known as the Sagebrush Rebellion Act. Nevada's legislators maintained that the huge expanse of federal lands within the state's borders (83 percent of the land in Nevada belonged to the federal government) put it at a disadvantage when compared to other states. They claimed the right to manage the 48 million acres of land then under the authority of the U.S. Bureau of Land Management (BLM), which constituted almost all the federal lands in the state. After passage of the act, Nevadans made no effort to seize control of the public lands. Instead, the state appropriated funds to initiate a lawsuit against the federal government.

Sagebrush rebels soon counted several victories throughout the region. Four western states enacted similar measures, with the Arizona legislature overriding Governor Bruce Babbitt's veto. In 1982 Alaskans showed their support for the so-called Tundra Rebellion when over 70 percent of voters supported efforts to take control of that state's BLM lands. Success was not guaranteed—Sagebrush rebels saw legislative or executive measures supporting their cause fail in seven western states—but the issue was clearly popular with western voters.

The movement gained a powerful boost in 1980 when president-elect Reagan declared himself to be a Sagebrush rebel. The president followed through on his rhetoric, nominating another "rebel," James Watt, for the position of interior secretary. Emboldened by the changed political climate in Washington, D.C., in May, 1981, senator Orrin Hatch of Utah again offered legislation requiring state management of BLM and Forest Service lands. As in the past, the proposal did not gain sufficient support in Congress. The failure of Hatch's legislation indicated that the Sagebrush Rebellion lacked support from key politicians. Although some government officials agreed with the westerners' complaints, others merely made campaign promises that they had no intention of fulfilling.

The Sagebrush Rebellion suffered from several weaknesses. First, the state initiatives offered a potent rallying point for the movement, but few politicians and observers took them seriously. The courts clearly would not approve measures that intruded upon the powers of the federal government. Land transfers had to take place at the federal level, and despite Reagan's stated enthusiasm for the rebellion, key members of his administration opposed the notion of state or federal government ownership of lands. They favored the sale of public lands to private concerns. In the early 1980's Sagebrush rebels allied themselves with environmentalists to thwart the privatization movement.

Observers also noted that the rebellion faced opposition from the very groups that it was intended to assist. In many western states ranchers complained about federal regulation but benefitted from the low cost of grazing on public lands. Mining companies also feared that a transfer to state control might raise their costs and threaten their mineral rights on public lands. Some western state governments estimated that the costs of managing the lands would exceed the income they generated while under state control. Thus, powerful western interests were arrayed against the Sagebrush Rebellion.

The movement was short-lived, lasting only from the late 1970's to 1983, when the debate over privatization effectively killed the movement. If it is assessed by its goals, the Sagebrush Rebellion failed because federal lands remained in the public domain. However, the movement changed the policy debate regarding public lands. It advanced the ethic of multiple-use management, which remained an important issue in public land policy during the 1980's and 1990's. It also provided public support for the Reagan administration's efforts to scale back environmental regulations.

Thomas Clarkin

SUGGESTED READINGS: Robert H. Nelson's *Public Lands and Private Rights: The Failure of Scientific Management* (1995) includes chapters on the Sagebrush Rebellion. William L. Graf's *Wilderness Preservation and the Sagebrush Rebellions* (1990) places the movement in the context of earlier conflicts over public lands in the western United States. R. McGreggor Cawley's *Federal Land, Western Anger: The Sagebrush Rebellion* (1993) offers an in-depth assessment of the movement. John D. Echeverria and Raymond Booth Ely edited *Let the People Judge: Wise Use and the Property Rights Movement* (1995), a collection of articles concerning the conflict over western lands during the 1990's.

SEE ALSO: Antienvironmentalism; Land-use policy; Wise-use movement.

St. Lawrence Seaway

DATE: opened April 25, 1959
CATEGORY: Land and land use

The St. Lawrence Seaway is an international waterway extending 3,769 kilometers (2,342 miles) from Lake Superior to the Atlantic Ocean, providing deep-draft ocean ships with access to the Great Lakes.

The St. Lawrence Seaway, a joint venture between the United States and Canada, is the world's longest waterway. It follows an ancient trade route known to Native Americans as "the river without end." The seaway uses lock systems to raise or lower ships to the appropriate water levels along the route. Each lock is at least 24 meters (80 feet) wide and 8 meters (26 feet) deep. The route from Montreal, Canada, to Lake Ontario contains seven locks. Another eight locks make up the Welland Ship Canal route from Lake Ontario to Lake Erie. The Soo Canals, which connect Lake Huron and Lake Superior, have four locks. The total lift from the mouth of the St. Lawrence River to Lake Superior is approximately 177 meters (581 feet).

The seaway was built before there was much interest in collecting baseline environmental data, making it difficult to assess the effects of operation since it opened on April 25, 1959. The massive project involved the excavation of more than 360 million tons of rock, construction of three dams, and completion of more than 105 kilometers (65 miles) of canals. The seaway can

The St. Lawrence Seaway, which opened on April 25, 1959, allows freshwater and oceanic ecosystems to intermix. Invasions of sea lampreys and zebra mussels have drastically affected the ecology of the Great Lakes. (National Archives)

accommodate ships up to 35,000 tons. The typical operating season is 270 days between April and December, with an average of 2,800 vessel transits. Ice-breaking equipment is used to enhance winter navigation. All tankers using the system must be double-hulled.

During the 1970's two sizable oil spills provided the impetus to develop an oil-spill contingency plan and to more closely monitor environmental impacts stemming from operation of the seaway. In 1989 the plan enabled a successful response after a tanker spilled 14,000 gallons of xylene, a volatile and hazardous solvent.

The seaway allows mixing of the Great Lakes ecosystems with oceanic and other ecosystems. Sea life lurking in cargo holds can enter the river and lake systems. Perhaps the most significant example of an intruding species is the zebra mussels, which arrived in the ballast water of an Eastern European ship during the 1980's and quickly spread throughout the Great Lakes region. Other invaders simply swim unaided into the Great Lakes from the Gulf of St. Lawrence, as in the case of sea lampreys.

The seaway also affects the riverine ecosystem through changes in river flow rates and replacement of natural channel features with canals. The St. Lawrence Seaway, with more than fifty associated ports, has spurred regional development, which has resulted in increased navigation on the river, recreational activities, shoreline development, tourism, point and nonpoint water pollution discharge, and loss of wetlands. Changes in operation of the seaway could have additional environmental consequences. Increased winter navigation would affect icing patterns, which in turn affect riparian habitat, archaeological resources, other natural resources, and the generation of power from dams. Pressure waves from ships disturb the bottom of the river and cause ice damage along the shoreline, including the shearing of plants caught in the ice. Other issues associated with increased operation range from potential effects on wildlife migration patterns to the consequences of shipping accidents.

Robert M. Sanford

SEE ALSO: Lake Erie; Oil Spills; Water Pollution; Zebra Mussels.

Sale, Kirkpatrick

BORN: June 27, 1937; Ithaca, New York
CATEGORY: Ecology and ecosystems

Ecologist and bioregionalist Kirkpatrick Sale's career as a writer and environmental activist has focused on political, economic, and environmental problems, all of which share the common theme of alienation from nature in urbanized cultures.

The first of Kirkpatrick Sale's books to address the interrelation of social and environmental ills was *Human Scale*, published in 1980. He attacked large-scale technology and urbanization because of their dehumanizing effects. He wrote that people are no longer in touch with themselves and nature. *Human Scale* embodies two major premises: Smaller systems are more efficient in producing social and economic outputs, and smaller systems are more humane, providing greater satisfaction for those involved in them.

Human Scale treats social and natural environments as interdependent entities. For Sale, creating a society on a human scale entails factors such as city size, architecture of buildings, food, health, education, energy, transportation, and waste disposal. He argues that all these human concerns have "economies of scale," namely, that quality and efficiency are maximized in smaller rather than larger units. For example, cities with populations under 100,000 are more livable and economically efficient. Small-scale energy resources—water, wind, and sun—enable communities to be less vulnerable to loss of power. Small, diversified farms producing foodstuffs for local communities have less overhead and market higher-quality produce. In addition, garbage is more easily recycled in small communities. According to Sale, human scale equals small size, enabling people to take direct control of their lives and environments.

In *Dwellers in the Land* (1991) Sale explores bioregionalism as an appropriate human scale. He defines bioregionalism as a way of life in which people of a natural region understand their land—its resources, limitations, and lore—and live their lives within the context of that place. Bioregionalism emphasizes self-sufficiency, stability, conservation, cooperation, and decentralization. The "dwellers in the land" become connected with their immediate environment because it is their primary sustenance.

Sale acknowledges that implementation of bioregional communities is a daunting task. One major obstacle is the practical reality that humans are defined not only by natural region but also by race, language, religion, class, and heritage. However, supporters of Sale would argue that bioregional principles enhance conservation and preservation of natural resources and provide an environment conducive to healthy social relationships.

In addition to his writing profession, Sale has extensive experience in political and environmental activism. He has served on the board of directors of Project Work, School for Living, and the American Schumacher Society. He cofounded the New York Green Party and has been involved with the Wetlands Restoration Project on the Lower Manhattan waterfront.

Ruth Bamberger

SEE ALSO: Bioregional Project; Ecology; Schumacher, Ernst F.; Social ecology; Sustainable development.

SANE

CATEGORY: Nuclear power and radiation

The National Committee for a Sane Nuclear Policy (SANE), a group of antinuclear activists and prominent scientists, was founded to focus attention on the effects of nuclear testing on the environment.

In April, 1957, prominent antinuclear activists and pacifists, led by *Saturday Review* editor Norman Cousins, met in Philadelphia; the ad hoc organization formed at this meeting evolved into SANE. In September, the group's officials adopted the name "The National Committee for a Sane Nuclear Policy" for the organization. The formation of SANE was announced in *The New York Times* on November 15, 1957, in an advertisement with the headline "We Are Facing a Danger Unlike Any Danger That Has Ever Existed." The advertisement, which was sponsored by such notables as theologian Paul Tillich, social critic Lewis Mumford, novelist James Jones, and humanitarian Eleanor Roosevelt, brought the group many new members and donations.

SANE was at first an informal venture meant to make the public aware of the dangers of nuclear tests, but it proved so popular that it became a permanent organization. The group's initial membership consisted largely of scientists, writers, and other professionals, but as the antitesting movement grew, thousands of other people joined. By mid-1958, SANE had approximately twenty-five thousand members and more than a hundred chapters.

Despite Cold War opposition, SANE and other groups pressed their campaign against testing. In August, the United States followed the lead of the Soviet Union and voluntarily suspended nuclear tests. The two countries further agreed to meet in Geneva, Switzerland, to begin negotiations for a test-ban treaty.

In the early 1960's, SANE came under congressional attack for its alleged harboring of communists. Investigations by SANE leaders established that some members did in fact have communist affiliations, and though congressional investigations eventually exonerated SANE's leadership of wrongdoing, the effects of the investigations on SANE were disastrous. Membership declined drastically, and several prominent members and sponsors resigned.

In 1961, the Soviet Union resumed the testing of nuclear weapons; SANE condemned the resumption of the tests and called for international protests. The organization also urged the United States to refrain from following the Soviet example, to no avail.

Such concerns became secondary during the Cuban Missile Crisis of October, 1962, when the world stood on the brink of nuclear war. In the aftermath, the test-ban treaty gained new life, and the Limited Test Ban Treaty was signed in July.

With the signing of the treaty, SANE's initial goal was partially achieved. SANE leaders subsequently chose to direct the group's energies to opposing the Vietnam War, but this approach proved divisive; in 1967, many key officials, including Cousins and the executive director Donald Keys, resigned from SANE. SANE's membership and influence subsequently declined precipitately, and in 1969 the group changed its focus to campaign against the construction of antiballistic missile systems. Yet a resurgence of the organization did not happen until the early 1980's, when the nuclear freeze movement gathered momentum. In the late 1980's, SANE merged with another organization, Freeze, to become SANE/Freeze, which resulted in the largest peace organization in American history.

Alexander Scott

SEE ALSO: Antinuclear movement; Nuclear testing.

Santa Barbara oil spill

DATE: January, 1969
CATEGORY: Water and water pollution

In January, 1969, Santa Barbara and other small communities along the Pacific coast were confronted by a devastating oil slick. The source of the polluted water was a drilling well in the Santa Barbara Channel that suffered a blowout.

Santa Barbara is an old Spanish mission town situated between the Pacific Ocean and the Santa Ynez Mountains in Southern California. Minor oil pollution at places along the coast is nothing new for residents of the city. Oil has been escaping from natural fractures in the ocean floor for thousands of years. Native Americans reportedly caulked their canoes with oil-based substances found near Coal Oil Point.

The site of the 1969 accident was Union Oil Company's oil-drilling platform A, less than 13 kilometers (8 miles) offshore in the Santa Barbara Channel. Gas-charged oil escaped from below the metal well casing (a pipe set in the well bore) before the blowout preventers could be closed. According to Joel Watkins and others in *Our Geological Environment* (1975), ocean water east of the platform "boiled" violently for several hours after the preventers were closed. Some reports suggested that the oil leaked to the surface along an unmapped fault. More than 11,000 tons of free oil reached the surface of the water and rapidly spread over an area of 200 square kilometers (80 square miles). An estimated 10,000 barrels of crude oil eventually reached the shoreline and coated gravel-sized beach rocks and rapidly infiltrated the sand. The light-colored Goleta Cliffs a short distance north of Santa Barbara were marked by a black band of gooey oil.

The spreading oil slick threatened marine life in the area, including porpoises, seals, whales, birds, and fish. Numerous oil-soaked birds perished, but many were saved at bird hospitals where the oil was removed with a solvent. U.S. Navy personnel expressed concern that the oil would harm porpoises near the base at Point Mugu, California.

Remediation commenced immediately and included burning the oil, application of chemical dispersants, skimming the oil, steam cleaning and vacuuming of beach areas, and spreading plant material along the shoreline. The most effective technique was the distribution of straw and other plant material along the beaches to absorb the oil. Many residents of Santa Barbara, including student volunteers from the University of California at Santa Barbara, worked many hours spreading and collecting the oil-soaked plant material. This material was taken away to be burned or buried.

As a result of the spill, Union Oil Company and three partners (Gulf, Mobil, and Texaco)—as well as the Interior Department of the federal government—were sued for damages by the state of California and several coastal communities. The claims ranged from $500 million to approximately $1.3 billion. The accident and ensuing pollution had major political repercussions. Drilling was temporarily halted in the channel, and more stringent regulations were imposed on oil companies drilling in offshore areas.

Donald F. Reaser

SEE ALSO: Mexico oil well leak; Ocean pollution; Oil spills.

Save the Whales Campaign

DATE: initiated in 1971
CATEGORY: Animals and endangered species

The Save the Whales Campaign is a sustained effort by a variety of people and organizations to keep whales safe by banning commercial whaling and ending the activities of pirate whalers and whale meat smugglers.

The unchecked commercial hunting of whales over several centuries in both open oceans and coastal waters caused the depletion of whale populations worldwide. Once harvested for their oil, whales later became a prize for their meat, considered to be a delicacy in Japan and a few other countries. Species after species approached extinction because of rampant overhunting.

The International Whaling Commission (IWC) was founded in 1946 to prevent further exploitation but instead presided over some of the worst excesses in whaling's history. In 1971 the Animal Welfare Institute (AWI), as well as several other organizations, launched the Save the Whales Campaign in an effort to end the harvesting of whales. Pursuing the theme of this campaign, the United Nations Environmental Conference in 1972 called for a ten-year moratorium on commercial whaling. In 1974 the IWC decided to regulate whaling according to the principle of maximum sustainable yield. Whenever a species of whale dropped below the optimal population for such a yield, the IWC instituted a ban on hunting that species so that the population could recover. In 1974 the blue whale, bowhead whale, and right whale populations reached low levels and were immediately protected.

Because of difficulties of obtaining reliable data and enforcing catch limits, in 1982 the IWC was encouraged by the Save the Whales Campaign to adopt an indefinite moratorium on commercial whaling, which took effect in 1986 as the International Whaling Ban. During the 1990's the philosophy of the Save the Whales Campaign expanded to include not only the conservation of the whale population but also the issues of animal rights and the aesthetic value of observing whales. Many people worldwide simply believe that it is wrong to kill and eat such large, unique animals, while the World Wildlife Fund and others have pointed out that whales must be preserved for observation because of their intrinsic value as mammals of great intelligence.

In opposition to the Save the Whales Campaign and the IWC, Japan, Norway, Russia, and Iceland resumed whaling of the minke species in the early 1990's, citing their rights to refuse specific IWC rulings. These countries based their decisions on information submitted by the Scientific Committee of the IWC stating that the minke population was large enough to absorb sustainable exploitation. However, a deoxyribonucleic acid (DNA) test of whale meat imposed by the IWC in 1995 revealed that Japan was also harvesting some fin and humpback whales. At its May, 1995, meeting, the IWC strongly censured the continued whaling activities of these countries and took serious steps toward stopping pirate whalers and the illegal trade of whale meat. Although reliable data are difficult to acquire, the Save the Whales Campaign has made a definite impact on the whale population. Many species appear to be recovering. For example, the bowhead whale was upgraded from endangered to vulnerable in 1995.

Alvin K. Benson

SEE ALSO: Greenpeace; International Whaling Ban; International Whaling Commission; Whaling.

Scenic Hudson Preservation Conference v. Federal Power Commission

DATE: 1965
CATEGORY: Preservation and wilderness issues

This U.S. Court of Appeals decision reversed a Federal Power Commission (FPC) order that had granted a license to Consolidated Edison Company (Con Ed) to construct a hydroelectric project on the Hudson River at Storm King Mountain to supply New York City's electric power needs.

In the early 1960's Scenic Hudson, a coalition of conservation and environmental groups, and the towns near Storm King Mountain in New York challenged Con Ed's license to construct a storage reservoir, a powerhouse, and transmission lines on the Hudson River in the region. However, the FPC denied their motions to intervene and reopen the hearings to consider evi-

A hiker sits on a hill above the Hudson River near Storm King Mountain, where the Consolidated Edison Company sought to build a hydroelectric plant during the 1960's. (Douglas Long)

dence on alternatives to the Storm King facility, the cost and practicality of underground transmission lines, and the feasibility of fish protection devices.

The U.S. Court of Appeals dismissed the FPC's argument that Scenic Hudson lacked standing to seek judicial review of Con Ed's license because it had made no claim of personal economic injury. Scenic Hudson clearly had economic interests to protect. The Storm King reservoir would inundate 27 kilometers (17 miles) of hiking trails, and the aboveground transmission lines would decrease the value of public land and tax revenues of private land and would interfere with community planning. Scenic Hudson also had noneconomic interests that would be affected by the Storm King project that were not barred from judicial review. The Court of Appeals held that the U.S. Constitution did not require an aggrieved party to have a personal economic interest, nor did the Federal Power Act (the FPC governing legislation), which permits "any aggrieved party" to obtain judicial review of an FPC order.

In terms of the licensing decision, the FPC had acknowledged that the "principal issue . . . is whether the project's effect on the scenic, historical and recreational values are such that we should deny the application." Since the FPC's standing argument was at odds with its recognition of the importance of noneconomic issues in its licensing decision, the court broadly read the statute to grant standing to organizations, such as Scenic Hudson, that had special interests in "aesthetic, conservational, and recreational" issues to act as private attorneys general to ensure that the FPC would protect the public interest. Since Scenic Hudson was an aggrieved party, the court read the statute's language, which gave the FPC the discretion to admit as a party "any person whose participation in the proceeding may be in the public interest," to create "an absolute right of intervention."

After the appellate court settled the standing issue, it found that the Federal Power Act gave the FPC a specific planning responsibility for the development of waterways for commerce, water power, and recreation. The statute's planning responsibility required the agency to weigh these three factors in making a licensing decision. Then the court broadly read the FPC's responsibility to consider recreational objectives to include "the preservation of natural resources, the maintenance of natural beauty, and the preservation of historic sights." In fact, the FPC regulations required a recreation plan.

The FPC, in granting Con Ed a license for the Storm King project, did not fulfill its statutorily mandated planning duties because it failed to develop a complete record. First, the FPC did not explore alternatives to the Storm King project, including the use of gas turbines or the purchase of power from interconnections with other electric utilities, nor did the agency require Con Ed to supply this information. Therefore, the agency was unable to determine whether Con Ed needed Storm King to meet its future power needs. Second, the FPC did not seriously consider the cost of underground transmission routes but accepted Con Ed's estimates on their cost even though the estimates had been questioned by its staff and hearing examiner. As a consequence, the agency was unable to weigh "the aesthetic advantages of underground transmission lines against the economic disadvantages." Third, there was sufficient evidence before the FPC on the danger of the Storm King project to fish to make further inquiries. Still the agency failed to make these inquiries and even dismissed as "untimely" the petitions of several groups to intervene and present evidence on Storm King's destructive impact on the spawning grounds for striped bass and shad and the unavailability of any powerhouse screening devices to protect fish eggs and larvae.

The Court of Appeals set aside the FPC's licensing order because the record had failed to support the agency's decision and then remanded the case with instructions to reexamine the alternatives to Storm King, the cost of underground transmission lines, and the fish spawning issue. In sum, the Scenic Hudson opinion was a bold declaration of judicial authority that granted environmental groups with noneconomic interests the right to intervene in federal regulatory agency licensing decisions and held a federal regulatory agency responsible for engaging in its own evidentiary inquiry, considering alternatives to granting a license, and weighing the environmental impacts of its decisions.

The FPC subsequently held hearings and issued Con Ed a license to build the Storm King project, which the federal Court of Appeals affirmed. However, the Scenic Hudson case was only the beginning of a two-decade controversy over the Storm King project. Further litigation on its threat to fish life expanded to include six other power stations along the Hudson River once the newly created Environmental Protection Agency conducted extensive hearings under the Federal Water Pollution Control Act amendments of 1972. The controversy was not resolved until 1981, when seventeen months of mediation led to Con Ed's agreement to forfeit its Storm King license and donate the site as a public park.

William C. Green

Suggested readings: Allan Talbot, *Power Along the Hudson: The Storm King Case and the Birth of Environmentalism* (1972), provides a valuable study of the case and its immediate aftermath. The subsequent controversy over Storm King and the details of the settlement are explored in Glenn A. Temple, "Peace at Storm King," *EPA Journal* 7 (1981), and Allan Talbot, *Settling Things: Six Case Studies in Environmental Mediation* (1983). Thomas Hoban and Richard Brooks, *Green Justice: The Environment and the Courts* (1987), provides a shorter study of the case with a set of study questions. The text of the case appears in the *Federal Reporter* 354 (1965).

See also: Dams and reservoirs; Hydroelectricity; Power plants.

Schumacher, Ernst Friedrich

Born: August 16, 1911; Bonn, Germany
Died: September 4, 1977; on a train en route to Zurich, Switzerland
Category: Philosophy and ethics

German-British economist Ernst Friedrich Schumacher became an important voice in environmental discourse when he published Small Is Beautiful: Economics as if People Mattered *in 1973.*

Published first in Great Britain, *Small Is Beautiful* found little following in England and Europe, but its U.S. publication in 1974 led Schumacher to become a guru to millions of environmentalists. His central theme, that people must replace materialistic values with nonmaterialistic ones, resonated with many Americans, especially college students. By the end of 1974 Schumacher had booked hundreds of lecture appearances, most of them in the United States.

Much of what is contained in *Small Is Beautiful* was foreshadowed by developments in Schumacher's life during the 1960's. At the time, Schumacher was the principal economic consultant to the National Coal Board in England. The German-born Schumacher had served as an adviser to British government officials during the 1940's. From 1946 to 1950 he had assisted economic planning in British-occupied West Germany, and his work there gained the attention of famed economist John Maynard Keynes, who recommended him to the National Coal Board in 1950.

Schumacher remained with the Coal Board for more than twenty years and served as its statistical director from 1963 to 1970. In the midst of this fairly routine career as an expert economist, Schumacher experienced a spiritual crisis that changed the focus of his life. While visiting Burma as an economic consultant in the early 1960's, Schumacher became enchanted with Buddhism. He began to speak of "Buddhist economics," by which he meant the combination of spiritual harmony and material well-being. Although he eventually determined that Buddhism was not applicable to Western culture (he became a Catholic), the influence of Buddhist philosophy is evident in *Small Is Beautiful.*

The main theme of *Small Is Beautiful* is that modern industrial production, with its emphasis on the most advanced technology, destroys the creativity and dignity of the worker. Schumacher therefore urges the use of intermediate or "appropriate technology" as a way of giving employees a greater sense of satisfaction. He advocates small-scale, regional industry rather than huge national or international corporations.

Throughout *Small Is Beautiful* Schumacher concentrates on the needs of people as opposed to an exclusive concern about the environment. Therefore, he is categorized as a social ecologist rather than a pure or "deep" environmentalist. Schumacher particularly encouraged this regional approach to job creation for developing countries. It would give people a sense of achievement and encourage them to protect their resources while gradually increasing a country's pool of skilled workers. In addition, it would slow the drift to urban areas, which leads to greater human misery and more pollution of the air and water. Schumacher's ideas proved unrealistic for Third World development, but they inspired greater regionalization of industry in the more environmentally conscious industrialized economies during the 1970's and 1980's.

Ronald K. Huch

SEE ALSO: Environmental economics; Renewable resources; Social ecology; Sustainable development.

Sea Empress oil spill

DATE: February, 1996
CATEGORY: Water and water pollution

The Sea Empress *oil tanker released approximately 79,000 tons of crude oil when it was grounded off the coast of Milford Haven in southwest Wales. It was the third-largest tanker spill in United Kingdom waters, causing adverse effects to a number of wildlife species and considerable difficulties for the fishing industry within the region.*

On February 15, 1996, the *Sea Empress* oil tanker ran aground upon a wave-exposed, current-scoured section of coastline near Milford Haven, Wales. Over the next six days, the vessel released more than 75,000 tons of North Seas light crude and about 400 tons of heavy fuel oil. Still leaking oil, the *Sea Empress* was recovered and towed into Milford Haven on February 21.

Weather and wave conditions that prevailed at the time of the spill facilitated the spread of oil

The Sea Empress *tanker lies aground off the coast of Wales. The damaged vessel released approximately 79,000 tons of oil into the ocean, negatively impacting local wildlife populations and fishing operations.* (AP/Wide World Photos)

slicks well beyond the immediate area of grounding. Heavy oil slicks drifted into Milford Haven, as well as north and south along the open Pembrokeshore coast. During the early weeks after the incident, oil was observed across a wide area of the Bristol Channel. More distant shores that were affected included those around Lundy Island and the southeast coast of Ireland.

Initially, the three main concerns were to establish the size of the area affected by oil, plan cleanup measures, and determine how badly individual shellfish, fin fish, and other wildlife had been contaminated. Shortly after the spill, a fishing exclusion order was applied to the affected region, banning the catching of any fish within a designated area. Some of the spilled oil was mechanically sucked up at sea. Between February 17 and 25, large amounts of chemical dispersants were used to break up the oil into small droplets in order to reduce the risk to the coastline and to birds at sea. Beaches were primarily cleaned using mechanical methods, but some dispersants were used to remove weathered oil from rocks next to selected beaches. The main recreational beaches were cleaned up by mid-April, allowing visitors to again enjoy them.

Since the fin fish were found to have little to no contamination, the ban on catching salmon and sea trout was lifted in May, 1996. The shellfish, however, were more heavily contaminated and recovered more slowly. Research studies assessing the impact and recovery for a range of key commercial fish species, particularly those that are important for food chains, was conducted between 1996 and 1998. By June, 1996, more than 6,900 oiled birds of at least twenty-eight species had been recovered dead or alive, and more than three thousand were cleaned and released.

Approximately one-third of the spilled oil evaporated from the sea surface, but because of a combination of natural and chemical dispersion, approximately 50 percent of the spill volume dispersed into the water column. The ultimate fate of this dispersed oil had yet to be assessed by the end of the 1990's. Water samples analyzed in 1997 showed low levels of total hydrocarbons, but the effectiveness of the cleanup operation needed further assessment.

Alvin K. Benson

SEE ALSO: *Amoco Cadiz* oil spill; *Braer* oil spill; *Exxon Valdez* oil spill; Oil spills; Tobago oil spill; *Torrey Canyon* oil spill.

Sea Shepherd Conservation Society

CATEGORY: Animals and endangered species

The Sea Shepherd Conservation Society is an aggressive group dedicated to the protection of marine life.

The Sea Shepherd Conservation Society was founded by Paul Watson, an early member of Greenpeace who had grown dissatisfied with that organization's unwavering commitment to nonviolence. Watson argued that the use of nonharmful force, especially against property, was often justifiable to protect animals and the environment. When he acted on his beliefs by physically interfering with the slaughter of baby seals in the Arctic, he was expelled from Greenpeace.

In August, 1977, therefore, he and several friends established Earthforce, a Vancouver-based organization dedicated to the use of "direct action" to protect animal life; the group's first undertaking was the production of a documentary film chronicling the slaughter of elephants in East Africa. Earthforce quickly encountered financial problems, and Watson turned for assistance to Cleveland Amory, a prominent writer who had founded the Fund for Animals in 1967. Amory gave Watson $120,000 to buy a retired fishing ship. With additional financial help from the British Royal Society for the Prevention of Cruelty to Animals, Watson outfitted the ship with supplies and a crew. The vessel was rechristened the *Sea Shepherd,* a name taken from the sixteenth century poem *The Faerie Queene,* and the organization also took the name.

The revitalized group then helped to inaugurate a new era of environmental activism, repeatedly intervening to protect baby seals and even ramming and sinking whaling ships and drift-net fishing trawlers. By the 1990's, the organization claimed more than fifteen thousand members

The Sea Shepherd Conservation Society's ship Sirenian, *a former U.S. Coast Guard cutter. The vessel is only one of several in the environmental organization's small fleet of ships used to combat seal and whale hunting.* (Reuters/Anthony Bolante/Archive Photos)

and was operating a fleet of vessels; it also had established its own intelligence agency. While critics portrayed the group's members as fanatics or terrorists, Watson reiterated that the actions of Sea Shepherd agents had never harmed a human being.

Robert McClenaghan

SEE ALSO: Amory, Cleveland; Animal rights; Commercial fishing; Drift nets and gill nets; Endangered species; Fish kills; Greenpeace; International Whaling Ban; Marine Mammal Protection Act; Save the Whales Campaign; Whaling.

Seabed disposal

CATEGORY: Waste and waste management

Debris, containers of toxic waste, and abandoned equipment dumped into the open sea settle on the ocean floor. Chemicals, heavy metals, and other harmful wastes are sometimes buried in the seabed. Both types of disposal disrupt the balance of the ocean ecosystem and may jeopardize ocean life.

The ocean covers more than 75 percent of the planet. People have long believed that the vast ocean waters would dilute, distribute, and render harmless any garbage, sewage, or other debris dumped into it. As space on land for dumps and landfills becomes more scarce, many have looked to the ocean and its tremendous seabed as an alternative for waste disposal. However, scientists and marine biologists have learned that the ocean can be harmed by waste disposal. Like the land, which is a network of many different biospheres, the ocean is a vast ecosystem serving a variety of marine life. For many marine creatures the balance is delicate, and any change in water quality, temperature, or food source is harmful.

In an effort to protect marine creatures and the balance of the ocean ecosystem, many countries have worked together to stop ocean pollution and open-ocean dumping. The International Convention for the Prevention of Marine Pollution (MARPOL) outlaws the dumping of plastic items anywhere in the ocean. It also regulates how far from shore other debris can be legally dumped. Much of this debris eventually settles on the seabed.

Scattered across the seabed are also containers of hazardous waste, chemicals, and radioactive waste. At one time scientists believed that the cold environment of the ocean floor was sufficient to dispose of sealed containers of such waste. They argued that this waste would not come into contact with other marine life, and radioactive waste would be safe in the extremely cold temperatures of the deep ocean. Dump sites for radioactive waste are located far north in the Atlantic Ocean and in the Arctic Ocean. After World War II several thousand tons of German chemical weapons, undetonated bombs, and other equipment were dumped in the waters off the coasts of Germany, Denmark, Norway, Sweden, and Poland. In 1972 an international agreement known as the London Dumping Convention banned ocean dumping of high-level radioactive waste. In 1994 this ban was expanded to cover low-level radioactive waste. Despite this agreement, Russia continued to dump large amounts of radioactive waste into the Barents and Kara Seas.

Research has shown that dumping of sealed containers of chemicals even in deep ocean waters poses threats to the environment. For example, several industrial waste sites were created in the middle of New England fishing grounds and used for twenty years. Though toxic and radioactive waste is no longer dumped there, the areas are still marked as hazardous because of the incredible staying power of these chemicals. Unlike organic matter, chemicals do not break down quickly over time. Despite warnings, fishermen still harvest fish, shrimp, and shellfish from these areas. In a joint effort among several state and federal agencies, seafood from these areas and the surrounding waters was collected and tested for dangerous levels of hazardous or toxic chemicals. Tests performed by the Food and Drug Administration (FDA) showed only trace contaminants in seafood—none had levels high enough to pose threats. When the anchor was pulled aboard at one location, however, the tip had traces of radioactive waste high enough to set off the sensors worn by the participants.

A video survey was taken of one site, known as the Foul Area, 29 kilometers (18 miles) off the coast of Boston, Massachusetts. The 3.2-kilometer (2-mile) expanse is officially named the Massachusetts Bay Industrial Waste Site. From 1953 to 1976 it served as a dumping ground for toxic and radioactive materials. The survey, taken in 1991, showed almost one hundred objects scattered across the ocean floor at eighteen separate sites. Of these objects, sixty-four were identified as cement containers. More than one-half of them had broken open over time.

Some scientists now advocate subseabed burial as an alternative to seabed dumping. Though proposed mostly for high-level radioactive material, subseabed disposal is argued to be safer than core-drilled burial on land. Proponents argue that the deep seabed is one of the most geologically stable places on Earth. They also argue that the sticky mud and clays found in the mid- and deep-ocean basins cause radioactive particles to cling to or bind with them, keeping them from migrating throughout the ocean waters. Tests conducted between 1974 and 1986 by an international team of scientists support these arguments. Opponents claim that retrieval of subseabed disposals would be more difficult and costly than those on land. They also question the types of containers to be used and how to safely transport the waste to the seabed. While proponents were making progress in addressing these concerns, research funding was cut off in 1986 to focus on land-based disposal. Though the London Convention has prohibited any dumping of radioactive waste at sea since 1994, subseabed disposal remains ambiguous, as does its effect on the ocean environment.

Lisa A. Wroble

Suggested readings: The issue of seabed disposal and dumping of heavy metals and chemicals into the ocean is addressed in *Environmental*

Hazards: Marine Pollution (1993), by Martha Gorman. The devastation of ocean life associated with human activities, including ocean dumping, is covered in *Ocean Pollution: Effects on Living Resources and Humans* (1996), by Carl J. Sindermann. The deep-ocean dumping of chemicals and its environmental effects is covered in *Endangered Species and Habitats: The Ocean* (1998), by Lisa A. Wroble.

SEE ALSO: Nuclear and radioactive waste; Ocean dumping; Ocean pollution.

Secondhand smoke

CATEGORY: Human health and the environment

Secondhand smoke consists of a mixture of smoke that has been exhaled by smokers and the smoke produced by the burning end of a cigarette, cigar, or pipe. Containing more than four thousand substances, forty of which are known carcinogens (cancer-causing agents), secondhand smoke is a strong respiratory irritant. When concentrated indoors, secondhand smoke becomes a significant pollutant that greatly reduces air quality.

It has been known since the early 1960's that smoke posed serious health risks to smokers. Numerous U.S. Public Health Service reports cited cigarette smoking as the largest single preventable cause of premature death in the United States. Research examining the effects of secondhand smoke, also termed environmental tobacco smoke (ETS) and passive smoke, followed later.

In 1986 the surgeon general of the United States reviewed the work of more than sixty physicians and scientists from the United States and elsewhere relating to secondhand smoke. A report published by the U.S. Department of Health and Human Services called *The Health Consequences of Involuntary Smoking: A Report of the Surgeon General* (1986) listed three conclusions: Involuntary smoking causes disease in nonsmokers; children exposed to secondhand smoke have an increased frequency of respiratory infections and respiratory symptoms, as well as a reduced rate of lung function capacity as the lung matures; and separating smokers from nonsmokers sharing the same airspace may reduce the exposure of nonsmokers to secondhand smoke but does not eliminate their exposure.

An examination of the contents of secondhand smoke reveals a selection of very strong chemicals. These chemicals are present in two forms: gases and particles. The major gaseous toxins in secondhand smoke include carbon monoxide, carbonyl sulfide, benzene, formaldehyde, hydrogen cyanide, and nitrogen oxides. The presence of carbon monoxide, which is also present in car exhaust, is a cause for particular concern. When inhaled, carbon monoxide interferes with the blood's ability to carry oxygen to the cells of the body. Red blood cells normally carry oxygen from the lungs to the cells. When carbon monoxide is present, it binds to red blood cells and makes them incapable of taking on the oxygen. The brain, heart, and other tissues do not get the oxygen they need.

The chemicals in secondhand smoke that are in particle form include tar, nicotine, and phenol. Tar is considered to be carcinogenic. Nicotine is toxic, and phenol promotes tumor growth. At least twelve other particulate chemicals in secondhand smoke are carcinogenic, including catechol, benz(a)anthracene, quinoline, cadmium, nickel, and polonium 210. So many cancer-causing agents are found in secondhand smoke that the U.S. Environmental Protection Agency (EPA) has classified it as a Group A carcinogen, which means that strong evidence of a cause-and-effect relationship in humans exists. In others words, there is strong evidence that secondhand smoke causes cancer in humans.

Being in the presence of secondhand smoke produces immediate consequences. Burning sensations in the eyes, nose, and throat; increased phlegm, heart rate, and blood pressure; and headaches and stomachaches are common reactions. When exposed to secondhand smoke over time, the risk of contracting the following problems increases: lung problems, cancer, heart attacks, and brain attacks (also known as strokes). Secondhand smoke is responsible for approximately three thousand lung cancer deaths annually in nonsmokers in the United

States and thirty-two thousand to forty thousand deaths from heart disease.

Infants and young children are especially susceptible to the dangers of secondhand smoke. Children whose parents smoke in the home experience a significantly higher risk of contracting lower respiratory tract infections and asthma. Bronchitis, tracheitis, and laryngitis are three acute respiratory illnesses that children of smokers suffer more frequently that do the children of nonsmokers. Children whose parents smoke also have more difficulty with chronic (persistent and ongoing) coughs and increased phlegm than do the children of nonsmokers. Children who suffer from asthma and are exposed to secondhand smoke show an increase in the severity of their symptoms and the frequency of their attacks.

The human and economic consequences of secondhand smoke are far-reaching. Exposure to secondhand smoke has led to increases in human illness, suffering, and medical expenses. Lost wages caused by secondhand-smoke-related illnesses also contribute to the economic costs. The overwhelming body of evidence pointing to the dangers of secondhand smoke for nonsmokers has led to a flurry of public outcry and new regulatory laws. As of August, 1997, 80 percent of U.S. employees worked for companies that had a smoking policy, and the Occupational Safety and Health Administration (OSHA) was in the process of reviewing a proposed rule that would ban smoking in the workplace except in specially ventilated, separate areas. In the late 1990's communities across the United States continued to struggle with the issue of smoke-free restaurants and community buildings.

Louise Magoon

SUGGESTED READINGS: For an in-depth overview of the research regarding secondhand smoke, see the U.S. Department of Health and Human Service publication *The Health Consequences of Involuntary Smoking: A Report of the Sur-*

Flight attendants and their lawyers celebrate in Miami, Florida, in October of 1997 after winning a lawsuit against the tobacco industry. Nonsmoking flight attendants sought damages for illnesses caused by secondhand smoke on airliners. (AP/Wide World Photos)

geon General (1986). Carol M. Browner's *Environmental Tobacco Smoke: EPA's Report* (1993), which is available on the Internet at http://www.epa.gov/docs/epajrnal/fall93/brown.txt.html, provides a concise description of the issue of environmental tobacco smoke. Although it is written in formal scientific language, Dr. James Pirkle, "Exposure of the U.S. Population to Environmental Tobacco Smoke," *Journal of the American Medical Association* (April 24, 1996), examines the extent of exposure to environmental tobacco smoke experienced by the American public. A small pamphlet produced by the Indiana State Department of Health, "Secondhand Smoke—It's No Joke," clearly describes secondhand smoke and is easy to read.

SEE ALSO: Indoor air quality.

Sedimentation

CATEGORY: Water and water pollution

Sedimentation is the deposition of particulate matter by wind, water, chemical precipitation, gravity, or ice.

Particulate matter accumulates as sediment through transport and subsequent deposition of materials. Transport mechanisms include wind, running water, gravity, and glaciers. Materials that have been dissolved and transported in solution may be deposited as sediment through chemical precipitation. Sedimentation is an ongoing process, and, in a natural system, there is a delicate balance between positive and negative effects.

Sedimentation takes place in low-energy environments. Particle transport distance is a function of particle shape, size, and mass, and the available energy for transport. High-energy transport mechanisms will transport larger particles than low-energy mechanisms, and less energy is required to transport small particles the same distance as larger particles. Sedimentation can result in a degradation of the quality of land or water, or a loss of use of these resources. Degradation can be measured in terms of the identifiable adverse affects on aquatic organisms, wildlife, and humans.

Sedimentation can have negative environmental effects as a result of natural processes or as a result of processes induced by humans. Negative effects from natural processes can result from landslides, mudslides, the migration of dunes in developed areas, and the in-filling of lagoons or lakes that might otherwise provide recreational or commercial opportunities. Natural degradation of water quality results from sedimentation in areas such as north-central Oklahoma, where large accumulations of salt have precipitated from saline groundwater to form the Great Salt Plains along the Salt Fork of the Arkansas River, thereby affecting the chemistry of the Arkansas River.

The discharge of dredged or fill material, industrial wastes, or other materials into a natural system can affect the chemical, physical, and biological integrity of the system. Industrial wastewaters are a major source of chemicals that may include highly stable organic compounds that are capable of accumulating in high concentrations through sedimentary processes. Both the sedimentation rate and the individual compound's rate of solubility in water will affect the concentration of accumulated materials.

Adult freshwater organisms are generally tolerant of the normal extremes of suspended solids, but the introduction of excess materials and resulting sedimentation will kill eggs, larvae, and insect fauna while altering the characteristics of the aquatic bottom environment. Studies have also shown that higher concentrations of suspended solids can interfere with the filter mechanisms of aquatic organisms and the ability of sight feeders to locate prey.

The positive effects of sedimentation can be seen through many natural processes. Beach rejuvenation and growth takes place as the ocean's waves and currents deposit materials in quiet-water environments. Similar land growth takes place in the quiet-water environments of lakes and rivers. This process has, in many instances, resulted in substantial land growth that has allowed development of commercial or recreational facilities. River floodplains are frequently rejuvenated with nutrients needed for agricultural land use through sedimentation. Salt harvested from brine ponds relies on the natural process of chemical precipitation and sedimentation. Sedimentary processes have also been responsible for creating accumulations of materials such as gold and platinum in river placer deposits.

Kyle L. Kayler

See also: Dredging; Erosion and erosion control.

Septic systems

Category: Waste and waste management

Septic systems are the most common form of onsite wastewater treatment used in the United States. Although they work well in isolation and in the appropriate soil environment, increased development in rural areas can cause environmental problems from improperly sited or separated septic systems.

The conventional septic system is a model of simplicity and economy. It consists of a large chamber—the septic tank—into which domestic wastewater flows and where anaerobic digestion of the waste occurs. Nondegradable or slowly degradable compounds fall to the bottom of the septic tank, where they accumulate as sludge. The effluent then flows by gravity to a distribution box, where it is distributed by pipes through one or more gravel-filled ditches buried in the soil. These ditches constitute the lateral field where wastes slowly percolate through the gravel and underlying soil. During leaching, pathogenic organisms and residual organic matter are consumed. The percolating water eventually finds its way to underlying groundwater.

One of the assumptions underlying use of septic systems is that there will be some pathogen reduction as the wastes are anaerobically digested and as they pass through the lateral field. This is normally the case. Anaerobic digestion reduces the pathogens in waste by about 10 percent, and up to a 99 percent reduction occurs as the pathogens pass through a biological mat that forms at the bottom of the lateral field. The biological mat is a rich collection of bacteria, fungi, and protozoa that consume harmful bacteria and viruses in domestic waste.

A second key assumption underlying septic system use is that they will be properly sited and maintained. The soil is generally an effective biological filter, so most septic systems are designed to have 45 to 60 centimeters (18 to 24 inches) of permeable soil beneath the lateral field before wastewater encounters groundwater and is diluted. Likewise, the nondegradable materials in sludge must periodically be removed and either disposed of in a landfill or further processed in a municipal wastewater treatment system. If not removed, the sludge eventually accumulates to such an extent that it reduces the wastewater residence time in the septic tank. This causes inadequately treated waste and some solids to flow into the lateral field, which is subsequently unable to adequately ensure proper wastewater treatment. The solids may also clog the leach field and cause wastewater to pond.

The dominant environmental concerns with septic systems are their failure to adequately reduce pathogens and their delivery of nutrients into groundwater. Domestic wastewater consists of graywater—water from showers, sinks, dishwashers, and clothes washers—and blackwater, which comes from commodes. Graywater is relatively nutrient and pathogen free. Blackwater, in contrast, is the major source of domestic wastewater that causes environmental concerns with septic systems because it is both nutrient rich and laden with potential human pathogens. Because blackwater is commingled with graywater in most septic systems, all of the water leaving a house must be thoroughly treated.

In isolation, septic systems work well if properly sited and maintained because there is usually adequate distance between the domestic wastewater source and receiving waters. However, as development finds its way to rural environments that are not served by either municipal sewage treatment or water facilities, the congregation of septic systems in a limited area often overwhelms the environment's capacity to effectively filter and dilute wastewater before it reaches drinking water wells or recreational water supplies.

Schematic Design of a Conventional Septic System

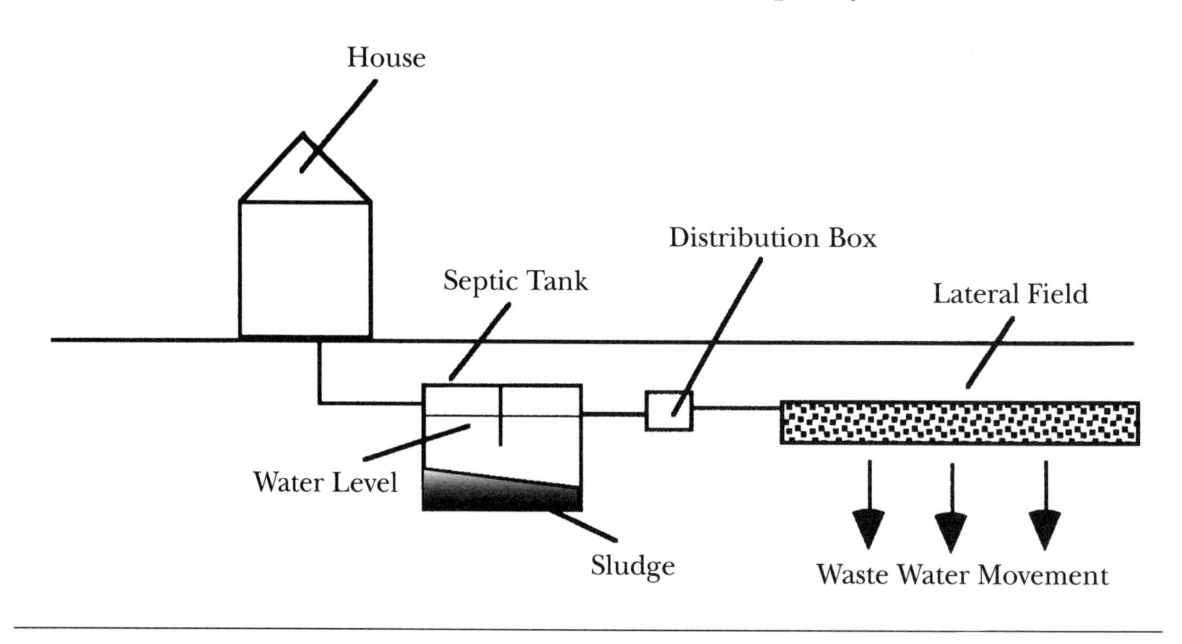

Properly locating septic systems is critical, and home sites may not always be located on soils that are suitable for conventional septic systems. The soils may be slowly permeable, the topography of the land may be too steep to permit an adequate lateral field, or there may be too little soil beneath the lateral field to allow proper filtration before impermeable layers and groundwater are reached. The consequence of these limitations to conventional septic systems is that soluble nutrients such as ammonium and phosphate, which come from the digestion of domestic wastes, may increasingly appear in underlying groundwater or shallow surface water and lead to eutrophication. Protozoan pathogens such as *Cryptosporidium* and *Giardia* may be inadequately treated or filtered. The potential for contamination by bacterial pathogens such as *Salmonella* and viral pathogens such as hepatitis A, B, and C is a greater concern because their escape from septic systems into drinking water or recreational water supplies represents a serious public health crisis.

As a result of these concerns with conventional septic systems, many states now require alternative forms of on-site wastewater treatment if either the soils or the density of habitation do not allow conventional septic systems to be used. These alternative treatments include wetlands, in which domestic waste from a septic tank first passes through a bed of vegetation before entering a lateral field. Aerobic digestion occurs in the wetland, and soluble nutrients are taken up by the wetland plants. Mound systems are often used where shallow soils exist. In a mound system, a raised bed of soil is constructed to which domestic waste effluent is pumped and then allowed to percolate. A variety of other on-site systems use percolation through sand or peat filters to help process wastewater and reduce the impact of the nutrients and pathogens in domestic waste.

Mark Coyne

SUGGESTED READINGS: Information on septic systems and their environmental consequences can be found in *So . . . Now You Own a Septic Tank* (1990), distributed by the National Small Flows Clearinghouse, and the Environmental Protection Agency's *Small Wastewater Systems: Alternative Systems for Small Communities and Rural Areas* (1980).

see also: Cultural eutrophication; Environmental health; Groundwater and groundwater pollution; Sewage treatment and disposal; Soil contamination; Waste treatment; Water pollution; Water quality; Water-saving toilets.

Serengeti National Park

Category: Preservation and wilderness issues

The Serengeti National Park is a designated World Heritage Site and biosphere reserve. The site of the world's largest ungulate migration, the park is threatened by human encroachment.

The Serengeti National Park, including the biologically important Ngorongoro Conservation Area, covers approximately 14,763 square kilometers of northeastern Tanzania; together with the neighboring 1,672-square-kilometer Maasai Mara Reserve in Kenya, it forms the continuous Serengeti-Mara ecosystem. The Serengeti (a Maasai term meaning "endless plains") is a complex system noted for unique biological diversity. It is characterized by an expansive plateau broken by hills called "kopjes." Altitudes in the park range from 1,170 meters to 1,900 meters (3,860 feet to 6,270 feet). Open grassland covers most of the park, but woodlands are found in wetter areas. Forests occur along the banks of the perennial Mara and Bologonja Rivers.

The Serengeti National Park was established in 1929 and attained its present conservation area in 1967. In 1972, the United Nations Education Science and Culture Organization (UNESCO) proposed the Serengeti ecosystem as the first World Heritage Site. In 1981, the park was designated a biosphere reserve, owing to its unique biodiversity.

The Serengeti-Mara ecosystem hosts a remarkable number of ungulate (hoofed) species and predators, including 1.3 million wildebeests, 200,000 zebras, 440,000 gazelles, 7,500 hyenas, and 2,800 lions. The ungulates cycle between the Maasai Mara and Serengeti areas in one of the largest and most spectacular migrations of large

The terrain of Serengeti National Park consists of a large plateau broken by hills. Although the park supports a large ungulate population, nearby human settlements threaten to destroy the ecology of the region. (Archive Photos)

mammals in the world. Migrants stay in the Serengeti during the wet months from December to June and move to the Maasai Mara in the dry months from July to October. The park is also the location of the Olduvai Gorge, the site of many of the most important hominid fossil finds.

The Serengeti ecosystem faces considerable challenges. Human settlements around the park have reduced it to a habitat island, decreasing the natural area by 40 percent. The Ngorongoro Conservation Area is threatened by increases in agriculture and overuse of water resources. Interaction between domestic animals and wildlife has caused diseases that have decimated wild dogs, and poaching is rampant. Wire snares are used to trap smaller animals, and guns and arrows are used to kill elephants and lions. A. R. E. Sinclair, an authority of the Serengeti, estimates that 200,000 animals are poached annually in the region. This illegal removal of animals has a major impact; the area's elephant population declined by 81 percent between 1970 and 1986, and the black rhino population was virtually wiped out. Moreover, the persistent killing of male lions has produced a female-biased sex ratio.

Efforts to reduce poaching in the park have had some effect, and populations of many threatened species are recovering. Measures to prevent the recurrence of other problems that have endangered the park in the past have also been proposed. Foremost is the involvement of local people in decision making through Integrated Conservation and Development projects, which allow local people to participate and benefit from wildlife while maintaining a sustainable population. Recent proposals include establishment of buffer zones, involvement of local community representatives, and controlled game cropping, and increases in vehicle and foot patrols in the park.

Joseph M. Wahome

SUGGESTED READINGS: A. R. E. Sinclair and P. Arcese's *Serengeti II: Dynamics, Management and Conservation of an Ecosystem* (1995) is a comprehensive overview of issues related to the Serengeti.

SEE ALSO: Biosphere reserves; National parks; Nature reserves; Poaching; Wildlife refuges.

Sewage treatment and disposal

CATEGORY: Waste and waste management

Wastewater consists of domestic and industrial effluent that is collected by a sewer system and conveyed to a facility where it is treated prior to release into the ground or, more usually, a surface watercourse. The proper disposal of wastewater is a critical parameter in environmental planning.

The Minoan civilization on the island of Crete near Greece had one of the earliest known sewers in the world (c. 1600 B.C.E.). A large sewer known as the Cloaca Maxima was built during the sixth century B.C.E. in ancient Rome to drain the Forum. The Romans also reused public bathing water to flush public toilets. London, England, had a drainage system by the thirteenth century, but effluent could not be discharged into it until 1815. Sewers were constructed in Paris, France, before the sixteenth century, but fewer than 5 percent of the homes were connected by 1893. In general, the widespread introduction of sewers in densely populated areas did not occur until the mid-nineteenth century.

Wastewater disposal systems consist of a system of pipes, a treatment plant, and an outfall to the ground or, more commonly, to a stream or the ocean. Older wastewater systems are generally combined: Domestic, industrial, and stormwater runoff are conveyed in the same pipe to a treatment plant. Although initially cheaper to build, combined systems are less desirable since most of the effluent must bypass the treatment plant during storms when street runoff rapidly increases. Newer wastewater systems are designed to be separate, with different pipes for wastewater and storm runoff.

About 60 to 75 percent of the water supplied to a community will wind up as effluent that must be treated and disposed. The remaining water is used in industrial processes, lawn sprinkling, and other types of consumptive use. Domestic sewage contains varying proportions of human excrement, paper, soap, dirt, food waste, and other substances. Much of the waste substance is organic and is decayed by bacteria. Accordingly, domestic sewage is biodegradable and capable of producing offensive odors. The composition of industrial waste varies from relatively clean rinse water to effluent that can contain corrosive, toxic, flammable, or even explosive materials. Therefore, communities usually require pretreatment of industrial effluent.

The organic material in sewage is decomposed by aerobic (oxygen-requiring) bacteria. However, the dissolved oxygen (DO) in water can be used up in the process of microbial decomposition. If too much organic waste enters the waterbody, the biochemical oxygen demand (BOD) can exhaust the DO in the water, thereby damaging the aquatic ecosystem. Indeed, most species of fish die if the DO falls below 4 milligrams/liter for extended periods of time.

The function of wastewater-treatment plants is to produce a discharge that is free of odors, suspended solids, and objectionable bacteria. Treatment processes are categorized as primary, secondary, or tertiary. Primary treatment is mostly mechanical, as it involves the removal of floating and suspended solids by screening and sedimentation in settling basins. It can remove 40 to 90 percent of the suspended solids and 25 to 85 percent of the BOD.

Secondary treatment involves biological processing in addition to mechanical treatment. One form of biological processing is a trickling filter, where wastewater is sprayed over crushed stone and allowed to flow in thin films over biologic growths that cover the stone. The organisms in the biologic growths—which include bacteria, fungi, and protozoa—decompose the dissolved organic materials in the wastewater. These growths eventually slough off and are carried to settling tanks by the wastewater flow. The other type of secondary treatment is the activated sludge process. In this procedure, flocs of bacteria, fungi, and protozoa are stirred in the wastewater with results that are about the same as trickling filters. Depending upon the efficiency of the plant and the nature of the incoming wastewater, both types of biological processes can remove 50 to 95 percent of the suspended solids and BOD. The efficiency of secondary treatment can be seriously lowered if the design capacity of the plant is overloaded with excessive effluent coming from stormwater runoff in combined sewers. This is one important reason why separate sewers, even though more expensive, are favored by public health officials. The biologic processes can also be severely impacted by toxic industrial waste, which can kill the "good" bacteria that are crucial in the treatment process. Accordingly, many communities require pretreatment of industrial wastes.

Tertiary treatment is the most advanced method and consequently the most expensive. It includes several procedures such as the use of ozone, which is a strong oxidizing agent, to remove most of the remaining BOD, odor, and taste; adding alum to remove phosphate; and denitrification. The final effluent from any treatment level is usually chlorinated prior to release.

In areas where population densities are fewer than about 2,600 people per square mile (1,000 per square kilometer), the cost of a sewer system and treatment plant are difficult to justify. Accordingly, septic systems are commonly used in low-density residential areas for wastewater disposal. Household effluent is piped to a buried septic tank, which acts as a small sedimentation basin and anaerobic (without oxygen) sludge digestion facility. The effluent exits from this tank into a disposal field, where aerobic biologic breakdown of dissolved and solid organic compounds occurs. In order to operate effectively, the soil must be of sufficient depth and permeability so that microbial decomposition can take place before the effluent reaches the water table. The Environmental Protection Agency estimates that 25 percent of the homes in the nation use septic systems.

Robert M. Hordon

SUGGESTED READINGS: *Water Supply and Sewerage* (1991), by Terrence McGhee, and *Environmental Engineering and Sanitation* (1992), by Joseph A. Salvato, are useful textbooks on waste management. Wastewater-treatment plants are covered by Syed A. Qasim in *Wastewater Treatment Plants: Planning, Design and Operation* (1994), and by Glenn M. Tillman in *Wastewater Operations: Troubleshooting and Problem Solving* (1995). A good reference for wastewater disposal in rural and suburban areas is *Wastewater Engineering Design for Unsewered Areas* (1986), by Rein Laak.

SEE ALSO: Environmental engineering; Sludge treatment and disposal; Water pollution; Water treatment.

Shanty towns

CATEGORY: The urban environment

Shanty towns consist of communities of poor people living in substandard housing. Such dwellings are usually built of flimsy materials on undesirable sites, exposing their inhabitants to many environmental hazards.

Millions of people throughout the world cannot afford housing that is safe, sturdy, and reasonably spacious. While inhabitants of poor housing in sparsely settled farm or forest regions run certain risks, environmental dangers multiply when many such houses huddle closely together in urban areas.

The most acute housing problems occur in Third World nations. Since the mid-twentieth century, millions of migrants from rural areas have crowded into these nations' cities. The jobs the cities provide are a powerful magnet for citizens whose customary farming life no longer sustains them. Indeed, their farmland has often been ruined by destructive processes such as deforestation or soil erosion. While the cities seem to promise a better life, most migrants arrive unable to afford the most basic housing with amenities. They are left to build their own shelter from whatever materials they can scavenge.

Statistics show the importance of shanty towns in housing these poorer residents. In Bogota, Colombia, 59 percent of the population lives in self-built housing, while 60 percent of the peo-

Shanty towns typically consist of flimsy structures built from found materials. Residents of such communities are exposed to a variety of environmental hazards, such as waterborne diseases from unsanitary drinking water sources. (Reuters/Archive Photos)

ple in Dar es Salaam, Tanzania, live in such dwellings. In Addis Ababa, Ethiopia, the number is close to 85 percent. It is estimated that in most Third World cities, 70 to 95 percent of all new housing is illegally built and held.

These houses occupy land unsuitable for other purposes. Shanty towns spring up on flood plains, steep hillsides, lots adjacent to contaminated industrial sites, and rail or highway right-of-ways. In Manila, Philippines, fifteen thousand squatters have lived for forty years on Smoky Mountain, a municipal dump whose refuse serves as a source of income. In Cairo, Egypt, 500,000 people live in a squatter community built amid ancient mausoleums. Shacks are constructed out of corrugated tin, packing crates, mud bricks made on-site, and similar materials.

Lack of sufficient drinking water and infrastructure to remove human and household wastes are the greatest environmental dangers in these communities. Few residents have water piped into their houses, so they must buy it from street vendors or manually carry it from a common spigot shared by many people. With either source, the expense or effort of obtaining it means households seldom have enough water for healthy day-to-day living. In some cases water from surface sources or shallow wells may be used; both are likely to be contaminated with biological or industrial wastes.

Without sewage treatment and disposal systems, disease vectors get cycled back into the immediate environment. On-site methods such as septic tanks or pit latrines, which are adequate for low-density conditions, break down in crowded settlements. Waterborne diseases such as dysentery, cholera, and infant diarrhea are common in shanty towns. Standing water and waste also shelter mosquitoes, which spread malaria. Dumped garbage attracts rats.

Industrial pollutants are another hazard. Many shanty towns are built next to factories. While this makes industrial jobs accessible to the inhabitants, most such sites generate contaminants. This is doubly true of factories in Third World countries, where public awareness and legal regulation of environmental toxins are much weaker than in the United States. The chemical disaster in Bhopal, India, in 1984 killed mostly low-income people who lived near the Union Carbide factory. Shanty town dwellers who live beside highways are subject to a constant barrage of fumes from motor traffic.

Many added environmental stresses arise in these communities. Families of four, five, or more people living in a single room readily exchange airborne infections. Some 60 percent of slum children in Kanpur, India, are estimated to have tuberculosis. The most dramatic threat, however, comes from natural disasters. Squatter settlements on bare hillsides, and houses built of mud or flimsy materials, are vulnerable to hurricane, flood, or earthquake damage.

Although their impact on the surrounding region is hard to separate from more general urban pollution, shanty town living conditions do affect a wider area. Fecal material and garbage from urban slums travel downstream in rivers for hundreds of miles. Yards of packed dirt and lack of drainage systems slow the absorption of rainwater into the soil. The costs of residents' excess illnesses and deaths, let alone time wasted in daily tasks like hauling water from a faraway spigot, are further drains on developing countries' limited resources.

Shanty towns are a rational response to housing needs by people who cannot afford standard houses. Merely bulldozing the shanties or forcing their residents out seldom works; such actions simply compound problems. Providing squatter villagers with a basic infrastructure of piped water and sewage disposal immediately upgrades their living conditions. It is also much cheaper, liter by liter, than the hauled water that slum dwellers buy. When such systems are provided, shanty town residents may further improve their homes with materials at hand. Economic, political, and cultural pressures on local governments often prevent such plans from going forward. It helps if shanty town residents first gain legal title to their dwellings and a modicum of political power. Given urban growth rates and economic problems in Third World nations, however, these problems may not be solved for many years.

Developed countries also have shanty towns, but they constitute a much smaller proportion of the total housing stock. Because these settlements are smaller and public health measures more widespread, environmental hazards to their residents are less overwhelming but still significant.

Emily Alward

SUGGESTED READINGS: *Squatter Citizen* (1989), by Jorge E. Hardoy and David Satterthwaite, provides a good overview of shanty towns worldwide. Daniel Litvin, "Living Dangerously," *The Economist* (March 21, 1998), discusses the problem of rapidly growing shanty towns in developing countries. Steve Lopez, "So Who's Crazy, Them or Us," *Time* (August 17, 1998), shows a contrarian view of a community of squatters in the California desert.

SEE ALSO: Environmental health; Environmental illnesses; Urbanization and urban sprawl.

Siberian pipeline

DATE: completed early 1990's
CATEGORY: Energy

The Siberian pipeline was constructed during the 1980's to transport natural gas from the Siberian Urengoi-Yamal gas fields, the largest in the world, to Western Europe.

After World War II Europe became increasingly dependent on oil-producing countries in the Middle East for its fuel needs. However, the Arab-Israeli conflicts of 1973 alarmed European countries and prompted them to seek other sources of energy. One alternative was to utilize larger amounts of natural gas; its use more than doubled during the late 1970's. Natural gas is cheaper than oil and coal, but transporting it requires enormous initial investments of capital and technology because pipelines are the only practical means of conveyance. Natural gas is moved along the pipes in liquid form by the use of strategically located compressing stations.

In 1980 West German chancellor Helmut Schmidt, after a visit to Moscow, announced a plan for the Soviets to build a Euro-Siberian gas pipeline. It would start at Urengoi, east of the Ural Mountains and near the Arctic Circle, and extend westward to West Germany, France, and Italy. Objections to building the pipeline came largely from Europe's North Atlantic Treaty Organization (NATO) ally, the United States. It was feared that if an armed conflict broke out, the Soviets would have an undue advantage over the Western Europeans.

The plan called for a 5,790-kilometer (3,600-mile) pipeline, four times the length of the controversial Trans-Alaskan pipeline, which had just been completed. Several small cities were built to accommodate construction workers. Preparations for extraction and construction of the pipeline, which began in the early 1980's, required the technical assistance, materials, and personnel provided by the potential European customer nations. In addition to West Germany, the largest purchaser, other countries were Austria, Belgium, France, Italy, the Netherlands, and Switzerland. As a result of the demise of the Soviet Union in 1991, the pipeline was completed and operated without adverse economic or political consequences.

Although successful as a political and commercial venture, the Siberian pipeline deserves attention because of its environmental impact. The resource-rich region of Siberia includes vast areas of tundra, taiga, and other fragile ecosystems, many of them in relatively pristine condition. As a result of the extraction process, building of roads, and related activities, significant damage occurred to the environment that will require centuries for nature to correct. Many of these environmental concerns were compounded by the collapse of the Soviet Union and the resulting economic chaos. With the new Russian republic in need of cash and with weak environmental laws, there is great concern that existing pipelines will not be properly maintained, thus creating the potential for considerable environmental damage. Several Siberian oil pipelines have burst, including a major oil seepage that occurred near the port of Archangel in 1994.

Thomas E. Hemmerly

SEE ALSO: Fossil fuels; Oil crises and oil embargoes; Trans-Alaskan pipeline.

Sick building syndrome

CATEGORY: Human health and the environment

Sick building syndrome (SBS) is an illness caused by exposure to indoor air pollutants during which more people than normal suffer from multiple symptoms of sickness for no apparent reason or with no identifiable cause. There is often no record of related or prior illnesses.

The average person spends 80 to 90 percent of the day indoors. The longer one spends in an affected environment, the more severe the SBS symptoms can become. SBS is often a problem in office buildings but can be evident in homes, schools, nurseries, and libraries. Since the 1970's, buildings have become more airtight in response to energy crisis concerns. Many heating, ventilating, and air-conditioning (HVAC) systems have been designed to recirculate indoor air rather than draw in fresh, filtered air from outside. Also, systems that bring in polluted outdoor air may contribute to building-related illnesses. A lack of ventilation is one primary cause of the trapping of natural and anthropogenic indoor pollutants and the onset of SBS.

The symptoms of SBS appear to increase in severity with time spent in the affected building and decrease or disappear with time away from the building. Symptoms include respiratory problems such as coughing; increased allergic reactions and sneezing; irritation of the eyes, throat, nose, and skin; and headaches. Some patients even experience severe fatigue and flulike symptoms. Not all individuals exposed to the same indoor contaminants experience similar symptoms, and some inhabitants feel no ill effects. Microenvironments may exist that cause inhabitants of offices sharing a common wall to experience unique symptoms. Many symptoms are similar to those of common allergies or illnesses, and sufferers may dismiss the problem without seeking help. A rise in SBS cases may be related to increased usage of synthetic building materials, greater stress and regimentation in the workplace, and the general increase in number of workers employed in office settings.

There is no one cause of SBS, but there are many sources of indoor pollution that may lead to SBS, including secondhand tobacco smoke, ozone and heat from photocopiers and computers, new carpet and furniture that off-gas volatile organic compounds (VOCs—compounds composed exclusively of hydrogen and carbon), asbestos, lead in paint, formaldehyde, microbials, respirable and inhalable particulates, dust mites, and gases such as carbon monoxide. Biological contaminants that are found where temperature and moisture levels are high are bacteria, pollen, mildew, and mold, all of which can increase SBS symptoms when inhaled. Computers seem to be a significant contributor to the onset of SBS because of increased heat production, which often results in the installation of air-conditioning systems that can exacerbate the problem. The computer display screen can also cause eye strain and headaches.

SBS should be distinguished from building-related illnesses caused by exposure to specific indoor contaminants that can be definitively diagnosed. Alveolitis, bronchospasm, rhinitis, and conjunctivitis can be caused by airborne allergens becoming lodged in the alveoli of the lungs, the bronchi of the lungs, the nose, or the eyes, respectively. Legionnaires' disease is associated with a bacteria commonly transmitted through a contaminated water source such as evaporative condensers or potable water distribution systems. SBS is rarely attributed to one specific exposure and is therefore more difficult to detect and treat.

Minor complaints by office workers should be accepted as serious, and investigation should begin to determine if indoor air quality is acceptable. Steps should be taken to identify the cause of a suspected outbreak of SBS. If no obvious breakdown of ventilation systems or major pollution sources are identified, additional precautionary measures should be taken. Filters should be cleaned, ventilation systems checked, humidity conditions lowered, and new carpets and paints allowed to off-gas and settle before spending time in the newly renovated areas.

Alternative solutions include the addition of common household plants that extract contaminants in the air through their leaves, particularly formaldehyde, benzol, phenol, and nicotine. A study conducted at the Botanical Institute at the University of Cologne in Germany showed that such hydroculture plants, or plants lacking soil, are efficient at absorbing pollutants and transforming 90 percent of the chemical substances into sugars, oxygen, and new plant material. These plants provide moisture to the air without the contribution of fungus spores. Another new treatment being considered is an indoor ecosystem composed of rocks, plants, fish, and microorganisms that inhales dirty air and exhales much cleaner air.

In the work environment, the health problems associated with SBS often lead to lost productivity among employees, high absenteeism, and the possibility of increased litigation. Scientists are unsure if increased incidences of cancer are related to exposure to contaminants in home and work environments. It is felt that avoiding prolonged exposure to indoor pollutants or reducing the number of indoor contaminants can add to the length and quality of one's life and lessen the symptoms and incidences of sick building syndrome.

Diane Stanitski-Martin

Suggested readings: A thorough overview of the causes, symptoms, and treatments of SBS can be found in "Sick-Building Syndrome," *The Lancet* 349 (April 5, 1997), by Carrie A. Redlich, Judy Sparer, and Mark R. Cullen. *Sick Building Syndrome*, published by the U.S. Environmental Protection Agency, Indoor Environments Division, Office of Air and Radiation, EPA Document 402-F-94-004 (April, 1991), discusses causes of SBS, describes building investigation procedures, and provides general solutions for resolving the syndrome. Thad Godish's *Sick Buildings: Definition, Diagnosis, and Mitigation* (1995) was written as a reference work and is intended for a variety of readers, including architects, those conducting indoor air quality investigations, public health and environmental professionals, and university students. It defines and discusses the issues related to SBS.

See also: Air pollution; Asbestos; Indoor air quality; Radon; Secondhand smoke.

Sierra Club

Date: established 1892
Category: Preservation and wilderness issues

The Sierra Club is one of the largest, oldest, and most influential conservation organizations in the world. Unique among large environmental groups because of its reliance on volunteer activists and its democratic structure, the Sierra Club pioneered many grassroots political techniques in its efforts to preserve wilderness and protect parks and other natural areas.

John Muir, amateur naturalist and writer, discovered for himself the wonders of Yosemite Valley in California's Sierra Nevada in 1868. He soon realized that such areas needed to be protected from development and resource extraction and used his writings and his friends to lobby for the creation of Yosemite National Park.

Muir conceived of the idea of an organization that would ensure the protection of Yosemite and the surrounding wildlands, and in 1892 the Sierra Club was born, with Muir as its first president. The club's stated purpose was

> to explore, enjoy, and render accessible the mountain regions of the Pacific Coast . . . and to enlist the support and cooperation of the people and government in preserving the forests and other natural features of the Sierra Nevada.

This statement set the stage for an organization that combined the recreational goals of a hiking and climbing club with political savvy and influence. Muir's idea was to build a constituency for nature by getting people out into the mountains and showing them areas that needed to be saved and explaining how they were endangered, a technique that has been successfully used by the Sierra Club for more than a century.

After the relatively easy success of the creation of Yosemite National Park, the Sierra Club's next major campaign was to be more difficult and ultimately unsuccessful. In 1907 a dam and reservoir to supply water to San Francisco was proposed for Hetch Hetchy Valley. Hetch Hetchy was within Yosemite National Park and was considered by many to be nearly equal to Yosemite

Valley in scenic grandeur. A long campaign to prevent the flooding of the valley included dissent within the club, pointing out the difficulties inherent in its strictly democratic structure. The battle was finally lost, and, through an act of Congress, the dam was built. Out of this failure, however, the Sierra Club gained experience in managing its own growing organization and building national support for its views by using media publicity and the network of other conservation organizations.

John Muir, sitting near his friend John Burroughs, helped establish the Sierra Club in 1892; it has since grown into one of the most influential environmental organizations in the world. (Library of Congress)

Muir died in 1914, one year after the Hetch Hetchy defeat, but the Sierra Club continued to grow in membership and influence. The outings program introduced people to the wonders of the Sierra Nevada and other wild areas as far afield as Montana and the Canadian Rocky Mountains. Club members developed mountaineering and rock-climbing techniques, as well as low-impact camping ethics. On the conservation front, the club was involved in campaigns to create Kings Canyon and Olympic National Parks, and prevent dams in Yellowstone and Glacier Parks.

In 1951 another proposed dam project pushed the Sierra Club to national prominence and influence. Another federally protected area, in this case Utah's Dinosaur National Monument, was to be the site for the Echo Park Dam and another dam as part of the Colorado River Project. Remembering the loss of Hetch Hetchy, the club vowed to fight harder this time and hired David Brower as its first executive director. Brower led the successful campaign against these dams with float trips on the river for influential politicians and members of the media, articles in national magazines, and a film called *Wilderness River Trail.* As part of the compromise that saved the canyons at Dinosaur National Monument, the Sierra Club agreed not to challenge other dams in the project. One of those, the Glen Canyon Dam in Arizona, flooded a canyon that turned out to be a magnificent slickrock gorge that Brower and others believed, in retrospect, should have been saved as well. This was another hard lesson learned by the Sierra Club, and one that made club members and leaders suspicious of compromises.

Brower led the Sierra Club for sixteen years, through a period of major expansion. He opened the membership rolls and actively sought new members from the public. He developed the publications program by putting the Sierra Club name on a series of calendars and "coffee table" books that combined beautiful photography with messages of preservation for specific areas. To prevent yet another dam, this time in the Grand Canyon, Brower used full-page advertisements in *The New York Times* to enlist the general public in a letter-writing campaign to dissuade Congress from authorizing the project.

The Sierra Club was also instrumental in creating the field of environmental law. Conservation organizations had a difficult time pursuing court cases because they were deemed to have no "standing," or no financial stake in the issue. In a case in New York State in the late 1960's, attorneys representing local conservation organizations and the Sierra Club won the right to get standing based on recreational, conservational, and aesthetic interests. In 1971 the Sierra Club created the Sierra Club Legal Defense Fund, a legally and financially distinct organization that represented the club, other conservation organizations, and individuals in environmental litigation.

By the end of its first one hundred years, its policies driven by the interests of its 500,000 members, the Sierra Club had taken on issues of water and air pollution, recycling, nuclear energy, population, and global warming. Through public education, grassroots letter-writing campaigns, lobbying, and litigation, the Sierra Club continues to affect environmental opinion and policy in the United States and, increasingly, the world.

Joseph W. Hinton

SUGGESTED READINGS: A detailed history of the Sierra Club, enlivened with beautiful photography, can be found in *Sierra Club: 100 Years of Protecting Nature* (1991), by Tom Turner. *The History of the Sierra Club, 1892-1970* (1988), by Michael P. Cohen, provides a more scholarly study, complete with policy documents and details of specific meetings. A volume specific to the early days of the club, Holway R. Jones's *John Muir and the Sierra Club: The Battle for Yosemite* (1965), paints a vivid picture of the life of the renowned naturalist and the beginnings of the environmental movement.

SEE ALSO: Brower, David; Echo Park Dam proposal; Environmental policy and lobbying; Glen Canyon Dam; Hetch Hetchy Dam; Muir, John; National parks; *Sierra Club v. Morton*.

Sierra Club v. Morton

DATE: 1972
CATEGORY: Preservation and wilderness issues

In 1972 the U.S. Supreme Court rejected the Sierra Club's attempt to challenge a Forest Service permit for the construction of a recreational complex in Mineral King Valley, a quasi-wilderness area located in the Sierra Nevada of California.

During the early 1970's the Sierra Club sued the U.S. Forest Service to prevent it from approving permits that would allow Walt Disney Enterprises to construct a $35 million complex of motels, restaurants, swimming pools, and ski trails in Mineral King Valley. Up to fourteen thousand visitors per day would gain access to the resort by using a 32-kilometer (20-mile) highway to be built, in part, through Sequoia National Park. The Federal District Court granted an injunction, but the Federal Court of Appeals reversed it. The Supreme Court did not consider whether the proposed development violated federal law, but whether the Sierra Club, as an organization with a special interest in the preservation of national parks and forests, had standing to challenge the federal agency's decision to issue the permits.

The Supreme Court had addressed the standing issue in *Association of Data Processing Service Organizations v. Camp* (1970), where it had held that people who sought judicial review of a federal agency's action under Section 10 of the Administrative Procedure Act had to claim that the agency had caused them injury in fact and that the injury was a harm within the zone of interests protected or regulated by statutes the agency was said to have violated.

In *Sierra Club v. Morton*, Justice Potter Stewart's opinion for the Court addressed only the injury in fact element of the Data Processing test. The Court accepted the Sierra Club's claim that noneconomic injury constituted injury in fact, that the "change in the aesthetics and ecology" from Mineral King's development "would destroy or otherwise adversely affect the scenery, natural and historic objects and wildlife of the park and would impair the enjoyment of the park for future generations." However, the Court rejected the Sierra Club's argument that it did not have to claim that its members would be adversely affected by the Mineral King development, because its longstanding concern for and expertise in environmental matters gave it standing as a "representative of the public." In denying standing, the Court held that an organization's sincere interest in an environmental problem, even if the interest is longstanding and the organization is highly qualified to speak on behalf of the public, is not enough to satisfy the injury in fact requirement. If it were, the Court feared that there would be "no objective basis on which to disallow a suit by any other bona-fide organization no matter how small or short-lived."

Justice William O. Douglas, in an eloquent dissent, argued that the case should have been entitled *Mineral King v. Morton* and that the Court should have designed a standing rule that would have permitted the Sierra Club to litigate the Forest Service use permit on behalf of the valley. Drawing upon and citing Christopher Stone's law review article "Should Trees Have Standing?", Douglas proceeded to sketch the broad outlines of an imaginative redefinition of standing. The law, he said, indulges a fiction that inanimate objects such as ships and corporations are people and may, therefore, be parties to litigation. So should it be with valleys, such as Mineral King, and with lakes, rivers, and forests. Who should speak for these inanimate objects and defend their rights? Congress, he argued, is "too remote . . . and too ponderous." Federal agencies, including the Forest Service, "are notoriously under the control of powerful interests." Only those who have an intimate relation with valleys, lakes, rivers, and forests, because they hike, fish, or "merely sit in solitude and wonderment," may speak for the values that these natural objects represent.

Justice Harry Blackmun's dissent was much more direct in his criticism of the Court's "practical" decision. Its "somewhat modernized" conception of standing, he argued, was not adequate to deal with the novel issues raised by the deteriorating state of the environment. He suggested two alternatives: Either approve the district court's decision on the condition that the Sierra Club amend its complaint to comply with the Court's standing rule, or redefine standing, as Justice Douglas had, to permit any bona fide environmental organization, such as the Sierra Club, to litigate on behalf of Mineral King. He did not fear, as the majority on the Court did, that an expanded definition would open a Pandora's box of litigation and had much greater faith in the ability to impose appropriate restraints on an "imaginative expansion" of standing.

Sierra Club v. Morton opened the federal courts to a wealth of environmental litigation because it settled the question left open by the Data Processing case about whether an injury to a noneconomic interest could provide the basis to challenge a federal agency decision. The Court further broadened its standing test in *United States v. Students Challenging Regulatory Agency Procedures* (1973) and *Duke Power v. Carolina Environmental Study Group* (1978) by allowing environmental groups to gain standing based on tenuous claims of causation between a proposed federal agency action and fairly speculative injuries. In *Lujan v. National Wildlife Federation* (1990), the Court tightened up standing and made it more difficult for environmental groups to gain access to federal courts and challenge federal programs by requiring them to allege that the specific lands involved were actually used by their members.

William C. Green

SUGGESTED READINGS: Christopher Stone, "Should Trees Have Standing? Toward Legal Rights for Natural Objects," *Southern California Law Review* 45 (1972), is the law review article upon which Justice Douglas drew in his dissenting opinion. The initial debate that the case generated over legal rights for natural objects appears in Martin Krieger, "What's Wrong with Plastic Trees?", *Science* 179 (1973), and Laurence Tribe, "From Environmental Foundations to Constitutional Structures: Learning Nature's Future," *Yale Law Journal* 84 (1975). Thomas Hoban and Richard Brooks, *Green Justice: The Environment and the Courts* (1996), provides a shorter study of the case with a set of study questions. The text of the Sierra Club case appears in the

United States Reports 405 (1972).

SEE ALSO: Kings Canyon and Sequoia National Parks; National parks; Nature preservation policy; Preservation; Sierra Club.

Silicosis

CATEGORY: Human health and the environment

Silicosis is a disabling condition of the lungs caused by the inhalation of crystalline silica.

Silica, or silicon dioxide, is one of the most common minerals. It occurs in a wide variety of natural and industrial settings. More commonly referred to as quartz, it is found in almost all rock and sand. Sandstone, for example, is primarily composed of silica. Thus, silicosis occurs most often as an occupational illness among workers in what are known as the dusty trades: foundry work, construction, and mining. Ancient Greek and Roman physicians recognized the high risk of lung diseases developing among quarrymen and miners, which was one reason criminals were often sentenced to work at those occupations. Still, researchers did not recognize silicosis as a distinct disorder until almost the twentieth century. Indeed, following the discovery of the tuberculin bacillus during the nineteenth century, some physicians disputed the existence of environmental disorders such as silicosis. Rather than accepting that a specific environmental factor, such as high levels of quartz dust, caused disabling lung conditions, many physicians blamed workers' ill health on a combination of unsanitary living conditions and bacterial infections. By the mid-twentieth century, however, health professionals recognized silicosis as a distinct occupational illness.

Physicians refer to silicosis as an environmental illness because the disorder is caused by exposure to silica rather than transmitted as an infectious disease. Although silica is not a toxic substance in itself, crystalline silica can scar sensitive lung tissue and lead to the development of fibrotic nodules. As these fibrotic nodules grow in size and number, the lungs stiffen, and breathing becomes more difficult. Depending on the level of exposure to silica, the onset of disabling and even fatal silicosis can occur within only a few weeks or months. More commonly, however, silicosis takes many years to develop. Sufferers of silicosis may eventually become disabled from diminished lung capacity and be at higher risk for developing other diseases such as tuberculosis or lung cancer. In some cases, the decreased flow of blood to the lungs caused by silicosis can cause the heart to enlarge. Although silicosis is becoming rarer in industrialized nations because of improvements in industrial hygiene, several hundred people still die in the United States each year from the effects of the disease.

In most industrialized nations, workers whose jobs expose them to high levels of crystalline silica are now required to wear protective masks and respirators. Some environmental ethicists are concerned that more stringent occupational health and safety standards in countries such as the United States and Canada would cause industry to move dusty, high-risk manufacturing operations to less-developed nations. Rather than improving working conditions at existing foundries in the American Midwest, for example, some corporations have chosen to close those facilities and build plants in countries where regulations are perceived as less burdensome, workers are less well educated and therefore unlikely to be aware of the risks associated with silica, and no system of workers' compensation insurance for occupational diseases exists.

Nancy Farm Männikkö

SEE ALSO: Environmental justice and environmental racism; Hawk's Nest tunnel disaster.

Silkwood, Karen

BORN: February 19, 1946; Longview, Texas
DIED: November 13, 1974; near Crescent, Oklahoma
CATEGORY: Nuclear power and radiation

Until her death in an automobile crash, nuclear industry worker and nuclear safety activist Karen Silkwood was practically unknown. After

Karen Silkwood, a nuclear safety activist who died in 1974 under mysterious circumstances, was portrayed by Meryl Streep (left) in the 1983 film Silkwood. (Archive Photos)

the accident, which occurred under mysterious circumstances, antinuclear activists saw her as a martyr to their cause.

Karen Silkwood, the daughter of Bill and Merle Silkwood, grew up in the Texas oil and gas fields around Nederland, Texas. She eloped with Bill Meadows after one year of study at Lamar College, where she was enrolled in a course of study leading to a degree in medical technology. The couple separated in 1972, and their three children remained with their father. Silkwood moved to Oklahoma to be near her parents and found work as a laboratory technician with the Kerr-Magee Corporation at their Cimarron plant in Crescent, Oklahoma.

Silkwood joined the Oil, Chemical, and Atomic Workers Union and was soon involved in actions protesting the company's lax health and safety procedures. One allegation that she made was that the company had falsified records to cover up missing nuclear material. Her concern in this area led to her being called to testify before the Atomic Energy Commission in Washington, D.C., in the early part of 1974. At this time Silkwood apparently agreed to clandestinely obtain film evidence of poor workmanship in the manufacture of the fuel rods.

In early November of 1974, plant monitoring equipment detected that Silkwood had become contaminated with a radioisotope. Further testing also showed that the contamination was present in her apartment. She was taken to the Los Alamos National Laboratory, where more refined testing was performed. It was determined that the level of the contamination was not serious and did not constitute a threat to her health.

On the night of her death, Silkwood left a union meeting supposedly carrying an envelope with proof of wrongdoing by Kerr-Magee. She was on her way to a meeting with Drew Stephens, a *New York Times* reporter and union representative. Her car crashed into a culvert alongside a dry, straight section of Oklahoma Highway 74.

Silkwood was killed in the crash, and the envelope was never recovered. The Oklahoma Highway Patrol concluded that she had fallen asleep while driving, but a private investigator concluded that her car had been forced off the road.

The resulting controversy, charges of a cover-up, and lawsuits provided a focal point for those concerned with safety in the nuclear power industry. Congressional hearings brought forth intriguing and bizarre stories but led to no definite conclusions. The Atomic Energy Commission confirmed that safety violations had occurred at the Cimarron plant, and it was eventually closed. After years in court, most of the questions surrounding Silkwood's death remained unanswered, but her father did win a large settlement from Kerr-Magee on behalf of her children. Her death and the events leading up to it were portrayed in the award-winning motion picture *Silkwood* (1983).

Kenneth H. Brown

SEE ALSO: Antinuclear movement; Hazardous and toxic substance regulation; Nuclear accidents; Nuclear regulatory policy; Power plants; Radioactive pollution and fallout.

Silver Bay, Minnesota, asbestos releases

DATE: began 1947; discovered 1973
CATEGORY: Human health and the environment

In 1947 the state of Minnesota granted the Reserve Mining Company of Silver Bay, Minnesota, permission to dump waste from its taconite processing plant directly into Lake Superior. Permission was granted in an attempt to revive mining in the Minnesota Iron Range.

Taconite is a low-grade iron ore used in the making of steel products. Rocks are crushed, and the ore is magnetically removed; the residual, or tailings, is industrial waste. In 1947 the Reserve Mining Company began dumping these tailings into a large chasm in Lake Superior at the rate of 67,000 tons per day.

On June 14, 1973, the Environmental Protection Agency (EPA) announced that high concentrations of asbestos fibers had been found in the drinking water of Duluth, Minnesota. The fibers were identified by Irving Selikoff, director of an environmental sciences laboratory in New York, as amosite, the same asbestos fibers that, when inhaled, were known to cause lung, stomach, and colon cancers after an incubation period of twenty to thirty years. Duluth took its drinking water directly from Lake Superior, which was thought to have the purest lake water in the world.

State and U.S. government experts charged that the Reserve Mining Company's tailings were the source for these fibers and that lake currents brought them into the water supply of Duluth and surrounding communities. At the time of the EPA warning, court action was under way to force Reserve to cease dumping its slurry into Lake Superior. Company executives protested that the fibers came into the lake naturally from eroding rocks located in tributary streams. They also contended that the "water scare" was a ploy to influence the court's decision.

With Reserve threatening to close if forced to find another dump site, Silver Bay residents, nearly all of whom owed their jobs to Reserve, strongly defended the company. Duluth's residents, while wanting Reserve to stop its dumping, generally reacted with equanimity when confronted with the EPA report, many noting that they had been drinking Lake Superior water for more than thirty years without ill effects. As no filtration system existed that could remove the fibers from the water, most Duluthians had no option except to continue to drink the water. Moreover, it was unclear whether ingesting the fibers had the same effect as inhaling them.

Numerous mortality studies were initiated by Selikoff and other scientists in 1973 to evaluate the impact of the fibers on Duluth's population. These studies failed to confirm that the water posed any serious threat to those who drank it. By 1975 the water scare dissipated and soon thereafter ceased to be an issue. After years of litigation, Reserve agreed to dispose of its taconite tailings on land. The company constructed a basin about 11 kilometers (7 miles) inland from Silver Bay to contain the tailings. In 1990 Reserve announced it was bankrupt and closed. Subsequently, another mining company reopened the Silver Bay taconite plant using the tailings basin built by Reserve.

Ronald K. Huch

SEE ALSO: Asbestos; Asbestosis; Clean Water Act and amendments; Drinking water; Water pollution.

Singer, Peter

BORN: July 6, 1946; Melbourne, Australia
CATEGORY: Animals and endangered species

Philosopher Peter Singer became a leading spokesperson for animal rights during the 1970's.

The son of Austrian Jewish refugees who moved to Australia in 1938, Peter Singer studied history and philosophy at the University of Melbourne in Australia and Oxford University in England. He began his academic career during the 1970's at Latrobe University in Australia and quickly established himself as one of that nation's premier philosophers. Later he became director of the Center for Human Bioethics at Monash University. Singer's appeal to the general public stems largely from his conviction that philosophers should apply theory to practical circumstances. It was this conviction that led to him becoming an outspoken advocate of animal rights.

In a series of books and articles, most notably *Animal Liberation: A New Ethics for Our Treatment of Animals* (1975), Singer protests that it is inappropriate to believe that animals exist for the use and benefit of humans. Singer bases his position on the fact that animals suffer pain and that

Activists block a street to protest the 1997 Labor Day pigeon shoot in Hegins, Pennsylvania. In his 1973 book Democracy and Disobedience *Peter Singer stated his belief that civil disobedience is acceptable when the rights of animals are at stake.* (AP/Wide World Photos)

humans can live well without inflicting this pain. His interest in this subject was inspired, in part, by comments written by early nineteenth century English utilitarian philosopher Jeremy Bentham, who argued that humans failed to grasp the suffering that animals endured.

Singer cites the use of animals in cosmetic screening as one example of unnecessary cruelty. He also questions the way that animals are treated by medical researchers, contending that much of the research could be satisfactorily accomplished without using animals. In *Animal Factories* (1980) Singer attacks agribusiness and insists that humans do not need to eat flesh to survive. His position on all these issues has brought strong criticism from farmers, scientists, sportsmen, and some economists. The controversy has given Singer the opportunity for many speaking engagements and media appearances.

Although not a proponent of extreme or violent measures to advance his cause, Singer, as he explains in *Democracy and Disobedience* (1973), considers civil disobedience acceptable in certain circumstances, and clearly the protection of animals is an acceptable circumstance. Singer also thinks it is important to live as he believes: He is a vegetarian and will not wear any clothing derived from animals.

With the assistance of other animal rights leaders around the world, including philosopher Tom Regan of North Carolina State University, Singer has dramatically raised the level of discourse in Western culture on how humans should relate to the natural world. Like most other environmentalists, Singer rejects as illegitimate the notion that nature exists for humans to exploit.

Singer has seen some positive results from the animal rights campaign. Most cosmetic companies have dropped the use of animals in their laboratory testing, and medical researchers are generally more respectful of the animals they use. On the other hand, sportsmen and those involved in producing animals for food and clothing have, for the most part, steadfastly resisted Singer's appeals.

Ronald K. Huch

SEE ALSO: Animal rights; Animal rights movement; Hunting; Ivory and ivory trade; Poaching; Whaling; Zoos.

Slash-and-burn agriculture

CATEGORY: Agriculture and food

Slash-and-burn agriculture is a practice in which forestland is cleared and burned for use in crop and livestock production. While yields are high during the first few years, they rapidly decline in subsequent years, leading to further clearing of nearby forestland.

Slash-and-burn agriculture has been practiced for many centuries among people living in tropical rain forests. Initially, this farming system involved small populations. Therefore, land could be allowed to lie fallow for many years, leading to the full regeneration of the secondary forests and hence a restoration of the ecosystems. During the second half of the twentieth century, however, several factors led to drastically reduced fallow periods. In some places such fallow systems are no longer in existence, resulting in the transformation of forests into shrub and grasslands, negative effects on agricultural productivity for small farmers, and disastrous consequences to the environment.

Among the factors that have been responsible for reduced or nonexistent fallow periods are increased population in the tropics, increased demand for wood-based energy, and, perhaps most important, the increased worldwide demand for tropical commodities during the 1980's and 1990's, especially in products such as palm oil and natural rubber. These last two factors have helped industrialize slash-and-burn agriculture, which was practiced for centuries by small farmers. Ordinarily, small farmers are able to control their fires such that they are similar to a small forest fire triggered by lightning in the northwestern or southeastern United States. However, the continued reduction in fallow periods, coupled with increased burning by subsistence farmers and large agribusiness, especially in Asia and Latin America, are resulting in increased environmental concern.

While slash-and-burn agriculture seldom takes place in temperate regions, some agricultural burning occurs in the Pacific Northwest of the United States, where it is estimated that

Declining Crop Yields with Successive Harvests on Unfertilized Tropical Soils

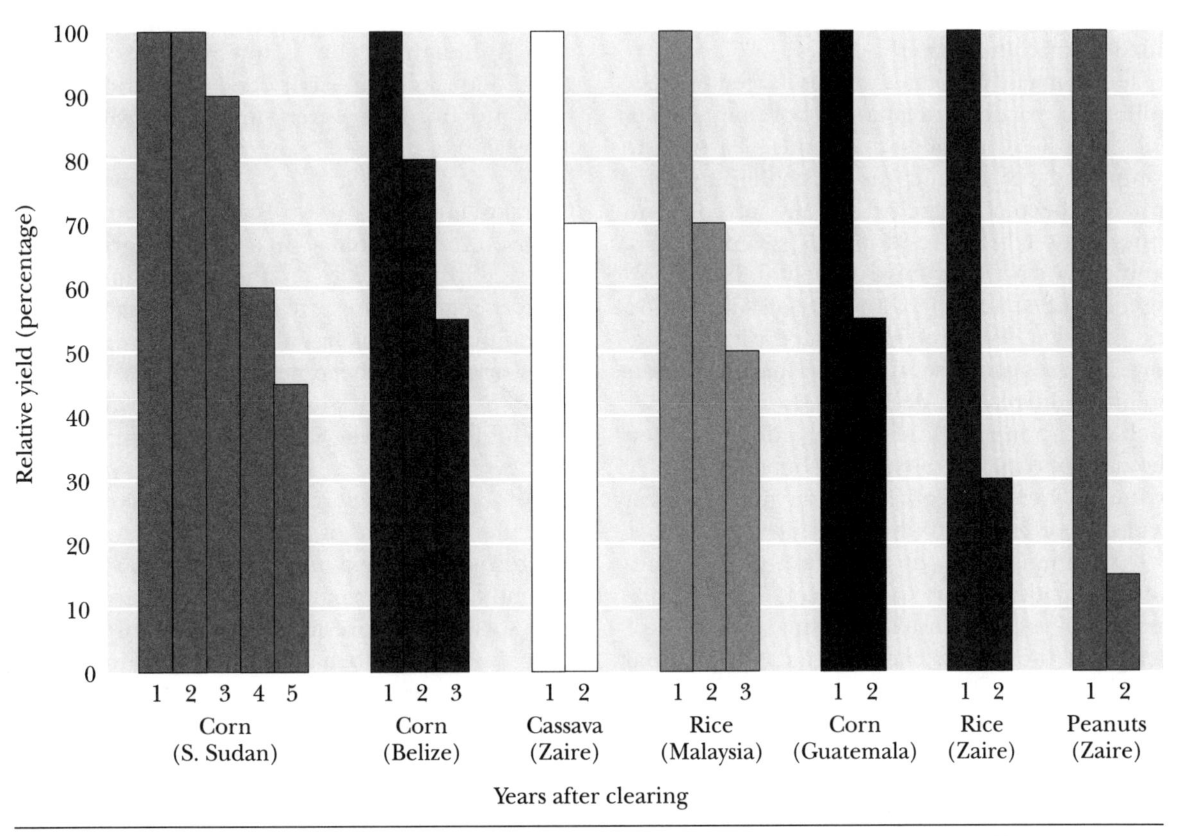

Source: Data adapted from John Terborgh, *Diversity and the Tropical Rain Forest.* New York: W. H. Freeman, 1992.

three thousand to five thousand agricultural fires are set each year in Washington State alone. These fires also create problems for human health and the environment.

Habitat Fragmentation

One of the most easily recognizable results of slash-and-burn agriculture is habitat fragmentation, which leads to a significant loss of the vegetation needed for the maintenance of effective gaseous exchange in tropical regions and throughout the world. For every acre of land lost to slash-and-burn agriculture, 10 to 15 acres of land are fragmented, resulting in the loss of habitat for wildlife, plant species, and innumerable macro- and microorganisms yet to be identified. This also creates problems for management and wildlife conservation efforts in parts of the world with little or no resources to feed their large populations. Fragmentation has also led to intensive discussions on global warming. While slash-and-burn agriculture by itself is not completely responsible for global warming, the industrialization of the process could make it a significant component of the problem, as more and more vegetation is fragmented.

Human Health

The impact of slash-and-burn agriculture on human health and the environment is best exemplified by the 1997 Asian fires that resulted from such practices. Monsoon rains normally extinguish the fires set by farmers, but a strong El Niño weather phenomenon delayed the expected rains, and the fires burned out of control for months. Thick smoke caused severe health problems. It is estimated that over 20 million people in Indonesia alone were treated for asthma, bronchitis, emphysema, and eye, skin, and cardiovascular problems as a result of the fires. Similar problems have been reported for smaller agricultural fires.

Three major problems are associated with air pollution: particulate matter, pollutant gases, and volatile organic compounds. Particulate compounds of 10 microns or smaller that are inhaled become attached to the alveoli and other blood cells, resulting in severe illness. Studies by the U.S. Environmental Protection Agency (EPA) and the University of Washington indicate that death rates associated with respiratory illnesses increase when fine particulate air pollution increases. Meanwhile, pollutant gases such as carbon monoxide, nitric oxide, nitrogen dioxide, and sulfur dioxide become respiratory irritants when they combine with vapor to form acid rain or fog. Until the Asian fires, air pollutants stemming from the small fires of slash-and-burn agriculture that occur every planting season often went unnoticed. Thus, millions of people in the tropics experience environmental health problems because of slash-and-burn agriculture that are never reported.

Soil and Water Quality

The loss of vegetation that follows slash-and-burn agriculture causes an increased level of soil erosion. The soils of the humid tropics create a hard pan underneath a thick layer of organic matter. Therefore, upon the removal of vegetation cover, huge areas of land become exposed to the torrential rainfalls that occur in these regions. The result is severe soil erosion. As evidenced by the impact of Hurricane Mitch on Honduras during 1998, these exposed lands can give rise to large mudslides that can lead to significant loss of life. While slash-and-burn agriculture may not be the ultimate cause for sudden mud slides, it does predispose these lands to erosional problems.

Associated with erosion is the impact of slash-and-burn agriculture on water quality. As erosion continues, sedimentation of streams increases. This sedimentation affects stream flow and freshwater discharge for catchment-area populations. Mixed with the sediment are minerals such as phosphorus and nitrogen-related compounds that enhance algal growth in streams and estuaries, which depletes the supply of oxygen that aquatic organisms require to survive. Although fertility is initially increased on noneroded soils, nutrient deposition and migration into drinking water supplies continues to increase.

Controlling Slash-and-Burn Agriculture

Given the fact that slash-and-burn agriculture has significant effects on the environment not only in regions where it is the mainstay of the agricultural systems but also in other regions of the world, it has become necessary to explore different approaches to controlling this form of agriculture. However, slash-and-burn agriculture has evolved into a sociocultural livelihood; therefore, recommendations must be consistent with the way of life of a people who have minimal resources for extensive agricultural systems.

Among the alternatives are new agroecosystems such as agroforestry systems and sustainable agricultural systems that do not rely so much on the slashing and burning of forestlands. These systems allow for the cultivation of agronomic crops and livestock within forest ecosystems. This protects soils from being eroded. Another possibility is the education of small rural farmers, absentee landlords, and big agribusiness concerns in developing countries to understand the environmental impact of slash-and-burn agriculture. While small rural farmers may not have the resources for renovating utilized forestlands, big business can organize ecosystems restoration, as has been done in many developed nations of the world.

Oghenekome U. Onokpise

SUGGESTED READINGS: J. Peterson, "Emissions from Wildland and Prescribed Fire," *Women in Natural Resources Journal* 19 (1998), discusses the pollution problems arising from both natural and human-made forest fires. L. M. Simons and M. Yamashita, "Indonesia's Plague of Fire," *National Geographic* 194 (1998), details the extensive effects of the 1997 Asian fires. W. D. Sunderlin, "Deforestation, Livelihoods, and Preconditions for Sustainable Management in Olancho, Honduras," *Agriculture and Human Values* 14 (1997), provides an in-depth case study of a community trying to find a balance between short-term economic needs and long-term survival. J. Terborgh, *Diversity and the Tropical Rain Forest* (1992), contains a good overview of tropical rain forests and the problems associated with slash-and-burn agriculture.

See also: Deforestation; Rain forests and rain forest destruction; Sustainable agriculture.

Sludge treatment and disposal

Category: Waste and waste management

Sludge is the sediment or residue left after the removal of water from sewage and industrial waste.

Most wastewater, whether from industrial discharge, storm drains, or sewage systems, goes through a process that separates the solids from the water. It takes place in up to three stages called primary, secondary, and tertiary treatment. Though tertiary treatment is desired before wastewater is discharged into lakes or oceans, secondary treatment is the minimum allowed for such sewage. Tertiary treatment involves polishing, or further treating, the liquid—or effluent—that was removed during the dewatering process.

Primary treatment involves collecting the wastewater in a sedimentation lagoon or clarifier. The water is allowed to settle so the solids and liquids will separate. The sediment that is left after the liquid is pumped out is called sludge. Sludge usually contains about 95 percent water. It is filtered to remove more water. This process is called dewatering.

The open lagoon method filters sludge through sand beds, allowing it to air dry. This may take several months. To accelerate the process and reduce offensive conditions, mechanical clarifying tanks are used at most treatment facilities. Rotating mechanical rakes move the settled solids to the center of the tank, where they are drawn off. This sludge is further dewatered before it is disposed of in a landfill or through incineration. Sometimes lime or other chemicals are added to increase the amount of solids that settle on the tank's bottom.

Secondary treatment involves the use of bacteria. There are several methods, including aerated lagoons, activated sludge treatment, digesting tanks, trickling filters, and oxidation ponds. Most involve supplying oxygen to the sludge so the microorganisms present will break down the organic contaminants in the sludge. Each method differs in the way in which the oxygen is supplied.

Aerated lagoons and oxidation ponds are similar. Both methods use large, shallow, open pits. Air and sunlight encourage bacteria and algae to grow. They work together to break down the organic matter: The bacteria consumes the organic matter, while the algae "feeds" on the sun and provides further oxygen for the bacteria to thrive. Sometimes oxygen is supplied mechanically, which allows the ponds to be smaller and still process the same amount of sludge as oxidation ponds. Sludge deposits are removed from the ponds on a regular basis through dredging.

A trickling filter is a large tank filled with stones, or a large plastic tank. Settled sewage is sprayed on top of the stones, or on the top and walls of the plastic tank. Water from the sediment trickles to the bottom of the tank, where it is collected, removed, and treated. Bacteria attack and metabolize the sediment that clings to the rocks or tank walls. Sealed digesting tanks use anaerobic bacteria, which work best without oxygen, to break down the organic matter in sludge.

The most common method for sludge treatment is the activated sludge process, which is an aerobic (with oxygen) biological system. Microbes that rely on air are used to help metabolize, or break down, organic waste. During the first stage of this treatment, sludge is mixed with settled sewage in an aeration tank. Large amounts of microorganisms are collected in the aeration tank and mixed with the semisolid slurry. Oxygen is added so the microorganisms will feed on the organic matter in the wastewater. Tiny particles of waste are consumed by the organisms. Large particles are broken down. After about twelve hours, the slurry in the aeration tank is pumped to a sedimentation tank. The sediment settles, and the organisms are returned to the aeration tank where they can consume more organic matter. Any remaining liquid is treated to remove nitrogen and phosphorous, which may cause excessive growth of plants or algae, before it is discharged in lakes or the ocean.

About 30 percent of the sludge is pumped back to the aeration tank to repeat the process. Recirculating the sludge and mixing it with fresh sewage sediment in the aeration tank is a key part of the activated sludge treatment process. Remaining sludge is dewatered using centrifuges, which spin the sludge to draw out water. Filter presses use belts or plates to squeeze out excess water. They create drier sludge cake.

The method used to dewater sludge depends on the disposal method. The more water that is removed from sludge, the less volume of sludge cake remains. Sludge cake needs to be very dry if it is to be incinerated. If the sludge cake is to be used as fertilizer, added to composting facilities, or disposed of in landfills, the water content can be higher. Chemicals are sometimes added to the sludge to encourage particles to clump together, which speeds the removal of water.

Separation of sludge from wastewater does not prevent toxins or other pollutants from remaining in the sludge, which means that toxins can enter landfills or crops. Bioremediation remedies this problem. Microorganisms known to consume inorganic particles, oil, and other toxins are added to the sludge. The Palm Beach County Solid Waste Authority in Florida has had great success using bioremedial treatment. They compost sludge and solid waste into a usable soil conditioner.

Lisa A. Wroble

SUGGESTED READINGS: Sludge treatment as part of the sewage treatment process is covered in *Flush: Treating Wastewater* (1995), by Karen Mueller Coombs. Sewage treatment and disposal of sludge in the ocean are covered in *Ocean Pollution* (1991), by Maria Talen. James D. Snyder describes the use of microbes and bioremediation in sewage treatment and toxic site cleanup in "Off-the-shelf Bugs Hungrily Gobble Our Nastiest Pollutants," *Smithsonian* (April, 1993). Bioremediation in the treatment of sewage and sludge is also covered in *Endangered Animals and Habitats: The Oceans* (1998), by Lisa A. Wroble. The Internet also provides many opportunities for further research on sludge treatment and disposal. Science Traveler International (http://scitrav.com/wwater/waterlnk.htm) provides links to many related sites as well as information on the process and the microbiology of sludge treatment and a background of terms.

SEE ALSO: Bioremediation; Composting; Sewage treatment and disposal; Waste management.

Smallpox eradication

CATEGORY: Human health and the environment

The eradication of smallpox by scientists working under the auspices of the World Health Organization represented the first time medical science was able to eliminate an infectious disease. Concern remains that eradication of a viral species may create an environmental niche into which other agents may enter.

While the precise origin of smallpox is unknown, it clearly was a disease of antiquity. Pharaoh Ramses V of Egypt (twelfth century B.C.E.) reportedly died of a disease resembling smallpox. The disease was prevalent in China and India for at least fifteen centuries before its appearance in the Middle East in the sixth century C.E. Crusaders returning from the Middle East spread smallpox throughout Europe during the twelfth and thirteenth centuries, while the disease was introduced to the Western Hemisphere by the Spanish under Hernán Cortés (1520). Reportedly, more than three million Aztecs died from smallpox within two years, opening Mexico to Spanish invaders. It is estimated that the annual death rate from smallpox in Europe during this period approached 400,000 people. Those who survived were often disfigured and scarred.

The first attempts at immunization against the disease were practiced by the Chinese. In a process called variolation, dried powder from smallpox crusts was inhaled. The procedure was taken to the Middle East during the seventeenth or eighteenth centuries, probably by Arab traders, where Lady Mary Wortley Montagu, wife of the British ambassador to the Ottoman Empire, became aware of the practice. Lady Montagu, herself a scarred survivor of the disease, had her young son successfully variolated in 1718. The British royal family soon heard of the practice and introduced it into the country several years later.

Edward Jenner tested the first smallpox vaccine. (Library of Congress)

While variolation was useful in producing immunization against smallpox, it remained a difficult and dangerous procedure. An alternative had long been practiced by British dairy farmers: the inoculation of people with material taken from the lesions that developed on the udders of cows infected with cowpox. In the 1790's, Edward Jenner, an English country physician, became aware of the practice and tested the procedure on himself and several volunteers. With the publication of his work *An Inquiry into the Causes and Effects of the Variolae Vaccinae* (1798), the use of vaccination quickly spread through both Europe and the Americas. Though smallpox was not eliminated, its incidence steadily declined over the next century. It was eliminated from the United States by 1949.

The key to global eradication was based on a characteristic of the virus that made it unlike most viral diseases: Humans represented the only reservoir for smallpox. Since both vaccination and recovery from infection resulted in lifelong immunity, once the chain of infection was disrupted, smallpox would cease to be an epidemic disease.

In 1950 a program was developed to eradicate smallpox in the Western Hemisphere. The program consisted of widespread immunization among susceptible populations; by 1958, the disease had been eradicated in most of the Americas, lending credibility to a Soviet proposal for the global elimination of the disease. Beginning in 1965 a program was developed to meet that goal. The program was based on the realization that immunization of 100 percent of the world's population was unrealistic. Rather, the goal was to detect and contain local outbreaks of disease, breaking any chain of transmission. Once the disease could no longer spread beyond local borders, any outbreak would die out.

In 1967 the incidence of smallpox approached an estimated ten to fifteen million cases in forty-six countries. Within ten years, however, the disease virtually ceased to exist; the last reported natural case was that of a young Somali in 1977. There was a single fatal laboratory-associated infection in 1978 at the Birmingham University Medical School in Great Britain, in which a medical photographer was infected by a virus being studied in an adjacent laboratory. On October 26, 1979, the World Health Organization announced the global eradication of the disease.

With the elimination of a viral species, concern remains that an environmental niche may be created that could provide a means for other viruses to enter the human population. Widespread infection by smallpox, in addition to general use of vaccination, had created a population

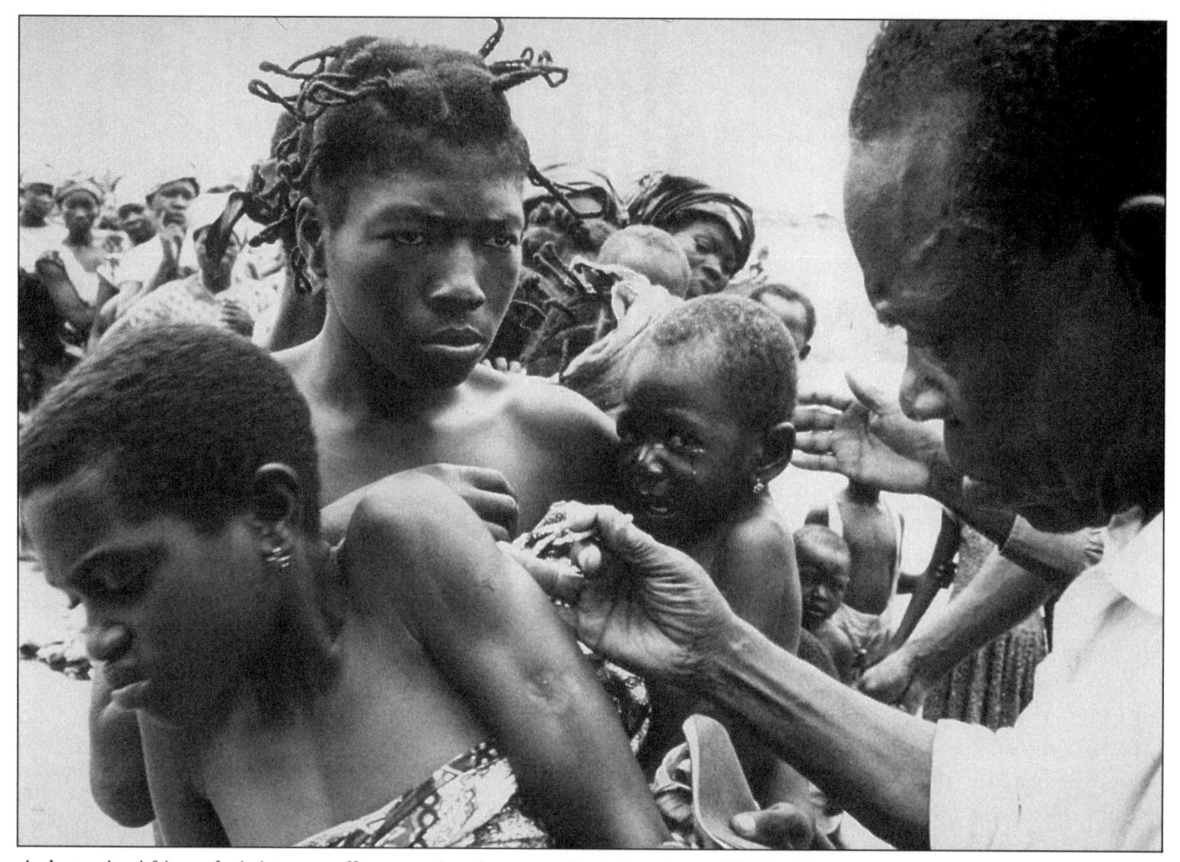

A doctor in Africa administers smallpox vaccinations to a line of patients. Worldwide vaccination efforts begun during the 1960's proved effective: The last reported natural case of smallpox occurred in Somalia in 1977. (Archive Photos)

Following the eradication of smallpox, the issue was raised as to whether to eliminate all laboratory stocks of the virus, which would mean the deliberate destruction of a species. By the mid-1990's, the only remaining virus stocks known to exist were stored at the Centers for Disease Control and Prevention in Atlanta, Georgia, and the Institute for Viral Preparations in Moscow, Russia. The deoxyribonucleic acid (DNA) genomes of several species of the virus were sequenced, but by the end of the 1990's no decision had been made as to disposition of the viral stocks.

that was resistant to infection by most other forms of poxviruses. While smallpox was species specific in only infecting humans, other viruses such as monkeypox or the whitepox viruses can infect a variety of primates, including humans. These viruses are neither as disfiguring nor as deadly as smallpox, and they are not transmitted as readily as smallpox, but combinations of poverty and social upheavals in poor Third World countries could conceivably facilitate their spread among human populations.

Richard Adler

SUGGESTED READINGS: The most complete description of the research that led to eradication of smallpox is found in "The Smallpox Story: Life and Death of an Old Disease," *Microbiological Reviews* (December, 1983), by Abbas Behbehani. An abbreviated article by the same author, "The Smallpox Story: Historical Perspective" is found in *American Society for Microbiology News* 57 (1991). Behbehani also authored a popular book on the subject, *The Smallpox Story in Words and Pictures* (1988). For those interested in a numerical discussion on the subject, *The Global Eradication of Smallpox* (1980), published by the World Health Organization, provides a clinical view of events.

SEE ALSO: Bacterial resistance and super bacteria; Genetically altered bacteria.

Smog

CATEGORY: Atmosphere and air pollution

Originally a blend of the words "smoke" and "fog," the term "smog" was coined to describe the severe air pollution that resulted when smoke from factories combined with fog during a temperature inversion. The word was later broadened to include photochemical smog—first noticed in the Los Angeles Basin—which resulted from sunlight acting on unburned hydrocarbons emitted from automobiles.

As one ascends upward from the earth's surface, the air temperature drops by about 5.5 degrees Fahrenheit every 1,000 feet. Temperature inversions occur when this normal condition is reversed so that a blanket of warm air is sandwiched between two cooler layers. The temperature inversion restricts the normal rise of surface air to the cooler upper layers, in effect placing a lid over a region. When air cannot rise, the air currents that carry pollutants away from their sources in cities stagnate, causing pollution levels to increase drastically. A combination of severe air pollution, prolonged temperature inversion, and moisture-laden air may result in a what has been termed "killer fog."

Several acute episodes of killer fog occurred during the twentieth century. One was in the Meuse Valley of Belgium. During the first week of December, 1930, a thick fog and stagnant air from a temperature inversion concentrated pollutants spewing forth from a variety of factories in this heavily industrialized river valley. After three days of such abnormal conditions, thousands of residents became ill with nausea, shortness of breath, and coughing. Approximately sixty people died, primarily elderly people or those with chronic heart and lung diseases. The detrimental effects on health were later attributed to sulfur oxide gases emitted by combusting fossil fuels concentrated to lethal levels by the abnormal weather. The presence of coal soot, combined with moisture from the fog, exacerbated the effect.

A second episode occurred in Donora, Pennsylvania, during the last week of October, 1948. Donora is situated in a highly industrialized river valley south of Pittsburgh. A four-day temperature inversion with fog concentrated the gaseous effluents from steel mills with the sulfur oxides released by burning fossil fuels. Severe respiratory tract infections began to occur, especially in the elderly, and 50 percent of the population became ill. Twenty people died, a tenfold increase in the normal death rate.

A third major episode occurred in London, England, in early December, 1952. At that time, many citizens burned soft coal in open grates. When a strong temperature inversion and fog enveloped the city for five consecutive days, residents began complaining of respiratory ailments. By the time the inversion had lifted, four thousand excess deaths had been recorded. In this case it was not only the elderly who were affected: Deaths occurred in all age categories. During the next decade London experienced two additional episodes: one in 1956, which claimed the lives of one thousand people, and one in 1962, which caused seven hundred deaths. The decline in mortality rates resulted from restricting the use of soft coal, with its high sulfur content, as a source of fuel. Sulfur oxide compounds are responsible for causing lung problems during these episodes; therefore, killer fogs have been more accurately renamed "sulfurous smogs."

A thick layer of smog obscures a view of the skyline of downtown Los Angeles. Photochemical smog produces ozone, which damages plants and causes respiratory problems in humans. (Ben Klaffke)

Photochemical smog, first noticed in the Los Angles Basin in the late 1940's, has been an increasingly serious problem in cities. Unlike sulfurous smogs, moisture is not part of the equation, and smoke-belching factories dumping tons of sulfur oxide compounds into the atmosphere are not required. Rather, photochemical smog results when unburned hydrocarbon fuel, emitted in automobile exhaust, is acted upon by sunlight. The Los Angeles Basin, hemmed in by mountains to the east and ocean to the west, has a high density of automotive traffic and plenty of sunshine. Varying driving conditions mean that gasoline is never completely consumed; instead, it is often changed into other highly reactive substances. Sunlight acts as an energy catalyst that changes these compounds into a variety of powerful oxidizing agents, which constitute photochemical smog. This type of smog has a faint bluish-brown tint and typically contains several powerful eye irritants. The chemical reactions also produce aldehydes, a class of organic chemical best typified by an unpleasant odor.

The complicated chemistry of photochemical smog also produces ozone, which is extremely reactive; it damages plants and irritates human lungs. Since ozone production is stimulated by sunlight and high temperatures, it becomes a particularly pernicious problem during the summer, especially during morning rush hours. Under temperature inversion conditions, the ozone created in photochemical smog can increase to dangerous levels. Ozone is highly toxic. It irritates the eyes, causes chest irritation and coughing, exacerbates asthma, and damages the lungs. Photochemical smog and ozone are now common ingredients in urban air. Although acute episodes of ozone-induced mortality are rare, concern is growing about the detrimental long-term consequences of the brief but repetitive exposures consistently inflicted upon commuters. It appears as though no curtailment of the problem will be possible in urban areas in the

United States without significant changes in transportation, strict limits on growth, and radical alterations in lifestyle.

George R. Plitnik

SUGGESTED READINGS: *Smog Alert: Managing Urban Air Quality* (1996), by Derek M. Elsom, discusses photochemical smog and what can be done to control it. Wyn Grant's *Autos, Smog, and Pollution Control: The Politics of Air Quality Management in California* (1995) explains the air pollution situation in California. Grant discusses why the pollution-control devices mandated by the state cannot hope to clean the air but can only keep the air quality from deteriorating further. An excellent overview that discusses the complicated relationship between energy use and its environmental effects is *Energy: Its Use and the Environment* (1996), by Roger A. Hinrichs. Chapter 8 details air pollutants, air-quality standards, and automotive emission control devices.

SEE ALSO: Air pollution; Air pollution policy; Automobile emissions; Black Wednesday; Catalytic converters; Clean Air Act and amendments; Donora, Pennsylvania, temperature inversion; London smog disaster.

Snail darter

CATEGORY: Animals and endangered species

The snail darter is a small fish whose critical habitat was threatened by the construction of the Tellico Dam on the Little Tennessee River by the Tennessee Valley Authority. The dam project was challenged by citizens using both the National Environmental Policy Act of 1969 and the Endangered Species Act of 1973.

The Tennessee Valley Authority (TVA) began building dams in the Tennessee Valley watershed in 1936. In 1960 the TVA had more than sixty dams in the region and focused on building the Tellico Dam on the Little Tennessee River. The Little Tennessee River and its valley were sacred to the Cherokee, used by more than three hundred farm families, and enjoyed by hundreds of canoeists and fishers as the region's only remaining stretch of natural river. In response to the proposed Tellico Dam, a citizens' coalition was formed in 1964. Although the local citizens were unable to stop the TVA from beginning the Tellico project, enactment of the National Environmental Policy Act (NEPA) in 1969 gave them a way to stop it. A court order stopped the construction of the dam until 1973, when the TVA produced a legally sufficient environmental impact statement.

Around the same time, a University of Tennessee professor discovered an endangered species of perch, the snail darter (*Percina tanasi*), living in the Tellico project area. The Endangered Species Act (ESA) prohibits federal actions that jeopardize the existence of endangered species or modify their critical habitat. Citizens opposed to the dam began administrative and court proceedings based on the ESA to stop the project. The Tellico citizens' group showed the court that the TVA did not properly value the river as a major recreational resource, the valley's rich agricultural lands, or the area's historic resources. The citizens pointed to the snail darter as an indicator species, a barometer of endangered human and economic values whose disappearance would signal devastating environmental degradation.

The U.S. Supreme Court decided that construction of the Tellico Dam violated the ESA and permanently stopped the completion of the project. However, Congress held meetings to consider the extreme nature of the law's application: a $100 million dam stopped because of a fish. These meetings found the Court's decision to be rational. Congress later created the Endangered Species Committee to authorize the extinction of certain species in compelling cases. The case of the Tellico Dam's snail darters was reviewed by this committee, which unanimously upheld protection for the endangered fish.

However, the legal victory for the snail darters was spoiled by political maneuvers. A senator and a congressman from Tennessee inserted language into an appropriations bill that overrode the Supreme Court's Tellico decision. Despite talk of a veto, the bill was signed into law. Consequently, the TVA finished the Tellico Dam and closed its floodgates on November 28, 1979. Although none of the estimated twenty thousand snail darters in the Little Tennessee River survived, small populations of the fish have been transplanted to and discovered in several downstream sites. Meanwhile, the TVA has been relatively unsuccessful in implementing its planned development for the flooded valley.

Michael D. Kaplowitz

SEE ALSO: Dams and reservoirs; Endangered species; Endangered Species Act; Endangered species and animal protection policy; *Tennessee Valley Authority v. Hill*; Tellico Dam.

Snyder, Gary

BORN: May 8, 1930; San Francisco, California
CATEGORY: Ecology and ecosystems

Poet and essayist Gary Snyder was one of the first writers to base his poetry, ethics, and spirituality around environmental ideas and values. He is one of the most influential figures in American nature writing.

Nature writer Gary Snyder, whose collection Turtle Island *won the 1975 Pulitzer Prize in poetry.* (Gary Snyder)

Gary Snyder grew up in a rural area outside Seattle, Washington, and started hiking in the mountains early in his life. The radical labor politics of the area laid the foundation for his critique of the dominant ideology and social structure of Western society, and the presence of American Indian and Asian culture in the Pacific Northwest attracted him to other traditions as constructive alternatives. In college he studied anthropology, followed by graduate study in Asian languages. He was a logger, mountain lookout, and merchant seaman; a love for work and an emphasis on the physical are key themes in his writings.

In the 1950's he moved to the San Francisco area, where he was associated with (but not a member of) the Beat movement. His early poetry was particularly influenced by Native American myth and Chinese nature poetry, which he translated in English. He studied Zen Buddhism in Japan during the late 1950's and early 1960's, developing a rich understanding of both the philosophy and practice of that tradition. Following his return to the United States in the late 1960's, he began emphasizing Native American wisdom along with Buddhism in his view of nature and the place of humans within it.

Snyder writes of the sacrality of natural processes, including predation and decomposition. He argues for the intrinsic value of all of nature and the importance of biodiversity, celebrating "the preciousness of mice and weeds." Thus he goes beyond the traditional romantic view of nature. He is known not only for his praise of wilderness but also for his exploration of its connection to the rich wilderness in the human mind. His view of the self and nature exhibits the Buddhist view of radical interrelationship, which combines a holistic vision of nature with an affirmation of the reality and value of individuals.

Snyder's poetry is suffused with a sensuous and mystical intimacy with nature, and he has proposed a shamanistic view of the poet's role as one who heals by bringing people into a close relationship with the natural world. Yet he also insists on attention to the practical details of living as a full member of one's bioregion. The interplay of culture and nature is seen as central to the development of a deeply rooted sense of place. Considered a major voice of deep ecology, his critique of Western society and ideology relates him to social ecology and ecofeminism as well. Snyder won the 1975 Pulitzer Prize in poetry for his collection *Turtle Island* (1974).

David Landis Barnhill

SEE ALSO: Biodiversity; Deep ecology; Ecofeminism; Environmental ethics; Social ecology.

Social ecology

Category: Philosophy and ethics

Social ecology is a philosophical movement whose adherents believe that the domination of nature by humans is derived from the domination of human society by the capitalist mode of production.

Social ecologists try, in many ways, to combine the environmental concerns of American preservationist John Muir with the economic concerns of German political philosopher Karl Marx. They argue that the domination of both disadvantaged people and the environment are derived from modern capitalist society and that both must be corrected together. All social ecologists look to the views of Karl Marx and Friedrich Engels for guidance, although they may differ in the policies they advocate.

There are two main theoretical approaches to social ecology: anarchist and socialist. Murray Bookchin is one of the leading proponents of anarchist social ecology. His view emphasizes the interdependence of humans and nonhuman nature. Hierarchy in society leads to the domination of some people by others and to the domination of nature by humankind. Bookchin and other anarchist social ecologists are suspicious of the state as an agent of domination. Ideally, they would do away with all hierarchy in society, which they contend would remove the hierarchy of exploitation of nature by humans. Social ecologists should thus work to eliminate the domination of nature by the material world by first taking into account social issues in order to address environmental problems. This view of social ecology is critical of mainstream environmentalism, which it sees as a mechanistic approach that treats nature as a resource for humans to use. It is also critical of some deep ecologists for being insensitive to social issues. Successful social ecology in this perspective emphasizes the achievement of small-scale communities in which local groups live in harmony with each other and the environment.

Socialist ecology is also rooted in the Marxian tradition, but instead of an emphasis on hierarchy and domination and the achievement of a utopian society modeled on nature, this school of thought calls for an economic transformation of society into an ecological socialist system. This variant is often the motivating force for much of the Green social and political movement. Particularly in the work of James O'Connor, socialist ecology starts from a Marxian framework of society but then incorporates the concepts of the autonomy of nature and ecological science. In an ecological socialist society, nature would be recognized as autonomous rather than as humanized and capitalized. The interrelatedness of all living organisms would be recognized, and socially and environmentally harmful means of production would be curtailed. Socialist ecology often emphasizes the importance of central planning and organizing the working class to achieve its aims of remaking first the state and then the human-nature relationship. An example of socialist ecology at work is integrated pest management, which is based on using biological rather than chemical means of controlling pests, a process that is less exploitive of nature and society.

Social ecology, no matter what its perspective, has much to offer anyone concerned about the environment. It also has many critics from a variety of perspectives. Deep ecologists, for example, criticize social ecology for its essentially human-centered perspective. Some ecofeminists contend that social ecology does not do enough regarding the domination of women. Social ecologists respond to the first criticism by maintaining that humans are part of the environment and that domination of humans is just as wrong as the domination of nature. Moreover, they argue that because the domination of nature is centered in the capitalist mode of production—a socioeconomic phenomenon—protecting nature must deal with capitalist, industrial society in a realistic fashion. Social ecologists also contend that dealing with the exploitation of humans deals with the exploitation of women.

A more telling criticism is that some socialist ecologists are overly caught up in Marxist rhetoric and are unable go beyond doctrine to offer workable alternatives to modern capitalist society. The deep ecologist argument that social ecology does not pay enough attention to environmental ethics would also appear to have some validity. Anarchist social ecology has a tendency to become utopian rather than practical in orientation, while socialist ecology is often doctrinaire and rigid. Both groups often expend a good deal of energy criticizing each other.

A far-ranging attack on social ecology comes from a market-based perspective. This approach, which social ecologists would argue is a mechanistic one that is ultimately exploitive of people and nature, argues that small-scale communal efforts at dealing with environmental issues are ineffective. Because modern society is based on property rights, environmental solutions must take property rights into account. Such critics argue that social ecology is too utopian and unable to deal with the problems facing modern society. One social ecologist response would be, "Yes, but that is why capitalist, industrial society needs to be remade rather than taking for granted a perspective that is, at base, exploitive of humans and nature."

Social ecology is both a philosophy and organizing principles for social reformers who are concerned about the environment. It aims to achieve an ecosystem in which humans and the rest of the natural world live in harmony in a nonexploitive setting. It has had some political success with the Green movement and provides trenchant criticism of modern society. However, both variants are treated with suspicion by many mainstream environmentalists.

John M. Theilmann

SUGGESTED READINGS: Murray Bookchin's work is fundamental to understanding the anarchist perspective. Among his more important books are *The Ecology of Freedom* (1982) and *Remaking Society* (1989). A good, but difficult, starting point for socialist ecology is James O'Conner, *The Fiscal Crisis of the State* (1973). Broad perspectives include Robyn Eckersley, *Environmentalism and Political Theory: Toward an Ecocentric Approach* (1992), and David Pepper, *Eco-Socialism* (1993).

SEE ALSO: Bookchin, Murray; Deep ecology; Ecofeminism; Environmental ethics; Green movement and Green parties.

Soil, salinization of

CATEGORY: Land and land use

Soil salinization is a process in which water-soluble salts build up in the root zones of plants, blocking the movement of water and nutrients into plant tissues. Soil salinization rarely occurs naturally. It becomes an environmental problem when it occurs as a result of human activity, denuding once-vegetated areas of all plant life.

Rainwater is virtually free of dissolved solids, but surface waters and underground waters (groundwater) contain significant quantities of dissolved solids, ultimately produced by the weathering of rocks. Evaporation of water at the land surface results in an increase in dissolved solids that may adversely affect the ability of plant roots to absorb water and nutrients.

In arid regions, evaporation of soil water potentially exceeds rainfall. Shallow wetting of the soil followed by surface evaporation lifts the available dissolved solids to near the surface of the soil. The near-surface soil therefore becomes richer in soluble salts. In natural arid areas, soluble salts in the subsurface are limited in quantity because rock weathering is an extremely slow process. Degrees of soil salinization detrimental to plants are uncommon.

Irrigating arid climate soils with surface or groundwater provides a constant new supply of soluble salt. As the irrigation water evaporates and moves through plants to the atmosphere, the dissolved solid content of the soil water increases. Eventually, the increase in soil salt will inhibit or stop plant growth. It is therefore necessary to apply much more water to the fields than required for plant growth to flush salts away from the plant root zone. If the excess water drains easily to the groundwater zone, the groundwater becomes enriched in dissolved solids, which may be detrimental.

If the groundwater table is near the surface, or if there are impermeable soil zones close to the surface, overirrigation will not alleviate the problem of soil salinization. This condition requires the installation of subsurface drains that carry the excess soil water and salts to a surface outlet. The problem with this method is that disposing of the salty drain water is difficult. If the drain water is released into surface streams, it degrades the quality of the stream water, adversely affecting downstream users. If the water is discharged into evaporation ponds, it has the potential to seep into the groundwater zone or produce a dangerously contaminated body of surface water, as occurred at the Kesterson Wildlife Refuge in California, where concentrations of the trace element selenium rose to levels that interfered with the reproduction of resident birds.

Robert E. Carver

SEE ALSO: Heavy metals and heavy metal poisoning; Irrigation; Kesterson National Wildlife Refuge poisoning; Runoff: agricultural; Soil contamination.

Soil conservation

CATEGORY: Land and land use

Soil conservation is the effort by farmers and other landowners to prevent the buildup of salts and fertilizer acids, as well as the loss of topsoil from wind erosion, water erosion, desertification, and chemical deterioration. Soil conservation involves the implementation of management practices to maintain soil quality and reduce pollution.

According to the United Nations Environment Programme, approximately 17 percent of the earth's vegetated land is degraded, which poses a threat to agricultural production around the world. The introduction of minerals, metals, nutrients, fertilizers, pesticides, bacteria, and pathogens suspended in topsoil runoff into waterways is a significant source of water pollution and is a threat to fisheries, wildlife habitat, and drinking water supplies. The introduction of soil particles into the air through wind erosion is a significant source of air pollution.

The Industrial Revolution of the nineteenth century and the population explosion of the twentieth century encouraged people to till new land, cut down forests, and disturb soil for the expansion of towns and cities. The newly exposed topsoil quickly succumbed to erosion from rainfall, floods, wind, ice, and snow. The Dust Bowl, which occurred in the Great Plains in the United States during the 1930's, is one example of the devastating effects of wind erosion.

Hugh Hammond Bennett, the so-called father of soil conservation, lobbied for congressional establishment of the United States Soil Conservation Service (approved in 1937) and the establishment of voluntary Soil Conservation Districts in each state. Bennett was named the first chief of the U.S. Soil Conservation Service in 1937. On August 4, 1937, the Brown Creek Conservation District in Bennett's home county, Anson County, North Carolina, became the first Soil Conservation District in the United States. Local landowners voted to establish the district by three hundred to one, proving that farmers were concerned about soil conservation. A reporter for the *Charlotte Observer* newspaper sought out the one negative voter; after having the program explained to him, he changed his opinion. By 1948 more than 2,100 districts had been established nationwide. They were eventually renamed Soil and Water Conservation Districts. There are more than three thousand such districts in the United States.

The U.S. Food Security Act of 1985 authorized the Conservation Reserve Program to take land highly susceptible to erosion out of production and required farmers to develop soil conservation plans for the remaining susceptible land. The Natural Resource Conservation Service estimates that the loss of topsoil was nearly cut in half, reduced from 1.6 billion tons per year to 0.9 billion tons. The European Community and Australia also adopted soil conservation measures during the 1990's.

Soil conservation practices include covering the soil with vegetation, reducing soil exposure on tilled land, creating wind and water barriers, and installing buffers. Vegetative cover slows the wind at ground level, slows water runoff, protects soil particles from being detached, and traps blowing or floating soil particles, chemicals, and nutrients. Because the greatest wind and water erosion damage often occurs during seasons in which no crops are growing or natural vegetation is dormant, soil conservation often depends on permitting the dead residues and standing stubble of the previous crop to remain in place until the next planting time. Annual tree-foliage loss serves as a natural ground mulch in forested areas. Planting grass or legume cover crops until the next planting season, or as part of a crop rotation cycle or no-till planting system, also reduces erosion.

Modern no-till and mulch-till planting systems reduce soil exposure to wind and rain, while plowing the land brings new soil to the surface and buries the ground cover. No-till systems leave the soil cover undisturbed before planting and insert crop seeds into the ground through a narrow slot in the soil. Mulch-till planting keeps a high percentage of the dead residues of previous crops on the surface when the new crop is planted. Row crops are planted at right angles to the prevailing winds and to the slope of the land in order to absorb wind and rainwater runoff energy and trap moving soil particles. Crops are planted in small fields to prevent avalanching caused by an increase in the amount of soil particles transported by wind or water as the distance across bare soil increases. As the amount of soil moved by wind or water increases, the erosive effects of the wind and water also increases. Smaller fields reduce the length and width of unprotected areas of soil.

Wind and water barriers include tree plantings and crosswind strips of perennial shrubs and 1-meter-high (3-foot-high) grasses, which act as wind breaks to slow wind speeds at the surface of the soil. The protected area extends for ten times the height of the barrier. In alley cropping, which is used in areas of sustained high wind, crops are planted between rows of larger mature trees. Contour strip farming on slopes, planting grass waterways in areas where rainwater runoff concentrates, and planting 3-meter-wide (10-foot-wide) grass field borders on all edges of cultivated or disturbed soil are additional methods for reducing wind speed and rainwater runoff and trapping soil particles, chemicals, and nutrients.

Buffers filter runoff to remove sediments and chemicals. Riparian buffers are waterside plantings of trees, shrubs, and grasses, usually 6 meters (20 feet) in width. Riparian buffers planted only in grass are called filter strips. Grassed waterways, field borders, water containment ponds, and contour grass strips are other types of soil conservation buffers.

Gordon Neal Diem

SUGGESTED READINGS: Soil conservation and related issues are discussed in the National Academy of Sciences' two-volume *Soil Conservation* (1986); *The Soul of the Soil: A Guide to Ecological Soil Management* (1986), by G. F. Wilson et al.; and *New Vegetative Approaches to Soil and Water Conservation* (1990), by M. Yudelman et al.

SEE ALSO: Agricultural chemicals; Air pollution; Clean Water Act and amendments; Conservation; Deforestation; Desertification; Dust Bowl; Erosion and erosion control; Fish kills; Groundwater and Groundwater pollution; Runoff: agricultural; Salinization; Sedimentation; Strip farming; Water pollution; Watersheds and watershed management.

Soil contamination

CATEGORY: Land and land use

Soils contaminated with high concentrations of hazardous substances pose potential risks to human health and the earth's thin layer of productive soil.

U.S. Environmental Protection Agency officials and visitors walk past a pile of decontaminated soil at a Superfund site in New Jersey. (AP/Wide World Photos)

Productive soil depends on bacteria, fungi, and other soil microbes to break down wastes and release and cycle nutrients that are essential to plants. Healthy soil is essential for growing enough food for the world's increasing population. Soil also serves as both a filter and a buffer between human activities and natural water resources, which ultimately serve as the primary source of drinking water. Soil that is contaminated may serve as a source of water pollution through leaching of contaminants into groundwater and through runoff into surface waters such as lakes, rivers, and streams.

The U.S. government has tried to address the problem of soil contamination by passing two landmark legislative acts. The Resource Conservation and Recovery Act (RCRA) of 1976 regulates hazardous and toxic wastes from the point of generation to disposal. The Comprehensive Environmental Response, Compensation, and Liability Act (CERCLA) of 1980, also known as Superfund, identifies past contaminated sites and implements remedial action.

Soils can become contaminated by many human activities, including fertilizer or pesticide application, direct discharge of pollutants at the soil surface, leaking of underground storage tanks or pipes, leaching from landfills, and atmospheric deposition. Additionally, soil contamination may be of natural origin. For example, soils with high concentrations of heavy metals can occur naturally because of their close proximity to metal ore deposits. Common contaminants include inorganic compounds such as nitrate and heavy metals (for example, lead, mercury, cadmium, arsenic, and chromium); volatile hydrocarbons found in fuels, such as benzene, toluene, ethylene, and xylene BTEX compounds; and chlorinated organic compounds such as polychlorinated biphenyls (PCBs) and pentachlorophenol (PCP).

Contaminants may also include substances that occur naturally but whose concentrations are elevated above normal levels. This includes such substances as nitrogen- and phosphorus-containing compounds. These substances are often added to agricultural lands as fertilizers. Since nitrogen and phosphorus are typically the limiting nutrients for plant and microbial growth, accumulation in the soil is usually not a concern. The real concern is the leaching and runoff of the nutrients into nearby water sources, which may lead to oxygen depletion of lakes. Furthermore, nitrate is a concern in drinking water because it poses a direct risk to human infants (blue-baby syndrome).

Contaminants may reside in the solid, liquid, and gaseous phases of the soil. Most will occupy all three phases but will favor one phase over the others. The physical and chemical properties of the contaminant and the soil will determine which phase the contaminant favors. The substance may preferentially adsorb to the solid phase. This may include either the inorganic (mineral) or the organic (organic matter) fraction of the soil. The attraction to the solid phase may be weak or strong. The contaminant may also volatize into the gaseous phase of the soil. If the contaminant is soluble in water, it will dwell mainly in the liquid-filled pores of the soil.

Contaminants may remain in the soil for years or wind up in the atmosphere or nearby water sources. Additionally, the compounds may be broken down or taken up by the biological component of the soil. This may include plants, bacteria, fungi, and other soil-dwelling microbes. The volatile compounds may slowly move from the gaseous phase of the soil into the atmosphere. The contaminants that are bound to the solid phase may remain intact or be carried off in runoff attached to soil particles and flow into surface waters. Compounds that favor the liquid phase, such as nitrate, will either wind up in surface waters or leach down into the groundwater.

Metals display a range of behaviors. Some bind strongly to the solid phase of the soil, while others easily dissolve and wind up in surface or groundwater. Polychlorinated biphenyls and similar compounds bind strongly to the solid surface and remain in the soil for years. These compounds can still pose a threat to waterways because, over long periods of time, they slowly dissolve from the solid phase into the water at trace quantities. Fuel components favor the gaseous phase but will bind to the solid phase and dissolve at trace quantities into the water. However, even trace quantities of some compounds can pose a serious ecological or health risk. When a contaminant causes a harmful effect, it is classified as a pollutant.

There are two general approaches to cleaning up a contaminated soil site: treatment of the soil in place (in situ) or removal of the contaminated soil followed by treatment (non-in situ). In situ methods, which have the advantage of minimizing exposure pathways, include biodegradation, volatilization, leaching, vitrification (glassification), and isolation or containment. Non-in situ methods generate additional concerns about exposure during the process of transporting contaminated soil. Non-in situ options include thermal treatment (incineration), land treatment, chemical extraction, solidification or stabilization, excavation, and asphalt incorporation. The choice of methodology will depend on the quantity and type of contaminants, and the nature of the soil. Some of these treatment technologies are still in the experimental phase.

John P. DiVincenzo

SUGGESTED READINGS: A good overview of the environmental health of soils can be found in *Soils and Environmental Quality* (1994), by Gary M. Pierzynski, J. Thomas Sims, and George F. Vance. *The Reuse and Recycling of Contaminated Soil* (1997), by Stephen M. Testa, includes a good introduction that provides an overview of the problem and the regulatory aspects of soil contamination. The remainder of the book is geared for the more advanced reader and discusses technical aspects of decontaminating soil. Donald L. Sparks, "Soil Decontamination," in *Handbook of Hazardous Materials* (1993), provides a good introduction to the sources of soil contamination and then gives a brief overview of the options for decontamination.

SEE ALSO: Agricultural chemicals; Bioremediation; Landfills; Pesticides and Herbicides; Polychlorinated Biphenyls; Soil conservation; Water Pollution.

Solar energy

Category: Energy

Solar energy technology uses the energy from the sun either directly—as in systems for heating water or air—or indirectly, by converting the solar energy to electricity or fuel. Diminishing supplies of fossil fuels and the pollution problems that result from their combustion have made renewable resources such as solar energy increasingly important.

The sun was essentially humankind's only source of energy before the Industrial Revolution. By the end of the twentieth century, direct and indirect solar sources were providing between 15 and 20 percent of the energy need worldwide and about 5 percent in the United States. Each day sufficient sunlight to equal the energy output of more than 22 billion barrels of oil (four times the 1979 U.S. oil consumption) falls on the United States. This energy is available to anyone who is able to collect, transform, and store it.

The greatest advantage of solar energy is its diversity. There are many different forms in which it can be used. Hydroelectric, wind, and wave power are all forms of solar energy, as is methane from swamps and animal digestive systems. Active and passive systems for heating air and water, as well as photovoltaic cells for conversion to electricity, are viable solar energy technologies. Since so many different methods are used in collecting and converting solar energy, it is considered a decentralized technology. Research is being carried out by major corporations and universities and by small start-up companies. The Solar Energy Research Institute (SERI) coordinates much of this research. Solar technologies fall into three broad areas: solar heating and cooling, fuels from biomass conversion, and solar-generated electricity.

Solar Heating and Cooling

The use of solar energy for heating and cooling is based on the simple fact that when an object absorbs sunlight it gets hot. The heat energy may be used in several ways: to provide space heating and cooling, to drive engines, or to heat water or other fluids. If this is accomplished by nothing other than appropriately designed and situated buildings and without moving parts, it is termed passive solar technology. Passive solar architecture was introduced about two thousand years ago by the Greeks and was a common feature in Islamic architecture.

The simplest way make use of passive solar heating in the Northern Hemisphere is to situate a building so that it faces south with its long axis running east-west. During the winter the sun is low in the sky and provides heating to the south-facing windows. In the summer, when the sun is high in the sky, most of the radiation falls on the roof. The windows and walls should be constructed to minimize heat transfer by conduction, convection, and radiation. Double-paned windows with a layer of air between the sheets of glass are effective at preventing conduction and convection. Glass is transparent to visible and infrared radiation, so sunlight enters and infrared radiation exits through the windows. During daylight hours more energy comes in than goes out. At night drapes or blinds can be closed to reduce the energy loss. There are coatings that can be applied to the windows to enhance their ability to reflect infrared radiation back into the room. Some buildings are designed so that the sun's heat creates convection currents, which draw cool air into the building through north-facing or underground ducts, while hot air is vented to the outside. Islamic architecture frequently used chimneys to vent this hot air. The Trombe wall, invented by French designer Félix Trombe during the 1970's, is a modern version of this feature.

Active solar heating involves heating a working substance, usually a liquid, and pumping it to where the heat is needed. Depending on the system used, the temperature increases of the fluid may be a few degrees or as much as a few thousand degrees. Such systems generally have tubes that are painted black or in thermal contact with a metal sheet that is painted black. Fluid that is circulated through the tubes is heated, then sent to a heat exchanger, where the heat energy is used directly or stored for future use. The technology is fairly simple, but the initial installation cost for most homeowners continues to be a stumbling block, particularly because the average time between moves for Americans is about five years.

BIOMASS CONVERSION

Biomass, the material of plants and animals, is the world's oldest energy source. Wood has been burned for heat and light since fire was first discovered. In some developing countries, cooking is still accomplished by burning fuelwood, crop residues, and cow dung. Even in the United States biomass is an important energy source, producing an amount of energy equal to that obtained from nuclear power. Annual world reserves of stored biomass have the energy content of all known fossil fuel reserves.

Although it is the simplest way to release energy, burning is not the most productive use of biomass resources. The material can be converted by biological or chemical processes to produce biofuels such as methane gas, ethanol liquid, or solid charcoal. The major ways of transforming biomass from energy crops are bioconversion (fermentation) and anaerobic thermal conversion. Fermentation processes have been used for centuries to produce beer, wine, and other alcoholic beverages. In Brazil a similar process distills ethanol from fermented cane sugar. This alcohol has replaced about 20 percent of that nation's gasoline use. During World War II automobiles and other vehicles in Germany were run using alcohol from potatoes. A major advantage is that the alcohol can be locally produced using whatever crop residues are available.

Besides burning the fuel in air to produce heat, there are three other thermal conversion processes. The first is pyrolysis, or heating without air. In this process wood is reduced to charcoal, which is lighter to transport and burns without fumes. The additional liquid and gaseous products are vented into the air in charcoal kilns and then collected and used. Gaseous residues can be used directly in gas turbines. Liquid products are refined for combustion. The second process converts the biomass into bio-oil, which can be burned or upgraded to diesel fuel or petrol. The third process is gasification, which produces either medium or low heating value gas. The two have different chemical characteristics but can be burned or used as starting materials in the chemical industry.

An increased use of biomass would result in social and political benefits, as well as advantages to the environment. Biomass can be locally pro-

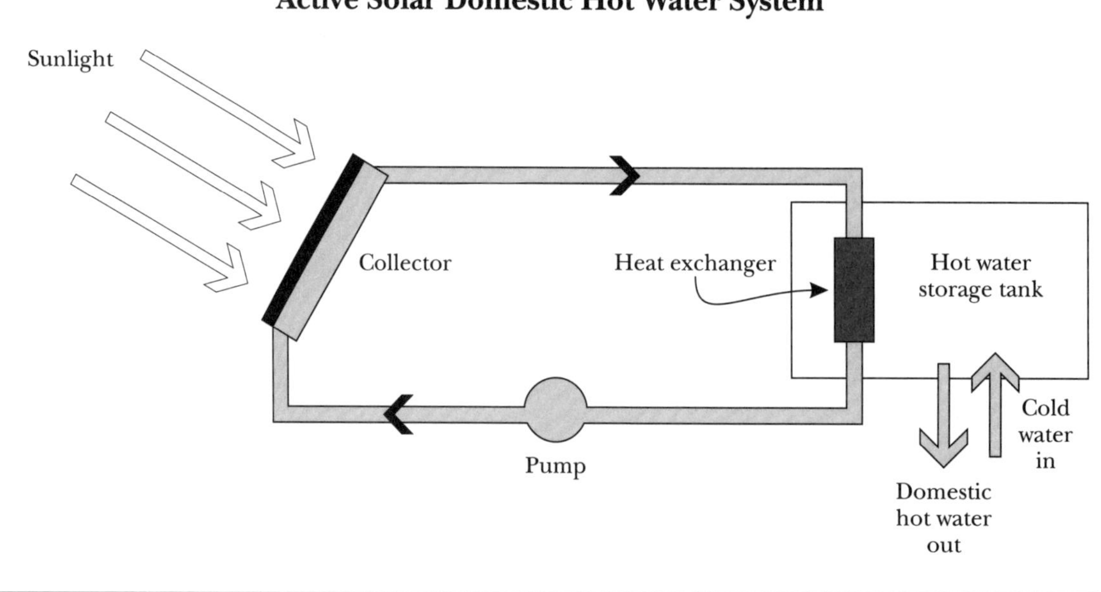

Comparison of Two Types of Solar Energy Collectors

Solar Photovoltaic Collector	Solar Thermal Collector
Converts solar energy directly into electricity for immediate use.	Collects heat from solar energy for conversion into electricity.
Electricity can be converted into heat for thermal use.	Heat is used directly.
Solar radiation of only a very small range of energy can be utilized.	Radiation of a wide range of energy can be used.
Requires additional storage devices that are costly and inefficient.	Some have built-in storage devices that are relatively inexpensive and efficient.
Ideal for micropower and small appliances.	Unsuitable for micropower and small appliances.

duced and used, avoiding the problems associated with concentrated supplies of fossil fuels. Vibrant rural communities founded on profitable employment should reduce the movement of rural population to urban areas. Some researchers have argued that global carbon dioxide emissions can be reduced by use of biomass within the commercial energy industry. The widespread transition is not without its problems, however. Wholesale conversion of croplands to fuel farms has the potential to endanger the world's food supply. Careful planning will be necessary if biomass is to play a more important role in the world's energy budget.

Electricity from the Sun

Solar cells, first developed in 1954 by scientists at the Bell Telephone Laboratories, are solid-state photovoltaic devices that convert sunlight into electricity. The earliest solar cells were made using single-crystal silicon wafers. Silicon is an important type of element known as a semiconductor, which has properties between conductors and insulators.

Electric conduction in silicon takes place by moving negative charges (electrons) and positive charges (holes). One way of accomplishing this is to add arsenic or phosphorous atoms—which have five outer-shell electrons—to the pure silicon, creating a semiconductor that has excess negative charge (n type). The addition of boron atoms with three outer-shell electrons creates a p-type semiconductor. Electric conduction occurs when p- and n-type slices are placed in close contact.

The simplest solar cell is a p-n junction sandwiched between two conductors. One of the tricks to fabricating efficient solar cells centers on the design of the top conductor. It should be large enough to capture the electrons but not so large that it blocks sunlight from passing into the center of the sandwich. Sunlight enters the top of the cell and excites electrons in the n-type layer, causing them to jump to the p-type layer, where they are captured and carried away to do work in the external circuit.

These early solar cells converted sunlight to electricity with about 1 percent efficiency and were expensive to produce. Since the 1950's photovoltaic technology has become more efficient and less costly. One of the newer types of cells uses amorphous silicon, in which silane gas is electroplated on the surface of the cell. They are less efficient but less expensive than the single-crystal silicon cells. Additional materials being considered for solar cells include gallium arsenide, cadmium sulfide, and gallium phosphide.

The efficiency of silicon cells depends, to a great extent, on the purity of the silicon. Working in vacuum and zero gravity, scientists can prepare exceptionally pure silicon on space shuttles. Transporting the materials into orbit and the finished product back down to Earth adds to its cost. An alternative proposal would leave the finished product in space, where the sun shines every day of the year. Although it is technically possible to build enormous solar cells in space and beam the energy down to Earth's surface in the form of microwaves, the concept is a controversial one. An orbiting solar power station would be very expensive, and the effects of microwave radiation passing through the earth's atmosphere are not yet clear.

Photovoltaics have been extensively used in space to power almost all satellites. On Earth, the range of consumer products with built-in photovoltaic cells is continually expanding. One of the advantages of photovoltaics is that they are modular: Individual units can be combined to produce outputs covering a wide voltage range. The technology is also transportable: Units can be set up wherever there is enough sunlight.

Grace A. Banks

SUGGESTED READINGS: *Sol Power* (1996), by Sophia Behling and Stefan Behling, provides a historical view of solar architecture and illustrates the effect solar energy has had on architecture and urban planning. *The Future of Energy Use* (1996), by Robert Hill, Phil O'Keefe, and Colin Snape, contains a chapter on solar energy set in a broad consideration of the future of energy use. A readable introduction to the topic of solar power may be found in *The Big Switch: Creating Jobs, Saving Money, and Protecting the Environment in the 21st Century* (1995), by Gavin Gilchrist. Ralph Nansen presents a plan for the development of solar satellites in *Sun Power: The Global Solution for the Coming Energy Crisis* (1995). For those interested in a more technical presentation of the topic, *Renewable Energy Resources* (1985), by John W. Twidell and Anthony D. Weir, is thorough but not overly mathematical.

SEE ALSO: Alternative energy sources; Alternatively fueled vehicles; Energy policy; General Motors solar-powered car race; Solar One; Sun Day.

Solar One

DATES: 1982-1988
CATEGORY: Energy

Solar One, which began operation in 1982, was a pilot plant that used large fields of tracking mirrors to focus sunlight onto central power towers for the generation of electricity. The Solar One facility proved that power towers work efficiently to produce utility-scale power from sunlight.

The first solar power tower system, Solar One, was built in the early 1980's near Barstow, California, by Southern California Edison, with the support of Sandia Labs, the Department of Energy, the Los Angeles Department of Water and Power, and the California Energy Commission. Rated at ten megawatts of power, Solar One could efficiently and cost-effectively store energy, making it unique among solar technologies. Solar One operated successfully from 1982 to 1988.

A power tower operates by focusing a field of thousands of mirrors (heliostats) onto a receiver located at the top of a centrally located tower. The receiver collects the sun's heat in a heat-transfer liquid. This liquid is used to generate steam for a conventional steam turbine, which then produces electricity at the base of the tower. The mirrors of Solar One reflected and focused sunlight onto a central tower that was nearly 100 meters (328 feet) tall. The tower's absorber panels were painted black and absorbed 88 to 96 percent of the incident light. In order to capture sun from the south, the field of mirrors was oriented mostly toward the north. Solar energy was focused onto six panels to preheat the water that traveled to eighteen superheat panels.

After leaving the superheat panels, the water was at 510 degrees Celsius (950 degrees Fahrenheit). The hot water was sent either to turbines, where it generated electricity at 35 percent efficiency, or to a heat exchanger, where it heated oil that was sent to a thermal storage tank and circulated through crushed granite. The stored heat could be drawn back from the tank through the heat exchanger to produce steam for the turbine. In addition, the thermal storage allowed a buffer system for periods of cloudiness so that the plant could keep operating through changes in weather conditions.

Solar One proved that the heat-transfer liquid cycle is reliable, that the system could meet expectations, and that thermal storage was cost-effective. Furthermore, the power tower system with energy storage showed a unique advantage over other solar power systems because it could supply power to the local utility during peak periods. In Southern California, these periods occur on hot, sunny afternoons and into the evenings during the air-conditioning season when power production is most valuable to the power company.

Based on experience with Solar One, which used water and steam as the heat-transfer liquid, solar engineers determined that power towers operate more efficiently using molten salt. The salt also has the further advantage of providing a direct, practical way to store heat. The concept of storing energy in molten salt and decoupling solar energy collection from electricity production formed the basis for Solar Two, which began operating in 1996. It was constructed from Solar One by converting the water and steam system to a molten salt system.

Alvin K. Benson

See also: Alternative energy sources; Energy policy; Solar energy; Sun Day.

Solid waste management policy

Category: Waste and waste management

Solid waste management policies are decisions made by communities about how to deal with the creation, accumulation, utilization, and eventual deposition of solid waste materials.

An array of large mirrors at the Solar One facility near Barstow, California. The mirrors reflect sunlight onto the central tower in the background, which collects the sun's heat and converts it to electricity. (California Edison)

There is no internationally recognized definition of solid waste. The Organization for Economic Cooperation and Development's definition excludes radioactive waste but includes hazardous waste. The federal government of the United States has adopted a definition that excludes most hazardous waste, except that included within municipal waste. Despite these definitional differences, the public policies adopted by most economically advanced nations have been reasonably similar. The purpose behind the definitions has been to identify levels of danger posed by waste materials so that adequate management regimes can be constructed. Materials defined as radioactive, hazardous, or toxic require more strict controls than do those defined as solid waste.

History

Though it has been said that in nature there is no waste, it would appear that there has always been waste in human communities. For many centuries, however, the amount of garbage generated and the dispersal of such materials were not recognized as a problem. Like many environmental issues, solid waste became problematic largely as a result of the increase in human population on the planet. As people began to live in more densely populated areas, the accumulation of solid waste could not be ignored. There is evidence that in the ancient cities of Africa and the Roman Empire people threw their solid waste on the floors of their dwellings. They lived in the midst of their waste, then built new streets and housing on top of the resultant mess. Eventually urban dwellers developed a variety of ways of removing their waste products from their living areas: cesspools, drainage systems, sewage systems, scavengers and collectors, and dumps located outside the densely populated areas.

Urbanization and industrialization led to greater accumulations of waste and to collective decisions regarding how to deal with those accumulations. In addition to the problems generated by high concentrations of people on relatively small land tracts, industrialization led to greater affluence, which correlates positively with solid waste generation. Municipal governments began to be expected to develop policies regulating waste disposal and to provide services to assist urban dwellers in dealing with their waste. In England, for example, the Poor Law Commission found in 1842 that a filthy environment promoted the spread of disease. This finding led to an increase in municipal sanitation services.

Sanitation services initially tended to operate on an "out of sight, out of mind" basis and focused on removing solid waste from densely populated areas. Collectors, scavengers, and recyclers gathered some of the waste and put it to other uses. Some waste was burned in incinerators, and some of the steam or electrical energy thus generated was put to use. Most of the solid waste, however, was dumped either into bodies of water or onto land.

Contemporary Trends

The United States leads the world in generation of solid waste; its per capita production is about twice that of other economically and industrially advanced nations. The People's Republic of China, the most populous nation in the world, generates less than one-third the total amount that the United States does.

In 1970 William Small brought attention to solid waste as an environmental health problem by calling it the "third" pollution, thereby distinguishing it from air and water pollution. However, it is now clear that solid waste is a multimedia environmental problem. In one sense there is no such thing as "disposal" of solid waste. Once solid waste is generated, it must be physically located in at least one of the three key environmental media: land, water, or air. When it is disposed of, solid waste remains in one of those three media. Thus, all solid waste management policies must decide how much waste will be allowed to be generated, how much will be redirected toward continued utility, how much will deposited into water, how much will be emitted into the air, and how much will be deposited on or into the land.

Although countries are attempting to reduce the volume of materials that are deposited in landfills, they are still the most common answer to the garbage problem. Around 70 percent of

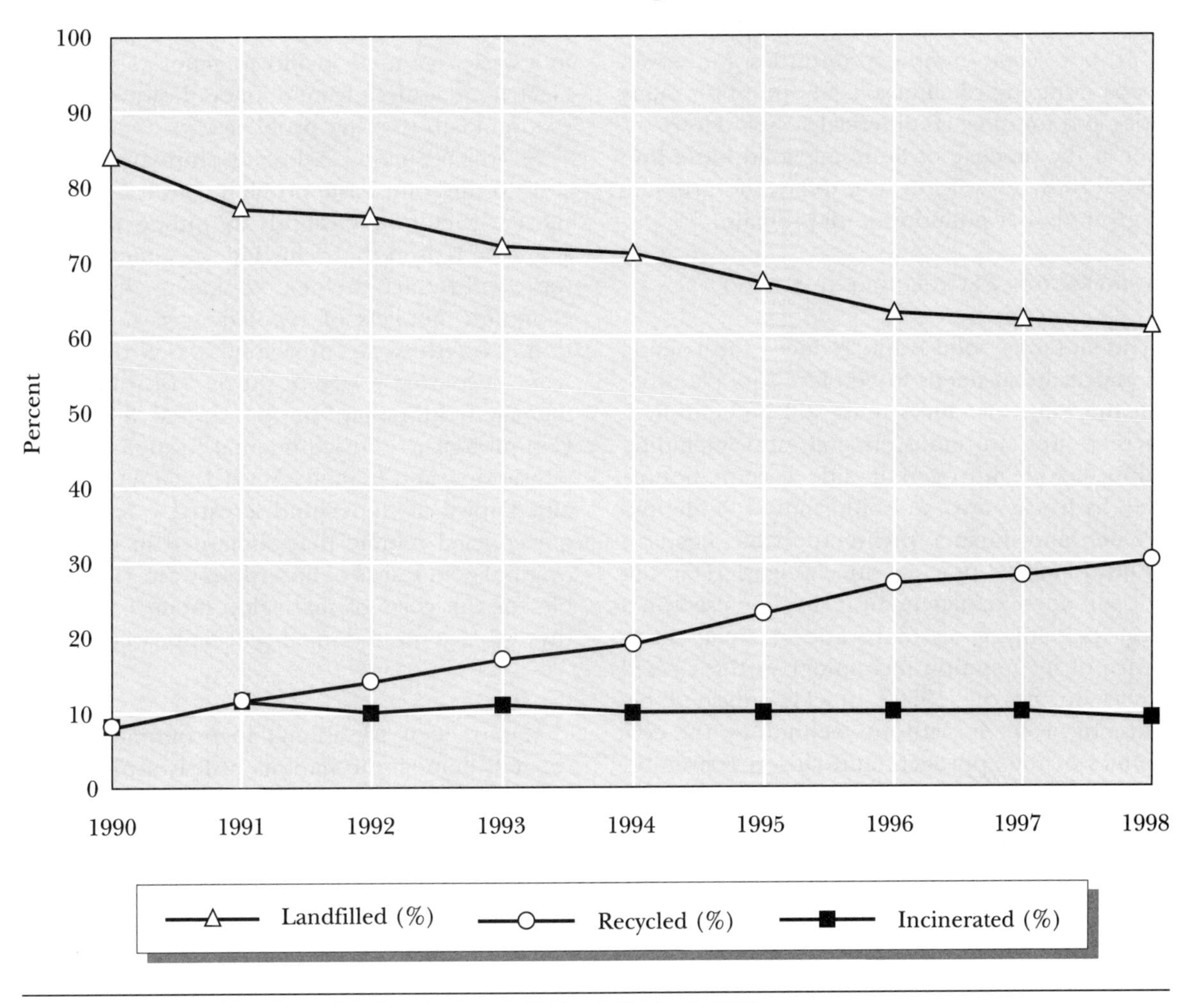

Source: Adapted from "The State of Garbage in America," *Biocycle* 39 (April, 1998).

the solid waste in economically advanced countries is deposited in landfills. Even when other solid waste management techniques are utilized, landfilling is often the final answer.

There are a wide variety of landfill technologies. The first landfills were nothing more than places to dump garbage, but approaches to landfilling have become more sophisticated over the years. During the twentieth century, town dumps were often replaced by municipal sanitary landfills; in advanced countries, sanitary landfills are being replaced by technologically sophisticated and carefully lined regional megalandfills. In 1988 the U.S. Environmental Protection Agency (EPA) predicted that three-fourths of the landfills then in existence would be closed by the year 2000. This development was fueled by community resistance to locating solid waste management facilities in their vicinity (the "not in my backyard," or NIMBY, syndrome) and the increasingly stringent technical requirements placed on landfills. The trend is toward fewer landfills, but the typical size of the newer landfills is much larger, the technology is more complex, and the cost is much higher.

At the beginning of the twentieth century, deposition of solid waste into bodies of water was seen as a good method of disposal, but it is no longer a popular option. There are numerous prohibitions against dumping solid waste into the ocean, such as the Ocean Dumping Act of 1974, but illegal dumping continues. Moreover, ocean dumping remains a legal option for many cities and for most commercial vessels. However, even if the practice of dumping solid waste into bodies of water were to stop today, the problem of groundwater pollution would remain.

Incineration, Recycling, and Source Reduction

Incinerating solid waste reduces the volume of material that needs to be placed in a landfill, but the ash it produces must still be landfilled. Incineration also emits air pollution, including sulfur dioxide, nitrogen dioxide, carbon monoxide, hydrocarbons, polychlorinated biphenyls (PCBs), and dioxins. Incineration has been an institutionalized practice since at least 1865, but it is still not a management technique used on a large percentage of solid waste. A small resurgence of incineration technology in the United States during the 1980's quickly subsided because of problems with the technology, the economics of the approach, and citizen resistance. Incineration is used more widely in Europe and Japan than in the United States.

If materials are not to be deposited in water or land, they must be reused. Resource recovery and recycling are politically popular aspects of solid waste management policy, but their contributions are limited in three ways. First, not all solid waste materials appear to have recycling potential. Second, the costs of recovery may be prohibitive. Third, even recovered and recycled products may eventually be deposited in a landfill.

One approach to recycling is the conversion of materials into another product. Cardboard and newspaper are sometimes recycled into other paper products. Aluminum cans may be recycled into other aluminum products or into other aluminum cans. Some plastic products can be recycled into other plastic products—for example, milk cartons into fibers for apparel or foam cups into plastic lawn furniture. Yard waste (grass, leaves, and branches) may be composted, but commercial compost facilities are experiencing difficulties in odor control. Solid waste may also be seen as an alternative energy source since a wide variety of materials may be burned in a waste-to-energy facility to generate steam or electricity; as already mentioned, however, incineration leads to other problems.

Source reduction is the most fundamental answer to the solid waste problem, but it is also the most difficult to establish as public policy. It stands to reason that reducing the generation of waste will reduce the need to manage it, but the economic 'impacts of regulating waste generation have thus far prevented waste reduction from becoming a significant part of solid waste management policy. In the United States the Comprehensive Environmental Response, Compensations, and Liability Act (CERCLA) of 1980, also known as Superfund, created a "cradle-to-grave" legal regime that attempted to hold the original generators of hazardous wastes responsible for the costs of managing them, even after they were deposited in a legally approved site.

Policy Making

It has been traditional in economically advanced, democratic nations to leave solid waste policy making to local governments. When solid waste was seen as a disposal problem or as a local health problem, it made sense to leave policy making at that level; but since solid waste was identified as an environmental problem, policymakers have been more inclined to recognize that, like other environmental problems, the impacts often extend beyond local political jurisdictions. Moreover, as the cost of managing solid waste has increased, the need for funding assistance from larger units of government has increased.

In the United States, for example, the federal government now plays a major role in developing solid waste management policy. Congress and the president have created legislation, the EPA and other regulatory agencies have promulgated regulations, and federal courts have set parameters within which state and local governments must remain. President Lyndon B. Johnson was a leader in bringing the federal government into solid waste policy making, initiating action that resulted in passage of the Solid Waste Disposal

Act in 1965 as an amendment to the Clean Air Act. In 1976 Congress passed the Resource Conservation and Recovery Act (RCRA), Subtitle D of which created the first national waste management program. In combination with other environmental laws, the RCRA changed the way solid waste was managed in the United States.

Federal court decisions have further diminished local government control over solid waste management policy making. *Philadelphia v. New Jersey* (1978) said that garbage was to be treated as any other commercial commodity under the Commerce Clause of the U.S. Constitution. This meant that state laws and local government ordinances that interfered with interstate commerce in solid waste would be found unconstitutional. Consequently, state and local governments are not allowed to restrict solid waste that was generated outside of their jurisdictions from entering into their jurisdictions. Moreover, the *Town of Clarkstown v. C. & A. Carbonne, Inc.* (1994) decision stated that local governments could not restrict solid waste that was generated within their jurisdictions from leaving. Without the ability to control entry into or exit from their jurisdictions, state and local governments are severely restricted in their ability to set solid waste management policy.

Another development in solid waste management policy is the trend toward privatization. Privatization is not a new approach to solid waste management, but at the end of the twentieth century it saw a resurgence in popularity. Legislative antipathy toward "command-and-control" regulatory approaches, combined with the legal system's treatment of solid waste management as a commercial activity rather than an issue of health and welfare, strengthened the role of private commercial enterprises in solid waste management.

Larry S. Luton

Suggested readings: *War on Waste: Can America Win Its Battle with Garbage?* (1989), by Louis Blumberg and Robert Gottlieb, provides an overview of waste disposal problems in the United States. *The Politics of Garbage: A Community Perspective on Solid Waste Policy Making* (1996), by Larry S. Luton, contains a close-up look at the process of making local solid waste policy decisions. *Garbage in the Cities: Refuse, Reform, and the Environment, 1880-1980* (1981), by Martin V. Melosi, puts urban garbage problems in a historical context. "The Viability of Incineration as a Disposal Option: Evolution of a Niche Technology, 1885-1995," *Public Works Management and Policy* 1 (1996), by Martin V. Melosi, provides an overview of the history of incineration. *The Third Pollution: The National Problem of Solid Waste Disposal* (1970), by William E. Small, brought the problem of waste disposal to the attention of the general public.

See also: Hazardous and toxic substance regulation; Landfills; Recycling; Superfund; Waste management.

Soviet nuclear submarine sinking

Date: October 3, 1986
Category: Nuclear power and radiation

An explosion and fire in a missile tube of a Soviet nuclear-powered submarine resulted in the deaths of four sailors and the sinking of the submarine, carrying two nuclear reactors and the warheads on sixteen missiles to the bottom of the Atlantic Ocean.

On October 3, 1986, the Yankee class Soviet nuclear submarine K-219 was cruising in the Atlantic Ocean, north of Bermuda, when an explosion of unexplained origin occurred in its fourth missile compartment. The explosion caused a leak in the compartment, and smoke from the missile fuel began to fill the submarine. The captain ordered the submarine to surface, but the explosion apparently breached the hull, and water began to leak into the vessel. Before abandoning the K-219, the captain ordered crewmembers to engage the shut-down mechanisms on the boat's two nuclear reactors; the control rods, which cause the nuclear reactions to cease, were manually inserted into the reactors. One crewmember died in the effort to insert these control rods, and three others died during the firefighting and evacuation efforts.

A U.S. Department of Defense photograph shows the damaged Soviet K-219 nuclear submarine off the coast of Bermuda in October of 1986. The vessel, along with two nuclear reactors and sixteen missiles, eventually sank to the bottom of the ocean. (AP/Wide World Photos)

The K-219 submarine sank into the Atlantic Ocean, taking its two nuclear reactors and the nuclear warheads on sixteen missiles to the seafloor. There was no evidence of immediate leakage of radioactive material from the wreckage. However, the long-term environmental consequences of the sinking are unknown. Once a nuclear reactor has generated power, some of its fuel is converted into a wide range of fission products. Some of these fission products are radioactive and are readily concentrated by living organisms. At some point, the corrosive action of the seawater will eat through the confinement housings of the reactors and the warheads, releasing radioactive pollution into the Atlantic Ocean. Depending on the leak rate and the local water currents, high concentrations of radioactive material could persist around the wreckage for months or years after the leakage begins. During that time, fish and other organisms may ingest substantial quantities of radioactive material. Eventually, because the radioactive material will be diluted into the vast volume of the Atlantic Ocean, the concentration of radioactive material will be reduced to the point where it constitutes a negligible hazard.

The K-219 is one of six nuclear-powered submarines that have sunk with fully fueled nuclear reactors onboard. The Soviet K-8 sank in the Bay of Biscay on April 8, 1970, and the Soviet K-278 sank in the Norwegian Sea on April 7, 1989. The Soviet K-27 developed a nuclear reactor problem on May 24, 1968, and was intentionally sunk in the Kara Sea in 1981, when Soviet officials decided that salvaging the submarine would be too costly. Two United States Navy submarines, the USS *Thrusher* (which sank on April 19, 1963, off New England) and the USS *Scorpion* (which sank on May 22, 1968, southwest of the Azores Islands) are both on the North Atlantic seafloor. Monitoring of the radiation around one or more of these sites could provide information on the potential long-term hazards at all six sites.

George J. Flynn

See also: Nuclear accidents; Nuclear weapons; Ocean pollution.

Soviet Plan for the Transformation of Nature

Date: 1948-1953
Category: Ecology and ecosystems

Soviet leader Joseph Stalin sought a quick remedy to severe food shortages following World War II by endorsing a dubious plan to transform nature, hoping for a dramatic increase in agricultural production. The plan was based on the quixotic notions of pseudobiologist Trofim Denisovich Lysenko, who rejected orthodox genetics.

Trofim Denisovich Lysenko and his followers, known as Lysenkoists, came into prominence during the 1930's. In 1932, at the International Congress of Genetics held at Cornell University, prominent Soviet biologist Nikolai I. Vavilov praised Lysenko's experiments with adapting grain and other plants to unfavorable climates by a mysterious process known as "vernalization" (preheating the seeds). Vavilov's praise for Lysenko was met with skepticism by other scientists, and eventually Vavilov also came to question Lysenko's results. After further inquiries, Vavilov became Lysenko's most outspoken critic in the Soviet scientific community.

Undaunted by rebukes from traditional scientists, the Lysenkoists made preposterous claims that wheat could be turned into barley, oats into rye, and oak trees into pine trees. The more bizarre the assertions, the more credibility they earned with Soviet Communist Party leaders. The Lysenkoists predicated all of their wild contentions on the conviction that orthodox genetics, and the whole notion of the primacy of genes, was a capitalist plot to hold back the advance of the Soviet Union. Environment, they insisted, could cause hereditary changes in plants. All that was necessary for the rapid improvement of Soviet agriculture was to assist crops in adjusting to different environments. There was no reason, therefore, that warm-weather crops could not successfully be grown in cold climates. Those who spoke against Lysenko's ideas found themselves charged with being anti-Communist. Many were removed from their posts, and some were purged from the party itself after 1935. Lysenko, meanwhile, rose quickly; in 1939 he replaced Vavilov as director of the Leningrad Plant Growing Institute. The following year, in a great irony, Lysenko was appointed head of the Genetics Institute. Shortly thereafter Vavilov was arrested and sent to Siberia, where he died in 1943.

During the period from 1941 to 1945, Stalin was preoccupied with the war effort, but when the fighting ceased, he turned his attention to rebuilding the Soviet Union. The country's greatest need was to increase agricultural production, and the Lysenkoists were promising great results with the abandonment of orthodox genetics. In July, 1948, Stalin and the Communist Party's powerful Central Committee gave the Lysenkoists an official endorsement. Party leaders liked the Lysenkoist notion of "proletarian" science, as opposed to the "degenerate, elitist" science practiced in the West.

On October 24, 1948, Stalin issued a decree for the Soviet Plan for the Transformation of Nature to begin. Using peasant labor, huge agricultural strips (60 meters, or 200 feet, wide) were to be established in western Russia. Lysenko's methods would be used to improve grain yields. In addition, millions of trees would be planted in the tundra regions to help ease the harshness of the climate. The trees were to be planted in clusters in the belief that some would survive by adjusting to their new environment. Those that survived would then reproduce and thus alter the Soviet terrain.

The plan also included transferring one breed of plant or tree to another by grafting and crossbreeding. By these methods it was believed that ordinary weeds could be turned into wheat. On January 1, 1949, Lysenko predicted that there would be a limitless growth in Soviet harvests. Soviet leaders proclaimed in 1950 that the plan was well on its way to success. In actuality, however, the opposite was true. In 1952 Stalin, embarrassed by reports of low crop production, attempted to deflect attention from the transformation of nature plan by introducing a massive scheme for the building of dams and canals. Stalin also permitted scientific criticism of Lysenko's methods. By the time of Stalin's death in March, 1953, Lysenkoism had fallen into disrepute.

With Stalin gone, the Soviet Plan for the Transformation of Nature was abandoned, and opposing scientific views were again recognized. Lysenko, however, continued to insist that he was right, and when Nikita Khrushchev emerged as Soviet leader in 1957, Lysenko made a brief return to prominence. Khrushchev, like Stalin, desperately sought a cure for his country's agricultural problems. This time, however, Lysenko's critics could not be silenced. Khrushchev's support for the discredited Lysenko contributed to the Soviet leader's removal from office in October, 1964.

The Soviet Plan for the Transformation of Nature produced no positive results for the Soviet Union. There were, however, negative consequences. Soviet biology was disrupted and set back by Lysenko's insistence that his critics be silenced and removed from their government posts. By the time Stalin realized that he had been misled by the Lysenkoists, the debacle could not be repaired in a short time. Moreover, other areas of Soviet science, including medicine, were also affected. A large number of individuals, who later became known as "harebrained" scientists, attempted to gain favor from the Communist Party. The party, clearlymore interested in proper ideology than proper science, all too often endorsed dubious scientific methods during the late 1940's. It was not until the 1970's that the Soviet Union fully rejoined the world of legitimate scientific inquiry.

Ronald K. Huch

SUGGESTED READINGS: For the best account of Lysenko's influence on Soviet science and the Soviet Plan for the Transformation of Nature, see David Joravsky's *The Lysenko Affair* (1970). Joravsky ably and clearly explains why the plan was doomed to failure. There are many good studies of the Stalin era, but three that help to put the transformation of nature plan into context are Ronald Hingley's *Joseph Stalin: Man and Legend* (1973), Roy Medvedev's *Let History Judge* (1971), and Adam B. Ulam's *Stalin: The Man and His Era* (1973).

SEE ALSO: Biotechnology and genetic engineering; High-yield wheat.

Space debris

CATEGORY: Pollutants and toxins

Human-made space debris largely consists of wreckage from destroyed spacecraft, which can cause damage to orbiting satellites. Natural space debris, such as meteoroids and cosmic dust, can affect the earth's environment. Huge amounts of dust thrown into the atmosphere from large impacts with asteroids and comets have the potential to modify the earth's climatic patterns.

Since 1957 more than thirty thousand objects have been launched into space. Most satellites successfully reached orbit, while others either exploded or failed once in orbit. Everything that reaches orbit becomes a satellite of the earth. Achieving a low-earth orbit requires a velocity of 17,500 miles per hour. Once in orbit, satellites are constantly under the pull of the earth's gravity and, in time, will slowly fall from orbit. The greater the distance from Earth, the longer the satellite will remain in orbit. Each object in orbit runs the risk of running into another object. The volume of space surrounding the earth is immense, and the chances of a collision between two objects are relatively low, except when they occupy the same orbit. Certain orbits are particularly desirable for communications and surveillance purposes. Various nations and commercial interests place their satellites into these strategic positions, thereby increasing the chances of collision.

Many different kinds of human-made space debris orbit the earth. In the past, many satellites were deliberately destroyed in weapons tests for antisatellite warfare. Other forms of space debris had a less dramatic origin, such as a glove from astronaut Ed White that slowly drifted away from his Gemini spacecraft. Objects ranging in size from spent rocket boosters to small chips of paint have the potential to damage other spacecraft. Each item adds to the ever-increasing number of human-made objects orbiting the earth. It is not the mass of the object that poses the danger, but its high velocity. Several space shuttles have been hit by microscopic particles that pitted their cockpit windows. The National Aeronautics and Space Administration (NASA) operates a program that tracks as many objects as possible in order to determine the potential danger of orbiting debris. The greatest danger is posed to the many communications satellites upon which the world depends.

Falling space debris is another hazard. Although the danger of a person being injured by a piece of a falling satellite is rather remote, it can happen. In 1979 large pieces of the Skylab space station survived its fiery plunge through the atmosphere and scattered debris all over western Australia. A Soviet satellite with a nuclear power source also survived reentry and scattered small amounts of radioactive material across Canada.

It has been estimated that 20 million bits of natural space debris collide with the earth each day. The majority of this is in the form of cosmic dust. These particles are so small that they do not even form meteors as they pass into the atmosphere. Most meteors that are seen come from particles the size of a pea, and sometimes the larger ones reach the earth's surface as meteorites.

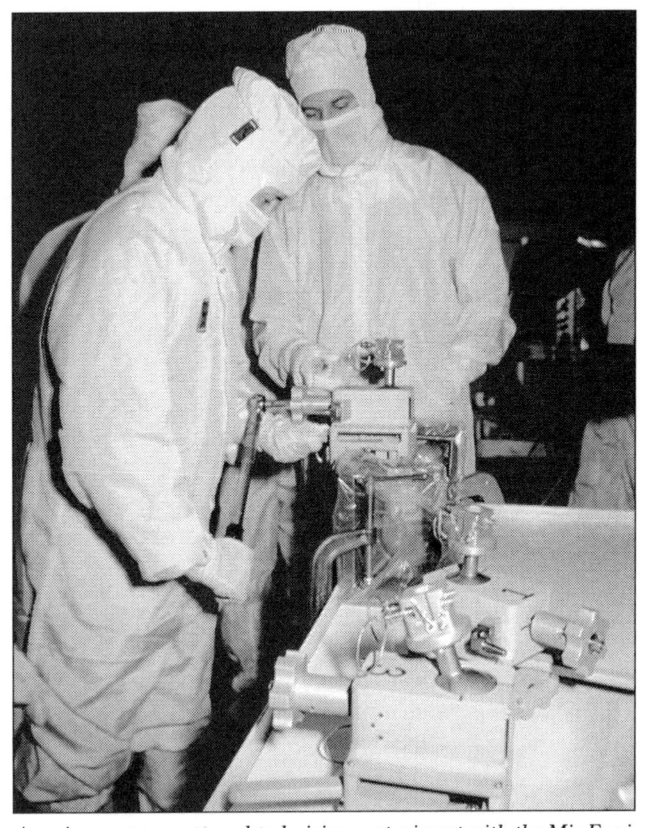

American astronauts and technicians experiment with the Mir Environmental Effects Payload (MEEP), a device designed to measure the amount of space debris that strikes the Russian MIR space station. (AP/Wide World Photos)

Occasionally an asteroid-sized object or comet collides with the earth. Such collisions have global implications. The impact destroys the comet and forms a huge crater. An enormous amount of gas and dust is carried into the atmosphere, creating a blanket of debris that blocks sunlight. Thus begins a nuclear winter, which can last anywhere from three months to three years. During this time most life forms will die as a result of the disruption of the food chain. Many scientists believe that such an event led to the mass extinctions of dinosaurs more than 65 million years ago.

Although giant impacts are one cause of nuclear winter, cosmic dust can produce the same effect. The solar system periodically runs into a cosmic dust cloud, thereby dramatically increasing the amount of dust that enters the atmosphere. A similar situation is also posed by periodic meteor storms. As the earth runs into the debris of old comets, the number of meteors that enter the atmosphere increases. The earth occasionally encounters a particularly dense region of comet debris. During such meteor storms, thousands of meteors can be seen each hour. The most notable is the Leonid meteor storm, which occurs every thirty-three years. Many scientists fear that increases in comet debris could knock out hundreds of satellites as they are hit by microscopic particles and greatly affect global communication.

Paul P. Sipiera

SUGGESTED READINGS: *Preservation of Near-Earth Space for Future Generations* (1994), edited by John A. Simpson, provides useful information on all aspects of space debris and how it can affect the future and the utilization of the space near Earth. *Artificial Space Debris* (1987), by

Nicholas L. Johnson and Darren S. McKnight, is a good companion volume that presents a thorough review of the problem of human-made space debris. *Cosmic Debris: The Asteroids* (1994), by Isaac Asimov, Greg Walz-Chojnacki, and Francis Reddy, presents good background information on asteroids, the accepted parent body for most meteorites. *Pollution in Space* (1995), also by Isaac Azimov, Greg Walz-Chojnacki, and Francis Reddy, provides a good overview of how humankind has affected its near-space environment.

SEE ALSO: Climatic change and global warming; Nuclear winter; Weather modification.

Spaceship Earth metaphor

CATEGORY: Ecology and ecosystems

The Spaceship Earth metaphor is used to compare planet Earth to a small spacecraft in a vast universe.

Environmentalists, economists, scientists, and others have used the Spaceship Earth metaphor to compare the systems needed for the survival and continuation of the diverse life forms on Earth to the systems needed on a spaceship to ensure the survival of the occupants. Major environmental systems needed for survival on Earth include air, water, food, shelter, economic goods and services, waste disposal, purification, recycling, adequate resources, security from violence, governmental justice, law, and order. Similar environmental systems needed on a spaceship include a sufficient air supply, clean water, adequate food, protection from outer space, adequate goods and services, effective waste disposal, pollution corrected by purification, recycling of used materials, and enforcement of appropriate safety rules and regulations by captain and crew.

In British economist Barbara Ward's book *Spaceship Earth* (1966), she indicated that she borrowed the comparison of Earth to a spaceship from the visionary American inventor R. Buckminster Fuller. To Ward, science and technology have created an intimate, spaceshiplike network of communication, transportation, and economic interdependence. This has resulted in a planetary fellowship, a close world community with the vulnerability of a spaceship. Inequities in power and wealth breed violence, which damages the general welfare in the crowded spaceship. Ward believed that the divisions fostered by conflicting belief systems should be replaced with planetary unity, rational rules for survival, and common world institutions, policies, and beliefs. Such a worldwide system of order and welfare is needed to avoid nuclear annihilation. Ward claimed that the current system of national sovereignty is a divisive tribal system that stands as a major barrier to the system of planetary loyalty, patriotism, and citizenship needed for adequate living on Spaceship Earth. The instinct to kill strangers from other tribes or nations must be replaced by the vision of a single community on a spaceship carrying a single human species.

Kenneth Boulding, longtime economics professor at Colorado University, took Ward's arguments and made them even more specific. He published an essay called "Economics of the Coming Spaceship Earth" in *The Environmental Handbook* (1970), edited by Garrett De Bell. Boulding labeled the open-earth economy of exploitation, extraction, consumption, depletion, pollution, and violence as the "cowboy economy." Boulding termed it a "fouling of the nest" system. In order to sustain the human species, Boulding advocated a closed "spaceman" economy in which the planet is seen as a single spaceship. To Boulding, maintenance of resources and conditions for long-term quality living should be top priorities on Spaceship Earth.

Lynn L. Weldon

SEE ALSO: Biosphere concept; Ecosystems; Sustainable development.

Speciesism

CATEGORY: Philosophy and ethics

Speciesism is the view that humans are the most important species in the universe, while other animal species are inferior and subservient.

Speciesism has deep roots in anthropocentric philosophies and theologies. Greek philosopher Plato was exclusively interested in human potential and regarded animals only as tools for human goals. French philosopher René Descartes believed animals were machines that were devoid of thought and feeling. Utilitarians have always prioritized human happiness over animal welfare. Even Immanuel Kant believed that humanity's duties were restricted to its own species.

Throughout history, subjugation of nonhumans was accepted and even institutionalized. Motivation for speciesism may come from many sources: fear of animals, religious dictates, survival, or sheer arrogance. Speciesism has been one of the most pervasive and persistent assumptions. Secular reasons for speciesism include personal prestige, sports and hunting, commerce, and economics. All religions, however, place some emphasis on fair treatment of nonhuman life. Most, however, and notably Christianity, give some speciesist privileges to humans. The Christian god created animals without souls for the purpose of service to humans, his best creation.

Critics view speciesism as unwarranted anthropocentric prejudice similar to sexism and racism. Generally, they contend that every species of sentient beings has the same rights to life, liberty, and the pursuit of happiness. The antispeciesist debate about this contention began in the 1970's in England and spread around the world. In 1977 the key philosophers of animal rights movements devoted an entire conference to speciesism at Trinity College in Cambridge, England. The outcome was the drafting of "A Declaration Against Speciesism." This charter became the basis for subsequent charters ratified in Great Britain and many other nations. The original document confirmed that there is evolutionary and moral kinship between humans and their "brother and sister animals." It expressed "total condemnation" on inflicting suffering on animals.

Antispeciesist philosophers find no moral, linguistic, cognitive, or divine basis for the superiority of humans, who share too many similarities with other sentient beings. Even making and using tools, language, and such moral sensibilities as altruism can be observed in animals. Antispeciesists such as Peter Singer appeal to moral instincts. Singer says that people are left with a startlingly simple realization that whatever is morally wrong and abhorrent for humans is probably wrong and abhorrent for nonhumans. Pain and suffering are bad and should be prevented or minimized regardless of race, sex, or species. The logic is simple and similar to a stripped-down version of the "universalization principle" that Kant used to test human morality. Singer's interspecies version of the Kantian principle is this: If it is wrong to torture and kill humans, then it is wrong to do it to nonhuman sentient beings. In the case of laboratory rats, for example, one should ask, "Would it be right to inflict severe electric shocks upon unwilling men and women?"

Almost all recent serious philosophical inquiries about speciesism conclude that the candidates for moral consideration include other nonhuman sentient beings. Antispeciesists maintain that speciesism is similar to sexism and racism. The debate is not easy to resolve and brings up many questions, including, "In what sense, if any, are people obligated to give equal moral standing to insects and rodents that, at times, wreak havoc on humankind?" Such questions are difficult to answer.

Chogollah Maroufi

SEE ALSO: Animal rights; Animal rights movement; People for the Ethical Treatment of Animals; Singer, Peter; Vegetarianism.

Stringfellow Acid Pits

DATES: operated from August, 1956, to November, 1972

CATEGORY: Human health and the environment

The Stringfellow site is a hazardous waste disposal facility in Southern California that was declared a top-priority federal Superfund site in 1983.

In the mid-1950's the state of California sought a location within a one-hour drive of Los Angeles that could serve as a dump for liquid industrial wastes. One promising site was Pyrite Canyon, located in a semirural area in the southern portion of the Jurupa Mountains near the community of Glen Avon in Riverside County. James Stringfellow, who operated a quarry in Pyrite Canyon, made an agreement with the state to allow waste disposal on a portion of his property. Disposal of liquid industrial wastes at the Stringfellow site commenced in August, 1956.

Over the next several years, roughly 128 million liters (34 million gallons) of liquid wastes were hauled to the 17-acre state-licensed dump site. These wastes—including acidic pickling liquor from steel plants and other wastes from metal finishing, electroplating, and dichloro-diphenyl-trichloroethane (DDT) production—were discharged into unlined evaporation ponds. Hundreds of private and public entities in California and Nevada contributed wastes to the facility during its operation.

In 1969 heavy rains flooded the Stringfellow facility, causing contaminated runoff from the overflowing disposal ponds to flow into Pyrite Creek and Pyrite Creek Channel, and from there into the community of Glen Avon. Subsequently, the facility operator worked with engineers to improve site integrity so the facility could continue operations. However, when chromium was detected in the well of the quarry office roughly 1 kilometer (0.6 miles) from the site, Riverside County revoked Stringfellow's special land-use permit. The operator closed the facility in November, 1972. Ownership of the site later reverted to the state of California because of the owner's failure to pay back taxes. Between 1975 and 1980, the state conducted studies to determine the type and extent of contamination and initiated cleanup efforts.

Groundwater beneath and downgradient from the site was found to be contaminated with volatile organic compounds such as trichloroethane and heavy metals such as cadmium, chromium, manganese, and nickel. Elevated levels of groundwater radioactivity were later determined to be naturally occurring and unrelated to the acid pits. Heavy metals were also found in the soil, along with pesticides, polychlorinated biphenyls (PCBs), and sulfates.

Public objection to the site was galvanized in 1978, when another season of heavy rains led the California Regional Water Quality Control Board to authorize a controlled release of 3 million liters (800,000 gallons) of wastewater from the site to avert flooding and uncontrolled discharge. As the contaminated water flowed through the streets of Glen Avon, residents experienced dizzy spells, breathing problems, and other health effects. Chronic illnesses spurred the community to file a personal injury lawsuit in 1984 against the state of California and two hundred companies that had contributed wastes to the site. One decade later, a settlement of more than $114 million was reached.

In 1980 the federal government became involved in site cleanup. The Stringfellow Acid Pits were made a Superfund site in 1983 when the site was included on the U.S. Environmental Protection Agency's National Priorities List. At the end of the 1990's, federal and state authorities were still working with a group of private parties to remediate the contamination. Total cleanup costs were projected to be between $250 million and $475 million.

Karen N. Kähler

SEE ALSO: Groundwater and groundwater pollution; Hazardous waste; Heavy metals and heavy metal poisoning; Soil contamination; Superfund.

Strip farming

CATEGORY: Agriculture and food

Strip farming involves growing crops in narrow, systematic strips or bands to reduce soil erosion from wind and rain; trap minerals, metals, fertilizers, pesticides, bacteria, and pathogens before they leave cultivated fields; provide habitat for wildlife; and improve agricultural production.

The origins of strip farming can be traced to the enclosure movement of postmedieval Great Britain. Landlords consolidated the small, fragmented strips of land farmed by tenant peasants into large block fields in an effort to increase agricultural production to meet the demands of growing human populations. Peasant plots were typically 1 acre in size: 220 yards, or one furlong (the distance a team of oxen can plow before resting), in length and 22 yards (the amount one team of oxen can plow in one day) in width. After enclosure, fields were 100 or more acres in size. Larger fields were more productive but were also more exposed to wind and water erosion and nutritional exhaustion.

As agricultural production gradually shifted to new lands in the Americas and colonial Africa, farmers continued to use large-field farming techniques and developed large-field plantations. By the early twentieth century, all readily tilled lands had been opened by the plow and were suffering the effects of water and wind erosion. Strip farming, also known as strip cropping, was developed as a soil conservation measure during the 1930's. During the 1960's strip farming became an important tool to prevent water and air pollution and improve wildlife habitat.

Wind erosion begins when wind velocity at 0.3 meters (1 foot) above soil level increases beyond 21 kilometers (13 miles) per hour. Soil moves by saltation and surface creep. In saltation, small particles are lifted off the surface, travel ten to fifteen times the height to which they are lifted, then spin downward with sufficient force to dislodge other soil particles and break earth clods into smaller particles. Surface creep occurs when particles too small to be lifted move along the surface in a rolling motion. The wider the field, the greater the cumulative effect of saltation and surface creep, leading to an avalanche of soil particles across the widest fields even during moderate wind gusts.

Water erosion begins when raindrops or flowing water detaches and suspends soil particles above the surface and transports them downslope by splash or runoff. Water ice crystals expand, then contract when melted, dislodging soil particles and making them available for both water and wind erosion. Water also leaches nutrients and chemicals from the soil, causing the soil to experience both nutrient loss and an increase in salts and acids.

The U.S. Department of Agriculture computes annual soil loss from agricultural and developed land using the formula A = RKLSCP. In this formula, A equals annual soil loss, R equals the amount of rainfall on the plot, K equals the erosion factor for the type of soil on the plot, L equals the length of the slope on which the plot is located, S equals the angle of the slope, C equals the type of crop or soil cover on the plot, and P equals the presence of management conservation practices such as buffers, terraces, and strip farming. Soil loss tolerances are developed for each plot. The tolerance is the amount of soil that can be lost without reducing productivity. Loss tolerances range from 1 to 5 tons per acre per year. Farmers and developers reduce soil losses to tolerance levels by reducing soil exposure to wind and rain and by utilizing conservation practices such as strip farming.

Strip farming reduces field width, thus reducing erosion. Large fields are subdivided into narrow cultivated strips. Planting crops along the contour lines around hills is called contour strip cropping. Planting crops in strips across the top of predominant slopes is called field stripping. Crops are arranged so that a strip of hay or sod (such as grass, clover, or alfalfa) or a strip of close-growing small grain (such as wheat or oats) is alternated with a strip of cultivated row crop (such as tobacco, cotton, or corn). Rainwater runoff or blown dust from the row-crop strip is trapped as it passes through the subsequent strip of hay or grain, thus reducing soil erosion and pollution of waterways. Contour or field strip cropping can reduce soil erosion by 65 to 75 percent on a 3 to 8 percent slope.

Cropping in each strip is usually rotated each year. In a typical four-strip field, each strip will be cultivated with a cover crop for one or two years, grain for one year, and row-crop planting for one year. Each strip benefits from one or two years of nitrogen replenishment from nitrogen-fixing cover crops such as alfalfa, and each strip benefits from one year of absorbing nutrient and fertilizer runoff from the adjacent row-crop strip.

Strip widths are determined by the slope of the land: The greater the slope, the narrower the strips. In areas of high wind, the greater the average wind velocity, the narrower the strips. The number of grass or small-grain strips must be equal to or greater than the number of row-cropped strips.

Terraces are often constructed to reduce the slope of agricultural land. At least one-half of the land between each terrace wall should be cultivated with grass or a close-growing crop. Diversion ditches are often used to redirect water from its downhill course across agricultural land. These ditches usually run through permanently grassed strips, through downhill grass waterways constructed across the width of the strips, and through grassed field borders surrounding each field.

Gordon Neal Diem

SUGGESTED READINGS: A description of historic land division and cropping systems is provided by John Fraser Hunt's *The Rural Landscape* (1998). The U.S. Department of Agriculture's *Universal Soil Loss Equation with Factor Values* is used to compute annual soil loss; the department produces a separate book for each state. The local Natural Resource Conservation Service and Soil and Water Conservation District distributes pamphlets on the techniques and benefits of implementing crop rotation and strip cropping systems.

SEE ALSO: Erosion and erosion control; Soil conservation; Sustainable agriculture.

Strip mining

CATEGORY: Land and land use

Strip mining is one of the most convenient and economical means of extracting natural resources from the earth. However, because the process involves the removal of huge amounts of soil, rock, and vegetation from a large area of the earth's surface, strip mining scars the landscape and promotes chemical pollution of both soil and water resources.

The use of strip mining, contour mining, and open-pit mining all essentially employ the same basic technique: rapid removal of surface material that overlies a valuable mineral resource. Strip mining is most often employed where a resource such as a coal seam lies fewer than 30 meters (100 feet) below the earth's surface. Contour mining is more applicable in a hilly region where coal is exposed in readily accessible layers. In the open-pit method, a valuable mineral resource is mined by using a technique that resembles quarrying, but on a much grander scale. A huge area is excavated to a specific depth. The mining process then moves first inward and then downward, creating a smaller but deeper base level. The process continues until all the valuable material has been removed or the pit is simply too deep to work. All three methods are considered relatively safe and quite economical when compared to shaft mining.

Coal is the primary earth resource that is removed by strip mining. Mining begins with site analysis and preparation. First, the extant of a coal deposit is determined by sinking a series of test holes. Once the deposit has been confirmed to be economically feasible, the land is prepared for mining. All surface vegetation is removed, and then dynamite is used to break up the overlying material. This overburden will later be removed by a huge dragline and smaller front-loaders. The dragline removes tons of overburden and deposits it in large piles called spoil banks. Often these spoil banks give the landscape a stark, moonlike appearance. A similar situation exists with contour mining. Huge front-loaders cut into the hillside to expose the coal seam and then deposit the overburden on the downward slope. This creates a potentially dangerous situation, depending upon the stability of the overburden on the slope.

Open-pit mining is more efficient for mining copper or iron ore since metal-bearing minerals are dispersed throughout various rock formations and are rarely found in concentrated form. Similar to strip mining, the open-pit method removes enormous amounts of ore-bearing rock, which is carried away for refining. The valuable metal leaves the refining process in concentrate form, leaving behind huge amounts of the ore material. This material has been crushed, dissolved in chemicals, and perhaps smelted, transforming the original ore material into a useless waste product. This waste material generally ac-

The Rabbit Creek strip mine in Nevada. Strip mining, although efficient, causes massive damage to the environment. In 1977 the United States government enacted legislation that regulates strip mining operations. (U.S. Geological Survey)

cumulates in large piles that will not support most kinds of vegetation. This is especially true in arid regions. Where water is plentiful, various chemicals can leach out of these waste piles and pollute nearby lakes, streams, and groundwater supplies.

Strip mining has an immediate impact on the environment since it completely modifies the surface conditions. The obvious effect is the loss of trees and other surface-covering vegetation. This is replaced by an enormous expanse of what appears to be a scarred wasteland of exposed rock and piles of overburden where little vegetation can grow. If left undisturbed, the mining pits quickly fill in from groundwater intrusion and rainfall. Sulfur leaching out of minerals such as pyrite and sphalerite combine with water to produce sulfuric acid. This acid mine drainage has polluted countless streams and even groundwater supplies in areas where extensive strip mining has occurred. Rainwater runoff cuts gullies into the vegetation-free spoil banks, exposing broken shale, which quickly turns into clay. In a contour-mining environment, spoil banks situated on hillsides or steep slopes can become waterlogged and move forward as giant mudslides. Quite often the miners live in towns at the base of the hills they are mining. This may lead to disaster, such as the 1966 mudslide that buried the town of Aberfan, Wales, and killed 144 people.

In 1977 the United States passed the Surface Mining Control and Reclamation Act, which established procedures for regulating surface mining and reclamation operations. The mining industry must now return the land, as much as possible, to its original condition. This involves leveling the spoil banks and grading the land. Finally, the land must be replanted with appropriate vegetation, and the chemistry of surface water must be continuously monitored. The act put an end to irresponsible mining operations. In addition, many coal companies have chosen to leave the traditional high-sulfur coal behind and mine the low-sulfur coal of the western United States. By mining low-sulfur coal, the coal industry has helped reduce the effects of coal pollution on both the atmosphere and water resources.

Paul P. Sipiera

Suggested readings: A good review of environment-related problems, along with potential solutions, is presented in *Healing the Planet* (1991), by Paul R. Ehrlich and Anne H. Ehrlich. Michael Silverstein, *The Environmental Economic Revolution* (1993), gives insight into how a balance between the needs of an industrial society can be met and still be in a reasonable balance with nature. Two titles, *Ancient Sunshine: The Story of Coal* (1997), by James B. Goode, and *Coal and People* (1997), by Shirley Y. Campbell, are both extremely helpful in giving readers a good insight into the background of coal, its uses, and how it affects society.

SEE ALSO: Environmental impact statements and assessments; Reclamation; Surface mining.

Sudbury, Ontario, emissions

CATEGORY: Atmosphere and air pollution

Nickel and copper sulfide ores were found near Sudbury in Canada during the 1880's. Throughout the twentieth century, sulfur dioxide released during the production of nickel and copper caused vast ecological devastation in the region. By the 1960's, Sudbury was known around the world for its acidified, lifeless lakes and its blackened, treeless landscape.

The process of smelting involves heating sulfur-containing metal ores in air to convert the sulfide to the oxide form. In the process, the sulfur is removed in the form of sulfur dioxide. Without environmental controls, this gas goes into the atmosphere, where it reacts with water and oxygen to form sulfuric acid. This acid rain damages foliage, lakes, and structures, and causes difficulty for anyone with respiratory ailments. The unnatural acid levels (measured at a pH of less than 3 in some cases) may also lead to the leaching of aluminum from soil into groundwater, which negatively impacts roots and aquatic life in streams and lakes.

In the late nineteenth and early twentieth centuries, a metal mining and processing industry took root in Sudbury, Ontario, Canada, just north of Lake Huron. Extensive logging took place in the region in order to furnish wood for the growing settlements and for fuel for the smelter operations. Trees were unable to grow back in the deteriorating environment. The denuded soil quickly eroded into waterways, adding to the devastation. At the same time, smelting emissions included airborne metal particulates. This led to toxic levels of nickel and copper in the surrounding soil and water.

Late in the 1960's the Canadian government began to respond to the growing worldwide environmental movement by ordering a reduction of sulfur dioxide and metal levels in the air around Sudbury. The companies responded by constructing 380-meter (1,247 feet) "superstacks," which acted to reduce local air pollution by spreading it over a larger area. During this period, the world was beginning to recognize the transboundary nature of pollution, particularly air pollution. Dangerously acidic conditions were found in a 50-kilometer (31-mile) radius of Sudbury and even farther in the direction of the prevailing winds.

Citizen outcry made it clear that improvements also had to occur inside the industrial plants. Among those implemented were the use of higher-grade ore, recycling of sulfur gases to make and sell sulfuric acid, and treatment of stack gases to remove residual acid and particulates. Sudbury has also become known for its remediation of lakes and soil by application of basic materials (such as lime) and for replanting and nurturing young trees and other plants. Over twenty years, soil and water pH levels rose measurably, and plant and aquatic life slowly began to return. In 1992 Sudbury received commendation at the Earth Summit in Brazil for its unprecedented clean-up, from which the rest of the world could learn.

Wendy Halpin Hallows

SEE ALSO: Acid deposition and acid rain; Air pollution; Air pollution policy.

Sun Day

DATE: May 3, 1978
CATEGORY: Energy

Sun Day was set aside by the United States government to increase awareness of solar energy and encourage the development of solar technologies.

During the 1960's and 1970's expanding demands for energy, increasing concern for environmental quality, and limited domestic capacity to meet energy demands with traditional fossil fuels brought the United States to the realization that renewable sources of energy, particularly solar energy, must be given a new priority. The urgency of the problem was dramatically impressed upon the leadership of the nation with the Middle East oil embargo in 1973. It became clear to the American public that while the oil embargo would eventually pass, the nation, and even the world, could never again operate under the assumption that the traditional dependence upon fossil fuels and other existing sources of energy could continue.

The need for a comprehensive program aimed at developing solar energy as a viable contributor to the future energy supply in the United States led to the creation of the Solar Energy Research Institute (SERI) in Denver, Colorado, in 1977 and the designation of May 3, 1978, as Sun Day. Solar energy awareness and development were emphasized throughout the week of May 1 through May 7, 1978. SERI provided technical support to the federal Sun Day Committee in their efforts to generate large volumes of information on solar energy for the public. A SERI-produced slide show on the technology and potential of solar energy was distributed throughout the nation to be shown at regular intervals in larger cities during the week. At the United States Customs House in Bowling Green, New York, solar energy displays were open to public view from May 3 to May 7.

On Sun Day, President Jimmy Carter visited SERI and gave an address on the future of solar energy in the United States, requesting that every federal government agency consider more ways to help solar energy become a part of everyday American life. Carter pointed out the importance of developing renewable and essentially inexhaustible sources of energy in the future, particularly placing new emphasis on the importance of solar energy in the country's coming energy transition. He concluded that the costs associated with solar power technologies must be reduced so that solar power could be used more widely and would help establish a cap on rising fossil fuel prices. In addition, Carter stated that he had just provided the Department of Energy with an additional $100 million for expanded efforts in solar research, development, and demonstration projects.

Following Carter's Sun Day address, a series of well-attended forums were conducted across the country. Participation involved the general public, congressional representatives, state and local government officials, industry, labor organizations, public utilities, and special interest groups. These public forums identified citizen groups interested in solar energy and helped in the development of national solar energy policies.

Alvin K. Benson

See also: Alternative energy sources; Energy policy; Solar energy.

Superfund

Date: Established 1980

Category: Human health and the environment

In 1980, the U.S. Congress passed the Comprehensive Environmental Response, Compensation, and Liability Act (CERCLA), better known as Superfund, a federal program designed to address the cleanup of severely contaminated sites that posed an immediate threat to human health or the environment. The Superfund includes a set of guidelines for assessing and cleaning sites, a system of applying legal and financial liability to responsible parties, and a fund to finance all or part of specific cleanup activities.

Superfund sites are those that have become contaminated from past industrial, commercial, and municipal activities. Sites include chemical manufacturing facilities, petrochemical plants, metal-related industries, and old landfills or waste dumps in which industrial wastes were indiscriminately mixed with household refuse. Chemicals of concern at such sites include arsenic, cadmium, chromium, lead, mercury, benzene, chlorinated dioxins, trichloroethylene, and polychlorinated biphenyls (PCBs).

Hazardous Waste Sites on Superfund Priority List, 1995

State	Total Sites	State Rank	Percent Distribution
Total	1,283		
United States	**1,270**		**100.00**
Alabama	13	30	1.02
Alaska	8	42	0.63
Arizona	10	38	0.79
Arkansas	12	32	0.94
California	96	3	7.56
Colorado	18	21	1.42
Connecticut	15	27	1.18
Delaware	19	20	1.50
District of Columbia	—	NA	0.00
Florida	55	6	4.33
Georgia	14	28	1.10
Hawaii	4	45	0.31
Idaho	10	38	0.79
Illinois	38	9	2.99
Indiana	33	12	2.60
Iowa	17	23	1.34
Kansas	13	30	1.02
Kentucky	20	19	1.57
Louisiana	17	23	1.34
Maine	12	32	0.94
Maryland	14	28	1.10
Massachusetts	30	13	2.36
Michigan	78	5	6.14
Minnesota	37	11	2.91
Mississippi	4	45	0.31
Missouri	22	18	1.73
Montana	9	41	0.71
Nebraska	10	38	0.79
Nevada	1	50	0.08
New Hampshire	17	23	1.34
New Jersey	107	1	8.43
New Mexico	11	36	0.87
New York	80	4	6.30
North Carolina	23	17	1.81
North Dakota	2	49	0.16
Ohio	38	9	2.99
Oklahoma	11	36	0.87
Oregon	12	32	0.94
Pennsylvania	103	2	8.11
Rhode Island	12	32	0.94
South Carolina	25	15	1.97
South Dakota	4	45	0.31
Tennessee	18	21	1.42
Texas	27	14	2.13
Utah	16	26	1.26
Vermont	8	42	0.63
Virginia	24	16	1.89
Washington	52	7	4.09
West Virginia	7	44	0.55
Wisconsin	41	8	3.23
Wyoming	3	48	0.24
Guam	2		
Puerto Rico	9		
Virgin Islands	2		

Source: U.S. Department of Commerce, *Statistical Abstract of the United States, 1996*, 1996. Primary source, U.S. Environmental Protection Agency.

Note: Includes both proposed and final sites on the National Priority List for the Superfund program.

In the past, many disposal activities were conducted with at least tacit permission from the relevant regulatory authority at the time. Prior to enactment of the Resource Conservation and Recovery Act (RCRA) of 1976, there were few, if any, national regulations addressing the proper management and disposal of the millions of tons of hazardous wastes annually generated in the United States. Many instances of inappropriate disposal of hazardous wastes were documented, including pouring of liquids into unlined ponds, placement of waste drums into unlined pits or trenches, abandonment of containers on the land, and uncontrolled incineration. Contamination has affected local soil, groundwater, surface water, air quality, and, ultimately, human health.

There are an estimated thirty-six thousand severely contaminated sites in the United States; however, some federal agencies claim the number to be much higher. As of November, 1996, there were 1,205 sites (1,054 nonfederal and 151 federal) on the Superfund National Priorities List (NPL). The Superfund pool, originally $1.6 billion, has grown to more than $15 billion. Monies to create the fund are primarily based on a tax on industries ("the polluter pays" principle), and a small portion is derived from individual income taxes. However, cleanup costs for uncontrolled and abandoned U.S. hazardous waste sites are estimated to exceed $350 billion.

Since the Superfund cannot remediate all sites, the U.S. Environmental Protection Agency (EPA) has formulated a system for the determination of potentially responsible parties (PRPs). Four classes of PRPs liable for the costs of investigation and cleanup at a Superfund site have been established: the current owner or operator of the facility, all previous facility owners or operators (at the time of hazardous materials disposal), anyone who arranged for disposal or treatment of hazardous substances, and transporters of the hazardous substances to the disposal facility. CERCLA allows a PRP that is liable to the government for cleanup costs to seek contribution from other PRPs. As a result, the payment process may drag on for years as PRPs seek out others who may offset some of the cleanup cost.

The Superfund declares that PRPs are liable for all costs of removal or remedial action, costs incurred by other parties, damages for destruction of natural resources, and costs of any health assessment conducted. Liability under CERCLA is joint, which means that costs for environmental assessment and cleanup may be distributed over all PRPs. This distribution is at the discretion of the EPA. No excuses for liability are accepted. Finally, liability is retroactive; in other words, a party can be held liable for cleanup costs even if all suspect wastes were disposed prior to the enactment of the Superfund law.

A CERCLA action at a contaminated site is designated as either a short-term removal action or a long-term remedial response. A removal action involves cleanup or other actions taken in response to emergency conditions or on a short-term basis. Actions may include installation of fences, evacuation of threatened populations, and construction of temporary containment systems.

The remedial response action is time-consuming and complex. Some of its basic elements include discovery of the affected site, preliminary investigation, ranking the hazard, selecting the cleanup remedy, preparing the official Record of Decision, detailing remedial design, remedial action, and project closure. Remedial designs range from the very simple to the highly complex. Designs are prepared on a case-by-case basis because each site's characteristics, history, and complexity of contamination are unique. Decisions on remediation methods are influenced by technical considerations, the degree of the present hazard of the site to local populations and ecosystems, and political considerations.

Since the enactment of the Superfund, the U.S. Congress, the EPA, and the American public have been concerned about the pace of cleanups. The average amount of time sites remain on the National Priorities List increased during the 1990's, as did cleanup completion times. Nonfederal projects averaged more than ten years from the time of listing to the time of site completion. Since the inception of Superfund, only about one hundred sites have been completely remediated. Progress on other sites has been delayed in court. Determination of the optimum choice of remediation plan may also delay the process. With the passage of amendments to Superfund in 1986 and 1991, the EPA introduced several initiatives to speed the overall cleanup process.

John Pichtel

SUGGESTED READINGS: A thorough review of Superfund court decisions appears in K. McSlarrow, D. Jones, and E. Murdock, "A Decade of Superfund Litigation: CERCLA Case Law from 1981-1991," *Environmental Law Reporter* (1991). An excellent overview of CERCLA regulations is provided in T. F. Sullivan, editor, *Environmental Law Handbook* (1997). Technologies for remediation of Superfund sites are presented in the U.S. Environmental Protection Agency's *Literature Survey of Innovative Technologies for Hazardous Waste Site Remediation* (1992).

SEE ALSO: Bioremediation; Environmental health; Environmental Protection Agency; Hazardous and toxic substance regulation; Hazardous waste; Landfills; Love Canal; Right-to-know legislation; Times Beach evacuation.

Superphénix

DATES: 1986-1997
CATEGORY: Nuclear power and radiation

The Superphénix, which operated in the Lyon area of France between 1986 and 1997, was the world's largest fast-breeder nuclear reactor. The facility was officially shut down in 1998.

After twelve years of construction, the Superphénix nuclear reactor, a 1,240-megawatt fast breeder, went into operation in Creys-Malville in the Lyon area of France in 1986. It operated at full power for the first time on December 9, 1986. Breeder reactors maximize the production of new fuel by using surplus neutrons not required to sustain the fission chain reaction to produce more fissionable fuels, such as plutonium. However, the Superphénix breeder was continually plagued by accidents and incidents during its twelve-year period of use, which resulted in it operating at full power for a total of only 278 days. The reactor's cooling system, which used liquid sodium, repeatedly suffered costly shutdowns because of leaks. In addition, low uranium prices undercut the value of the plutonium fuel produced by the Superphénix.

In 1994 it was decided to convert the Superphénix from a breeder into a burner of plutonium and to use the facility only as a research tool. Consequently, on July 11, 1994, the reactor license was changed from a power reactor to a

Cows graze near the Superphénix fast-breeder nuclear reactor near Lyon, France. The reactor, which faced numerous technical problems and protests from environmental groups, was closed down in 1997 after only twelve years of intermittent use. (Reuters/Robert Pratta/Archive Photos)

research reactor for the demonstration of burning nuclear waste in breeder reactors. The Superphénix was closed temporarily in December, 1996, for repair, maintenance, and reconstruction, with the plan to restart it in June, 1997. Based on procedural grounds, however, the reactor license was canceled in February, 1997. On June 6, Europeans Against Superphénix, a confederation of 250 environmental and antinuclear groups, demanded that the Superphénix be shut down permanently. Subsequently, on June 19 newly elected French prime minister Lionel Jospin announced in his general policy statement to the Parliament that operations at the Superphénix breeder reactor would be discontinued for economic reasons. Running the Superphénix had cost France billions of dollars, with only approximately six months of electricity in return. The final announcement of closure came on February 2, 1998.

The dismantling of the Superphénix was scheduled to begin in 2005 and was predicted to cost around $1.76 billion. However, since France's huge nuclear power industry generates 80 percent of the country's power, many groups have lobbied for the Superphénix to reopen. In Creys-Malville residents have protested the shutdown of the Superphénix, worried over losing about 1,300 Superphénix-related jobs in a town of twelve thousand people. In the late 1990's politicians began discussing a plan for decommissioning the reactor, job conversion, and protection of the environment, as well as a new economic plan for the region. Dismantling the reactor would actually double the employment at Creys-Malville for a period of five to six years.

The closure of the Superphénix made necessary a review and revision of the French breeder and plutonium recycling programs. Serious questions needed to be answered about what to do with the new and spent nuclear fuel. The shutdown struck a serious blow to the French breeder program and raised questions about breeder programs in other countries, particularly Japan, India, and Russia.

Alvin K. Benson

SEE ALSO: Antinuclear movement; Nuclear and radioactive waste; Nuclear power; Nuclear weapons; Radioactive pollution and fallout.

Surface mining

CATEGORY: Land and land use

Surface mining involves the removal of overlying material to reach valuable mineral deposits. This method of mining is used throughout the world. Environmental problems associated with surface mining include air pollution, water pollution, and land-use disruption.

The extraction of mineral resources from the earth's surface is conducted in several ways, with surface mining, dredging, and hydraulic mining being the most common. Surface mining is a current term that has replaced strip mining as a description of the process of removing overlying material to reach minerals. Surface mining has been practiced since prehistoric times wherever a mineral element is close enough to the surface to extract it by removing the overlying material.

The development of large-scale earthmoving machinery in the late nineteenth century enhanced the development of surface mining by enabling miners to move larger volumes of overlying material and work at greater depths. The change in technological power from steam to internal combustion engines and then to electrical power also made surface mining more efficient. Therefore, surface mining is used as a means of acquiring a variety of minerals. Coal, phosphate rock, shale, limestone, and clay are some of the minerals commonly mined in this way.

In practice, surface mining involves the removal (stripping) of the overlying material, called overburden, to expose the desired mineral. This initial removal is usually done with draglines. Once exposed, the mineral is then broken up with explosives and loaded onto vehicles for transportation away from the mining site. The loading operation is usually carried out with power shovels or front-end loaders. Transportation is often initially done by trucks since they are more mobile and flexible for operation in the mining area. The mined mineral may then be transported longer distances by railroad or river barge, or, in some cases, may be mixed with water and pumped through a pipeline to its ultimate destination.

Mining for bituminous coal is generally the best known example of surface mining. In the United States, bituminous coal mining using surface methods is extensive in the Appalachian, Eastern Interior, Western Interior, and Rocky Mountain coal fields. On a global scale, surface mining of coal is also practiced in Germany, Russia, Poland, China, and Great Britain.

Nearly all aspects of the mining and transporting process have the potential for raising environmental issues. During the mining process, there is extensive destruction of the surface environment. In addition to the visual impact, such mining is noisy and produces considerable dust. The negative impact in the immediate mining area is, therefore, of both an aesthetic and environmental nature. Other problems resulting from surface mining include the loss of productive surface land uses, such as agriculture and forestry, from the local tax base. Such land-use changes may, in turn, have a negative impact on wildlife because of habitat removal, loss of cover, and disruption of migration paths.

Debris produced during the mining process may also create problems away from the site itself. Such debris, when carried away by runoff following rainfall, becomes a serious source of sediment. The sediment, when carried into the surface stream systems, clogs the channels and results in flooding. Small particles cause a deterioration of the aquatic environment by diminishing the quality of the water for fish and other users of the water. Sediment-choked stream valleys may also kill trees and other streamside vegetation by changing the water level.

In some cases surface mining exposes toxic materials, which can also contribute to water quality decline. This is especially true if such substances as marcasite and iron pyrites are exposed to the air and water during the mining process. These minerals are often found adjacent to coal, and, when so exposed, they combine to form acid mine drainage. This drainage may be toxic to aquatic life and streamside wildlife and may cause a deterioration in structural facilities such as dams and bridge supports.

Environmental disruption became so evident in coal mining areas of the United States by the mid-twentieth century that laws began to be passed to regulate the mining activity. Statewide laws were most common, although some counties and townships did enact regulations at a local level. Ohio, Pennsylvania, and Illinois were among the first states to enact legislation. The early laws were not very demanding in their requirements, but as environmental benefits began to result from the initial legislation, the laws became more stringent over time.

Federal interest was raised because surface mining is so widespread in the United States and because the problems associated with surface mining for coal extended, in some cases, across state boundaries. In August of 1977, the federal government became involved in controlling the problems resulting from surface mining with the passage of the Surface Mining Control and Reclamation Act of 1977. This act identified areas of concern and established a federal enforcement office. It also provided for state mining and mineral resources research. The act gave attention to abandoned mine lands and "control of the environmental impacts of surface coal mining." To control the environmental impacts, a permit system was established, as was a reclamation plan requirement to ensure that, following the mining process, the land would be returned to a productive use. Environment performance standards were also established by the law. The national standards became a minimum level of reclamation to which mining companies were required to adhere. States are free to develop more stringent standards and requirements for postmining land uses.

Jerry E. Green

SUGGESTED READINGS: A general review of surface mining in an environmental context is provided by G. Tyler Miller in *Resource Conservation and Management* (1990). Details regarding legal requirements for reclamation at the national level are contained in Public Law 95-87, August 3, 1977. A review of postmining land uses in the United States is covered in Jerry Green, *International Journal of Environmental Education and Information* (1989).

SEE ALSO: Reclamation; Restoration ecology; Strip mining.

Sustainable agriculture

Category: Agriculture and food

Sustainable agriculture is the practice of growing and harvesting crops in a manner that has minimal impact on the environment.

Most twentieth century agricultural practices were based upon continued economic growth. This practice demonstrated dramatic increases in production but had a negative impact on the environment through the losses of plant and animal habitats, depletion of soil nutrients, and an increase in pollution of water supplies. The concept of sustainable development is based on using renewable resources and working in harmony with existing ecological systems. The World Commission on Environment and Development phrased the concept of sustainable development as being able "to meet the needs of the present without compromising the ability of future generations to meet their own needs." Sustainable agriculture strives to manage agricultural activities in such a way as to protect air, soil, and water quality, as well as conserve wildlife habitats and biodiversity.

Problems Caused by Agriculture

Water pollution is one of the most damaging and widespread effects of modern agriculture. The runoff from farms accounts for over 50 percent of the sediment damage to natural waterways, and the chemicals and nutrients associated with this runoff in the United States are estimated to cost between $2 billion and $16 billion per year to clean. Heavy application of nitrogen fertilizers, insecticides, and herbicides has raised the potential for groundwater contamination. Feedlots that concentrate manure production lead to further groundwater contamination. Several of the most commonly used pesticides have been detected in the groundwater of at least one-half of the states in the United States. In addition, growing highly specialized monoculture crops, which requires a heavy reliance on agricultural chemicals, has depleted the natural organic nutrients that were formerly rich in North American topsoils.

Research has found that many of the farm-based chemical agents, pesticides, fertilizers, plant-growth regulators, and antibiotics are now found in the food supply. These chemicals can be harmful to humans at moderate doses, and chronic effects can develop with prolonged exposure at lower doses. Further, widespread pesticide use has been shown to severely stress other animals, including bee populations. Pesticides have often caused resurgences of pests after treatment, occurrences of secondary pest outbreaks, and resistance to pesticides in the target pest.

Because of these growing problems, many American farmers are turning to sustainable agriculture. The U.S. federal government has offered guidance for this transition through the Sustainable Agriculture Farm Bill, passed by Congress in 1990. The bill provides that sustainable agriculture, through an integrated system of plant and animal production practices, can, over the long run, meet human food and fiber needs, enhance environmental quality and natural resources, make the most efficient use of nonrenewable resources, maintain economic viability of farm operations, and enhance the quality of life for farmers and society.

Water and Soil Conservation

Water is one of the most important resources for agriculture and society as a whole. In the western United States, it is an important factor in allowing arid lands to produce crops through irrigation. In California, limited surface water supplies have caused overdraft of groundwater and the consequent intrusion of salt water, which causes a permanent collapse of aquifers. In order to counteract these negative effects, sustainable farmers in California are improving water conservation and storage methods, selecting drought-resistant crop species, using reduced volume irrigation systems, and managing crops to reduce water loss. Drip and trickle irrigation can also be used to dramatically reduce water usage and water loss while helping to avoid such problems as soil salinization.

Salinization and contamination of groundwater by pesticides, nitrates, and selenium can be temporarily managed by using tile drainage to remove water and salt. However, this often has adverse affects on the environment. Long-term solutions include conversion of row crops to production of drought-tolerant forages and the restoration of wildlife habitats.

One of the most important aspects of sustainable agriculture is soil conservation. Water runoff from a field having a 5 percent slope has three times the water volume and eight times the soil erosion rate as a field with a 1 percent slope. In order to prevent excessive erosion, sustainable farmers can leave grass strips in the waterways to capture soil that begins to erode. Contour plowing, which involves plowing across the hill rather than up and down the hill, helps capture overland flow and reduce water runoff. Contour plowing is often combined with strip farming, where different kinds of crops are planted in alternating strips along the contours of the land. As one crop is harvested, another is still growing and helps recapture the soil and prevent the water from running straight down the hill. In areas of heavy rainfall, tiered ridges are constructed to trap water and prevent runoff. This involves a series of ridges constructed at right angles to one another. Such construction blocks direct runoff and allows water to soak into the soil.

Another method of soil conservation is terracing, in which the land is shaped into level shelves of earth to hold in the water and soil. To provide further stability to soil, soil-anchoring plants are grown on the edges of the terraces. Terracing, although costly, can make it possible to farm on steep hillsides. Some soils that are fairly unstable on sloping sites or waterways can require that perennial species of grasses be planted to protect the fragile soil from cultivation every year.

Sustainable agriculture tries to match crops and livestock to the topography, soil characteristics, and climatic conditions that exist at a given farm. The selection of crops should be well suited to the existing soil and site conditions as well as being resistant to known pests in the area.

Livestock and Animal Manure

Livestock such as ruminant animals (sheep, cattle, and goats) can be raised on rangeland, pasture, cultivated forage, cover crops, shrubs, weeds, and crop residues. The breeds that have lower growth and milk production potential can adapt better to environments with sparse or seasonal forage. Growing row crops on level soil and growing pasture on steeper slopes can help reduce soil erosion. Putting pasture and forage crops in rotation can help improve soil quality. Allowing ruminants or other farm animals to graze allows the pastures to be fertilized naturally. Farmers can take advantage of animal manure by using portable fencing to make the animals graze one area or strip of pasture all the way down before they are moved to another strip of the field while the first strip of pasture recovers.

Farmers also use green manure—crops that are raised specifically to be plowed under—to introduce organic matter and nutrients into the soil. Green manure crops help protect against erosion, cycle nutrients from lower levels of the soil into the upper layers, suppress weeds, and keep nutrients in the soil rather than allowing them to leach out. Legumes such as sweet clover, ladino clover, and alfalfa are excellent green manure crops. They are able to extract nitrogen from the air into the soil and leave a supply of nitrogen for the next crop that is grown. Some crops, such as beans and corn, can cause high soil erosion rates because they leave the ground bare most of the year. One way sustainable farming is combating this is by leaving crop residues on the land after harvest. Residues help reduce soil evaporation and even excessive soil temperatures in hot climates. Many farmers choose to use cover crops rather than residue crops. Which cover crop to use depends on which geographical area farmers live in and if they wish to control erosion, capture nitrogen in the fall, release nitrogen to the crop, or improve soil structure and suppress weeds.

Cover Crops

When planting crops with high nitrogen requirements, such as tomatoes or sweet corn, a cover crop such as hairy vetch or clover is well suited to the needs of the crop. Both cover crops decompose and release nutrients into the soil within one month. To fight erosion, a farmer might chose a rapid-growing cover crop, such as rye. Rye provides abundant groundcover and an extensive root system below the soil to stop erosion and capture nutrients. Alfalfa, rye, or clover can be planted after harvest to protect the soil and add nutrients and can then be plowed under at planting time to provide a green manure for the crop. Cover crops can also be flattened with rollers, and seeds can be planted in their residue. This gives the new plants a protective cover and discourages weeds from overtaking the young plants. Use of natural nitrogen also reduces the risk of water contamination by agricultural chemicals.

Sustainable agriculture emphasizes the use of reduced tillage systems. There are three reduced tilling systems that sustainable farmers use to disturb the soil as little as possible. Minimum till involves using the disc of a chisel plow to make a trench in the soil where seeds are planted. Plant debris is left on the surface of the ground between the rows, which helps further prevent erosion. Several sustainable planting techniques help prevent soil erosion. Conser-till farming uses a coulter to open a slot just wide enough to insert seeds without disturbing the soil. No-till planting involves drilling seeds into the ground directly through the ground cover or mulch. When mulch is still in place, a narrow slit can be cut through the cover or crop residues in order to plant the new crops.

Crop Rotation and Monoculture

Planting the same crop every year on the same field can result in depleted soils. In order to keep the soil fertile, nitrogen-depleting crops (such as sweet corn, tomatoes, and cotton) should be rotated every year with legumes, which add nitrogen to the soil. Planting a winter cover crop, such as rye grass, protects the land from erosion. Such cover crops will, when plowed under, provide a nutrient-rich soil for the planting of a cash crop. Crop rotations improve the physical condition of the soil because of variations in root depth and cultivation differences.

In nature, plants grow in mixed meadows, which allow for them to avoid insect infestations. Agricultural practices that use monoculture place a great quantity of the food of choice in easy proximity of the insect predator. Insects can multiply out of proportion when the same crop is grown in the field year after year. Since most insects are instinctively drawn to the same home area every year, they will not be able to proliferate and thrive if crops are rotated and their crop of choice in not in the same field the second year.

Crop rotation not only helps farmers use fewer pesticides but also helps to control weeds naturally. Some crops and cultivation methods inadvertently allow certain weeds to thrive. Crop rotations can incorporate a successor crop that eradicates the weeds. Some crops, such as potatoes and winter squash, work as cleaning crops because of the different style of cultivation that is used on them. Pumpkins planted between rows of corn will help keep weeds at bay.

Integrated Pest Management

Most sustainable farmers use integrated pest management (IPM) to control insect pests. Using IPM techniques, each crop and its pest is evaluated as an ecological system. A plan is developed for using cultivation, biological methods, and chemical methods at different timed intervals. Although effective, profitable, and safe, the IPM techniques have been widely adopted only for a few crops, such as tomatoes, citrus, and apples.

The goal of IPM is to keep pest populations below the size where they can cause damage to crops. Fields are monitored to gauge the level of pest damage. If farmers begin to see crop damage, they put cultivation and biological methods into effect to control the pests. Techniques such as vacuuming bugs off crops are used in IPM. IPM encourages growth and diversity of beneficial organisms that enhance plant defenses and vigor. Small amounts of pesticides are used only if all other methods fail to control pests. It has been found that integrated pest management, when done properly, can reduce inputs of fertilizer, lower the use of irrigation water, and reduce preharvest crop losses by 50 percent. Reduced pesticide use can cut pest-control costs by 50 to 90 percent and increases crop yield without increasing production costs.

Toby Stewart and Dion Stewart

SUGGESTED READINGS: *Alternative Agriculture* (1989), by the Board on Agriculture National Research Council, provides pages of detailed information on sustainable agricultural methods. *Grow It* (1972), by Richard W. Langer, is an easy-to-read, hands-on book explaining how to use and apply sustainable farming methods to both farm and garden. *Sustainable Agriculture: Task Force Report* (1994) lists the goals and strategies for initiating sustainable agriculture in the United States.

SEE ALSO: Agricultural revolution; Alternative grains; Green Revolution; Integrated pest management; Organic gardening and farming.

Sustainable development

CATEGORY: Ecology and ecosystems

Sustainable development meets the consumption needs of the current generation without compromising the ability of future generations to increase their economic production to meet future needs. Environmental benefits arise as a consequence of changes in human attitude and behavior, technology, and resource utilization.

According to the 1987 United Nations World Commission on Environment and Development, also known as the Brundtland Commission, humanity has the ability to make development sustainable—to ensure that it meets the needs of the present without compromising the ability of future generations to meet their own needs. Sustainable development is a process of change in which the exploitation of resources, the direction of investments, the orientation of technological development, and institutional change are all in harmony and enhance both current and future potential to meet human needs and aspirations. The commission envisions the possibility of continued economic growth, population stabilization, improvements in global economic equity between rich and poor nations, and environmental improvement, all occurring simultaneously and in harmony. Since publication of the Brundtland Commission report, sustainable development has become the dominant global position on the environment, ecology, and economic development.

Sustainable development is a normative philosophy, or value system, concerned with equal distribution of the earth's natural capital among current and future generations of humans. Sustainable development promotes three core values. First, current and future generations should each have equal access to the planet's life-support systems—including Earth's gaseous atmosphere, biodiversity, stocks of exhaustible resources, and stocks of renewable resources—and should maintain the earth's atmosphere, land, and biodiversity for future generations. Exhaustible resources, such as minerals and fossil fuels, are used sparingly and conserved for use by future generations. Renewable resources, such as forests and soil fertility, are renewed as they are used to ensure that stocks are maintained at or above current levels and are never exhausted.

Second, all future generations should have an equal opportunity to enjoy a material standard of living equivalent to that of the current generation. In addition, the descendants of the current generation in underdeveloped regions are permitted to increase their economic development to match that available to descendants of the current generation in the industrialized regions. Future development and growth in both developed and underdeveloped regions must be sustainable.

Finally, future development must no longer follow the growth path taken by the currently industrialized countries but should utilize appropriate technology. Development should also limit use of renewable resources to each resource's maximum sustained yield, the rate of harvest of natural resources such as fisheries and timber that can be maintained indefinitely through active human management of those resources.

Weak sustainability requires that depletions in natural capital be compensated for by increases in human-made capital of equal value. For example, the requirements for weak sustainability are met when a tree (natural capital) is cut for the construction of a frame house (human-made capital). However, if the tree is cut and cast aside in a land-clearing project, the requirements for

A housing development in southern Florida encroaches upon the Everglades ecosystem. Advocates of sustainable development insist that the needs of a growing population must be balanced with the need to maintain biodiversity for future generations. (AP/Wide World Photos)

weak sustainability are not met. Strong sustainability requires that depletions of one sort of natural capital be compensated for by increases in the same or similar natural capital. For example, the requirements for strong sustainability are met when a tree is cut and a new tree is planted to replace it, or when loss of acreage in equatorial rain forests in Brazil is compensated for by an increase in the acreage of temperate rain forests on the Pacific coast of North America.

Sustainable development is promoted through a combination of public policies. First, to the extent possible, elements in the earth's support system are assigned monetary values in order to make the economic and financial calculations that are necessary to ensure that the requirements of weak sustainability are met. Second, economic development in the underdeveloped world is shifted away from high-resource-using, high-polluting patterns of Western development and toward more sustainable or "appropriate" patterns. Suggested appropriate technologies include solar energy, resource recycling, cottage industry, and microenterprises (factories built on a small scale). Third, objective and measurable air, water, and resource quality standards are established and enforced to ensure that a continuing minimum quality and quantity of natural capital is maintained and that certain stocks of natural capital are protected through the establishment of wilderness areas, oil and gas reserves, and other reserves. Finally, each individual human adopts a personal commitment to a sustainable lifestyle, thus making a minimal personal impact on the earth's natural capital.

Environmental improvement results from the changes in resource utilization. For example, reductions in use and waste of natural capital reduces the environmental impact of resource extraction industries such as strip mines, and waste disposal industries such as incinerators. Environmental quality standards and maintenance of biodiversity leads to implementation of antipollution and ecosystem restoration efforts.

Gordon Neal Diem

SUGGESTED READINGS: *Agenda 21* (1992), the report of the United Nations Earth Summit, discusses implementation of sustainable development and reduction of wealth disparities between rich and poor nations. The World Bank's *World Environment Report* (1992) discusses ways environmental management and economic development can proceed together, while their *Monitoring Environmental Progress* (1995) proposes a measure of sustainable national well-being to rival more established measures of development such as gross national product. Lai Lee, *Compass and Gyroscope: Integrating Science and Politics for the Environment* (1993), describes early efforts to implement sustainable development and ecosystem rehabilitation in the Columbia River Basin. John Dryzek, *The Politics of the Earth: Environmental Discourse* (1997), compares sustainable development to other contemporary development concepts. John Bowers, *Sustainability and Environmental Economics* (1997), provides an economic analysis of sustainable development. Daniel Sitarz, *Sustainable America: America's Environment, Economy and Society in the 21st Century* (1998), presents plans to protect the environment, build the economy, and provide for social equality.

SEE ALSO: Brundtland, Gro Harlem; Earth Summit; *Global 2000 Report, The*; Intergenerational justice; Sustainable agriculture; Sustainable forestry; United Nations Environmental Conference.

Sustainable forestry

CATEGORY: Forests and plants

Sustainable forestry is a system of forest management that relies on natural processes to maintain a forest's continuing capacity to produce a stable and perpetual yield of harvested timber and other benefits, including recreation, wildlife habitat, and forest-related commodities.

Forest management in the United States first became an issue in 1827 when the Department of the Navy and President John Quincy Adams saw the need for a continuous supply of mature timber for ship construction. In the 1860's the American Association for the Advancement of Science first discussed the need for sustained-yield forestry. In 1878 the Cosmos Club, a Washington, D.C., club of intellectuals, proposed the wise use of natural resources for the greatest good, for the greatest number, and for the longest time, establishing the foundation for the conservation movement. The first national forest reserves were established by the U.S. government in 1891, and the first selective logging and marketing of U.S. government timber reserves occurred in 1897. Clear-cutting was the general method of timber harvesting. Continued clear-cutting during the twentieth century deforestated private and Forest Service lands, leading to concerns about soil erosion, water pollution, loss of wildlife habitat, and the sustained availability of forest resources.

Forest science developed the high-yield forestry plantation tree farming system in the 1930's. By the 1960's ecological concerns had led to restoration forestry, which emphasized human intervention to reconstruct forest ecosystems and return forests to baseline conditions that existed before clear-cutting or plantation planting. By the 1980's new understandings concerning the complexity of forest ecosystems led to an emphasis on perpetually sustaining existing forest resources rather than relying on human efforts to reconstruct forests.

Sustainable forestry is an alternative to clear-cutting, the standard logging practice. Clear-cutting removes all timber in one harvest that usually occurs no more than once every sixty to one hundred years. Both mature and immature trees are removed in one process. Logging roads are cut into the forest so heavy machinery can remove all trees from a large area, usually about 100 acres at a time. Road construction and clear-cutting lead to soil erosion, topsoil and nutrient loss, silting and pollution of waterways, the loss of wildlife habitat, and the loss of recreational benefits. Repeated cycles of growth and clear-cutting erode soil nutrition, destroy plants, animals, and microorganisms in the ecosystem necessary for healthy forest growth, and reduce the value of future harvests.

Sustainable forestry is also an alternative to monoculture plantation forestry. Plantation forestry requires active human intervention to plant tree seedlings, control disease and pests, and nurture the timber stand to maturity. Plantations usually feature a grid planting of a single tree species, with all trees maturing simultaneously. The lack of species and age diversity makes tree plantations unsuitable for wildlife habitat or recreation and makes trees susceptible to disease and pests. Monoculture plantations also deplete species-specific minerals and other nutrients in the soil, reducing its future productivity.

Sustainable forest management techniques seek a perpetual high yield of timber and pulpwood while maintaining biological diversity and natural forest ecosystems and permitting forests to restore their vitality through natural processes, such as foliage decomposition and fire.

Sustainable forestry maintains a balance between natural environmental stresses and the human needs for timber, pulpwood, recreation, and a variety of harvested forest products. In spite of the effort to maintain this balance, various sustainable forestry methods often tend to favor either ecosystem maintenance or high timber yields.

Sustainable forestry with an ecosystem emphasis is the discipline of repeated thinning of natural tree stands to sustain a mixed-age, mixed-species forest that is naturally perpetuated by seeds from the mature trees. The forest is periodically thinned, usually every twenty years, to provide a steady income to the forest owners, permit the remaining trees to reach their full maturity, and provide space for new seedlings to grow. When the timber stand reaches full sustainable maturity, immature trees are continuously harvested for pulpwood, and mature trees over one hundred years of age are continuously harvested for high-quality lumber. Natural processes promote the health of the forest and revitalize the forest soil. Diversity in both age and species makes the forest a suitable habitat for a variety of forest-dwelling species and human recreation. The forest is able to quickly recover from natural disasters, fires, or drought.

A logging company sign shows the timetable for reharvesting a plantation area. Sustainable forestry, which relies on efforts to maintain ecosystem biodiversity, is an alternative to such monoculture forestry. (Jim West)

Sustainable forestry with an emphasis on timber yield divides the forest into subplots, then manages each subplot to produce two sequential high-yield plantation crop cycles of eighty years each before permitting the plot to grow to maturity in a third four-hundred-year cycle. The third cycle permits the forest soil to restore its vitality and produces an old-growth forest suitable for wildlife and eventual timber harvesting. Once fully implemented, this system ensures that each forest has subplots at each stage of growth and harvesting, from newly planted plots to old-growth plots with trees at or near four hundred years of age.

Gordon Neal Diem

SUGGESTED READINGS: Various forestry techniques, including clear-cutting, sustainable forestry, and restoration forestry, are described in John Berger's *Understanding Forests* (1998). The economics and social desirability of sustainable forestry are discussed in Chris Maser, *Sustainable Forestry: Philosophy, Science and Economics* (1994). The Institute for Sustainable Forestry book *Working Our Woods: An Introductory Guide to Sustainable Forestry* (1996) describes successful examples of sustainable forests, emphasizing forester Craig Blencowe's work in California. The U.S. Department of Agriculture Forest Service General Technical Report PNW-OTR-319, "A Framework for Sustainable Ecosystem Management," provides technical guidance for implementing sustainable practices.

SEE ALSO: Forest management; Logging and clear-cutting; National forests; Renewable resources; Restoration ecology; Sustainable development; Wise-use movement.

Synthetic fuels

CATEGORY: Energy

Synthetic fuels are solid, liquid, or gaseous fuels that do not occur naturally. They are normally produced from abundantly occurring natural resources such as coal, tar sands, oil shale, and biomass.

One of the main objectives of producing synthetic fuel is to eliminate sulfur and nitrogen from the fuel compound, thereby creating an environmentally clean energy source. Oxides of nitrogen and sulfur dioxide are among the most undesirable of common air pollutants. Sulfur dioxide is one of the major causes of acid rain, which is created when gases such as sulfur dioxide combine with water vapor in the atmosphere to form sulfuric acid. Similarly, oxides of nitrogen produce nitric acid. These acids fall back to earth as rain and are detrimental to aquatic life as well as botanical life. Synthetic fuel manufacturers thus strive to eliminate these pollutants, as well as others such as carbon monoxide, hydrocarbons, particulates, and photochemical oxidants, from the fuel supply.

PRINCIPLES OF SYNTHETIC FUEL MANUFACTURE

Liquid and gaseous synthetic fuels are normally manufactured by transforming naturally occurring carbonaceous raw material, using a suitable conversion process. The techniques employed include hydrogenation, devolatilization, decomposition, and fermentation. The principle aim in the manufacture of synthetic fuel is to achieve a low carbon-to-hydrogen atomic mass ratio, or a high hydrogen-to-carbon atomic ratio, whenever possible. This results in a clean-burning fuel that releases by-products that are harmless to the environment. For example, pure methane (CH_4), with a molecular weight of 16, has a high hydrogen-to-carbon ratio of 4:1. Methane gas is a common component that is absorbed into coal. The gas can be released by fracturing the coal and exposing it to low pressures. Coal-bed methane is one of the cleanest-burning fossil fuels; the by-products of burning it are simply carbon dioxide and water. Synthetically generated substitute natural gas is more than 90 percent methane. Natural gas (of which methane is the chief constituent) has a hydrogen-to-carbon ratio of approximately 3.4:1, which is also quite high. The ratios for liquefied petroleum gas and for naphtha lie between 2:1 and 3:1. (In comparison, the ratios for gasoline and fuel oil are less than 2:1. Bituminous coal has one of the lowest values, with ratio of much less than 1:1.)

Coal Gasification and Liquefaction

Although coal is among the most abundant natural energy sources, it is also among the dirtiest. The composition of this solid fossil fuel is a major disadvantage; it consists of about 70 percent carbon and about 5 percent hydrogen, translating to a highly undesirable carbon-to-hydrogen mass ratio of 14:1. Coal-burning power-generating stations thus spew out large quantities of gases that are harmful to the environment. Despite the use of such emission-reduction devices as electrostatic precipitators, the levels of pollutants emitted by coal-burning plants remain high. Techniques such as coal gasification and coal liquefaction yield synthetic fuels that are safer for the environment.

The process of coal gasification involves making coal react with steam at very high temperatures (in the range of 1,000 degrees Celsius). This process produces synthetic gas. Three types of synthetic gas are in common use. Low-calorific-value gas (also called "producer" gas) is used in turbines. Medium-calorific-value gas (also called "power" gas is used as a fuel gas by various industries. High-calorific-value gas (also called "pipeline" gas) is a very good substitute for natural gas and is well suited to economical pipeline transportation. Pipeline gas contains more than 90 percent methane; as a result, it has a high hydrogen-to-carbon ratio.

The process of coal liquefaction is employed to generate a liquid fuel with a high hydrogen-to-carbon ratio; it is also used to obtain low-sulfur fuel oil. Several methods are employed to accomplish coal liquefaction, including direct catalytic hydrogenation, indirect catalytic hydrogenation, pyrolysis, and solvent extraction. All of these methods produce fuels that are much safer for the environment than the original coal.

Tar Sands and Oil Shale

Naturally occurring tar sands contain grains of sand, water, and bitumen. Bitumen, a member of the petroleum family, is a high-viscosity crude hydrocarbon. A method known as hot water extraction is used to procure bitumen from tar sands. The bitumen is subsequently upgraded to synthetic crude oil in refineries. Synthetic crude oil (also called "syncrude") is similar to petroleum and can be obtained by coal liquefaction as well as from tar sands and oil shale.

Large deposits of tar sands are found in Alberta, Canada; the United States has huge reserves of oil shale in Utah, Wyoming, and Colorado. Oil shale is probably the most abundant form of hydrocarbon on earth. Oil shale is a sedimentary rock that contains kerogen, which is not a member of the petroleum family. A popular method known as "retorting" is used to produce oil from shale. The process involves the method of pyrolysis, which reduces the carbon content in the raw hydrocarbon by distillation. Because of the low cost of petroleum, however, it has not been possible to produce shale oil economically.

Biomass Fuels and Gasohol

Like oil and coal, biomass is derived from plant life. Oil and coal, however, are considered non-renewable resources, as it takes vast periods of time for geological processes to produce them naturally. Because biomass consists of any material that is derived from plant life, it is produced in far shorter spans—a hundred years or less—and is thus considered renewable. Wood is the most versatile biomass resource; farm and agricultural wastes, municipal wastes, and animal wastes are also considered to be biomass. Biomass can be processed in a variety of methods. Fermentation, for example, yields ethanol, or ethyl alcohol (sometimes called "grain alcohol"). Other processes used to convert biomass into fuels include combustion, gasification, and pyrolysis.

Gasohol is a mixture of gasoline and small quantities of ethanol. The mixture burns cleaner than conventional gasoline; however, it can cause damage to plastic and rubber materials used in automobile engines. In the United States, therefore, the Environmental Protection Agency (EPA) permits the addition of only 10 percent ethanol by volume to gasoline to create gasohol. Methanol, or methyl alcohol (also called "wood alcohol") can also be combined with conventional gasoline to produce cleaner fuel; however, the EPA limits the amount of methane in such mixtures to 3 percent.

OTHER METHODS

A nonpolluting rocket fuel based on alcohol and hydrogen peroxide has been developed by U.S. Navy research engineers at China Lake, California. The Navy's nontoxic homogeneous miscible fuel (NHMF) can be modified and used to drive turbines, which in turn drive alternators that produce electricity. Further developments of this fuel may permit its use in automobiles. During World War II, moreover, Germany produced synthetic fuels in large quantities to meet its energy demands, employing coal gasification and also creating diesel oil and aviation kerosene using a reconstitution process; this process is still in use in many places.

Although the present abundance of natural petroleum limits the economic competitiveness of most synthetic fuels, the finite nature of the world's oil supply virtually ensures that synthetic fuels will become increasingly important energy sources. Thus, the U.S. Department of Energy and governmental agencies in many other countries provide funding to encourage research into the creation of less expensive, environmentally safe, and renewable synthetic fuels.

Mysore Narayanan

SUGGESTED READINGS: Information on the production and use of synthetic fuels may be found in J. Douglas's "Quickening the Pace in Clean Coal Technology" (*EPRI Journal*, January-February, 1989); *Environment: 98/99* (1998), by John L. Allen; John M. Fowler's *Energy and Environment* (1975); Richard L. Bechtold's *Alternative Fuels Guidebook* (1997); *Alternative Motor Fuels* (1996), by Maureen Sheilds Lorenzetti; and *The Energy Sourcebook* (1991), edited by Ruth Howes and Anthony Fainberg,

SEE ALSO: Alternative energy sources; Alternative fuels; Alternatively fueled vehicles; Biomass conversion; Refuse-derived fuels.

T

Tansley, Arthur G.

Born: August 15, 1871; London, England
Died: November 25, 1955; Grantchester, England
Category: Ecology and ecosystems

Ecologist and environmental pioneer Arthur G. Tansley coined the term "ecosystem" and published views concerning natural processes that have become central to ecological theory. He was also instrumental in the founding of the Nature Conservancy.

Arthur G. Tansley was the only son of George Tansley, a businessman and teacher at the Working Men's College, and Amelia Lawrence. Tansley was educated at University College, London, and Trinity College, Cambridge, where he studied natural science. He joined F. W. Oliver as a professor of botany at University College and worked on fernlike plants. He married one of his student collaborators, Edith Chick, in 1903 and had three daughters who had careers in physiology, architecture, and economics.

Tansley founded *The New Phytologist* journal in 1902 and edited it for thirty years. He directed about one dozen botanists known as the British Vegetation Committee, who published *Types of British Vegetation* in 1911. He formed the British Ecological Society in 1913, became its first president, and edited the society's *Journal of Ecology* from 1917 to 1938. In an essay on the use and abuse of vegetation concepts and terms published in the July, 1935, issue, Tansley coined the word "ecosystems," although the general underlying ideas of ecosystems had existed prior to 1935. Tansley stressed the integration and interdependence of succession (a continuous process of vegetation change), animal life, organic and inorganic matter, and climate and soil, with change usually occurring in a gradual manner. The climax is the highest stage of integration and the nearest approach to perfect dynamic equilibrium. However, equilibrium, or stability, is never quite perfect.

Tansley is known for emphasizing ecology as an "approach to botany through the direct study of plants in their natural conditions." Since plants exist in communities, ecologists should concern themselves with the structure of communities. The study of a habitat should include a study of its parts, such as green plants, herbivores, carnivores, fungi, bacteria, dead and organic matter, solar energy, water, oxygen, carbon dioxide, nitrogen, heat, respiration, and nutrient losses.

Tansley did not personally direct many research students and has been criticized for avoiding experimentation while concentrating on description, comparison, and synthesis in the sphere of ecological theory. His views have been central to most British and American ecological theory.

From 1907 to 1923 Tansley was a lecturer in botany at Cambridge. During that time he became interested in psychology and published *The New Psychology and Its Relation to Life* in 1920. He went to Vienna, Austria, and studied with Sigmund Freud in 1923-1924. Like the evolutionary naturalists Charles Darwin and Alfred Russel Wallace, Tansley had a side that also sought answers from a spiritist or unconscious realm. He was appointed Sherardian Professor of Botany at Oxford in 1927, a post he held until 1937. Two years after retirement, he published his best-known book, *The British Islands and Their Vegetation* (1939). He entered the field of public policy when Great Britain was planning post-World War II conservation. He was instrumental in the founding of the Nature Conservancy in 1949 and served as its first chairman from 1949 to 1953. He was knighted in 1950.

Oliver B. Pollak and Aaron S. Pollak

SEE ALSO: Ecosystems; Nature Conservancy.

Taxol

CATEGORY: Human health and the environment

Taxol is a potent cancer-fighting drug originally derived from the bark of the Pacific yew tree, a small- to medium-sized understory tree that occupies Pacific coastal forests from southwestern Alaska to California.

Development of taxol as a drug began in 1962 with the collection in Washington State of the reddish-purple bark of the Pacific yew tree (*Taxus brevifolia Nutt*) by Kurt Blum, then a technician with the National Cancer Institute (NCI). The NCI was employing a "shotgun" approach to cancer research: A wide variety of plant parts of various species were being screened for anticancer activity. Thereafter, several scientists, including Monroe Wall and M. C. Wani at Research Triangle Institute in North Carolina, and Susan Horwitz and Peter Schiff of the Albert Einstein College of Medicine in New York, recognized the potential of taxol and became intensely interested. After years of delay, Bristol-Meyers Squibb pharmaceutical company continued tests and production, but on a larger scale. By the late 1980's taxol had become the drug of choice, despite its high cost, for the treatment of a wide range of cancers, but especially ovarian and breast cancer. It arrests the growth of cancer cells by attaching to their microtubules, thus preventing cell division.

In spite of taxol's prominence as a success story in the "herbal renaissance" of the twentieth century, several problems involved in production and medicinal use have persisted. For one, the cost of taxol treatment has been prohibitive for many who desperately need it. The large amount of bark required (all the bark from a one-century-old tree yields only enough taxol for a 300-milligram dose) raised fears among conservationists that continued harvesting could threaten the species. While occurring over a wide area, the tree exists only in relatively small numbers. Furthermore, it is a slow-growing species that rarely reaches a height of more than 18 meters (60 feet); stripping the bark kills the tree.

Several means of producing taxol without the destruction of wild yew trees have been proposed. Attempts have been made to produce taxol from tissue cultures. Plantations of the Pacific yew tree could be established, but it would take years before they would be productive. Attempts to identify other *Taxus* species that may contain taxol have been only marginally successful. In 1993 the Bristol company announced that it had found a semisynthetic method for producing taxol that does not require yew bark. Taxol-like compounds have been found in extracts from needles of the European yew tree (*Taxus baccata*) and those of several yew shrub species. An important advantage is that needles can be harvested without killing the trees or shrubs. Similar compounds have also been found in a fungus that grows on *Taxus* species.

Thomas E. Hemmerly

SEE ALSO: Biodiversity; Logging and clear-cutting.

Tellico Dam

DATE: completed November 29, 1979
CATEGORY: Preservation and wilderness issues

The Tellico Dam project involved the construction of a hydroelectric dam on the Little Tennessee River near Knoxville, Tennessee. Controversy over the transformation of a river valley into an artificial lake brought national attention to the conflict between wilderness preservation and human development.

As early as 1936, the Tennessee Valley Authority (TVA) had made plans to build a dam across the Little Tennessee River to facilitate navigation below Fort Loudoun, but the project was vetoed in 1942 because of a scarcity of steel. The project remained a high priority until 1959, when the TVA undertook a thorough study of how the region would be affected by the dam. The cost of the dam was also compared to the benefits it would create, which included electricity as well as employment opportunities and recreational sites. Although the cost was estimated to be about equal to the possible benefits, the TVA decided that building the dam would be economically feasible and beneficial to the area.

The federal money appropriated for the dam's construction had to be approved by the president of the United States. On October 17, 1966, the TVA received $3.2 million to begin building the dam in 1967, with the completion date estimated to be 1970 or 1971. The final cost of the dam was actually $120 million, with a completion date of November 29, 1979. The delay and extra cost were caused by many factors, including inflation and the diversion of federal money to support the Vietnam War.

Besides economic factors, however, the construction of the dam was also delayed during the 1970's because of legal action taken against it by the Cherokee Nation, local residents, and environmentalists. The Little Tennessee Valley, which would be flooded upon completion of the project, contained many Cherokee historical sites, including sacred burial grounds and ruins of the Seven Towns, which were the center of the Cherokee Nation before the Cherokees were sent to reservations in Oklahoma and North Carolina. Local residents claimed an interest in maintaining homes that had been in families for generations.

Prime farmland would also be lost once the valley flooded. Environmentalists pointed out that the dam was unnecessary since, compared to the TVA's total electrical output, the dam would put out very little electricity. Within 96 kilometers (60 miles) of Tellico, twenty-four major dams already existed. The recreational opportunities of a new lake, such as boating, swimming, and fishing, were trivial compared to the greater wilderness activities associated with an untamed river near the Great Smoky Mountains National Park. Building the dam would restrict the last free-flowing stretch of the Little Tennessee River.

After the discovery of an endangered species of fish called the snail darter in the Little Tennessee River in 1973, environmentalists filed suit in 1977 to halt construction of the dam because it would destroy the fish's habitat. However, a special law was passed to exempt the TVA from complying with the Endangered Species Act of 1973, and Tellico Dam went into operation in January of 1980.

Rose Secrest

SEE ALSO: Dams and reservoirs; Hydroelectricity; Snail darter; Tennessee Valley Authority; *Tennessee Valley Authority v. Hill.*

Tennessee Valley Authority

DATE: established 1933
CATEGORY: Energy

The Tennessee Valley Authority (TVA), established by the Tennessee Valley Authority Act of 1933, was created as a federal corporation that was authorized to generate, transmit, and sell electric power through municipal distributors and rural electric cooperatives. In 1936 the Supreme Court upheld the constitutionality of this act and, in 1939, upheld the right of the TVA to supply and sell electric power.

Established in 1933, the TVA was, by 1945, the largest electric utility in the United States. In addition to its original hydroelectric capability, the TVA has also invested in thermal-electric and nuclear power plants, giving it the capacity to annually produce 53 billion kilowatt-hours of electricity and provide electric power to more than 7 million customers.

The Tennessee River stretches for 1,049 kilometers (652 miles) and drains most of the state of Tennessee and parts of six other states. The TVA provided for the maximum development of the Tennessee River basin, which has been referred to as the most comprehensive environmental program in history. In addition to providing inexpensive and abundant electric power, the TVA was designed to control the floods that periodically devastated the basin and increase the navigation potential of the Tennessee River. With its system of locks and dams, the river is now navigable from Knoxville, Tennessee, to its junction with the Ohio River at Paducah, Kentucky. Other goals of the TVA were to institute good conservation practices and agricultural programs, improve air and water quality, attract industry and commerce, increase resource development, and generally increase the quality of life for the residents of the region. During World War II and throughout the Cold War, the TVA provided the energy for aluminum processing factories and uranium enrichment facilities.

To accomplish its purpose, the TVA built a total of twenty-nine dams on the Tennessee River and its tributaries; the largest is Kentucky Dam, and the highest, at 146 meters (480 feet), is Fontana Dam. The TVA dams are of two types: high dams with large reservoir capacities constructed on the tributaries to provide flood protection and electric power generation, and low, broad dams on the Tennessee River designed to control navigation. To construct the first sixteen TVA dams between 1933 and 1944, the federal government purchased or condemned 1.1 million acres of land and moved fourteen thousand families.

Total TVA assets are estimated to be as high as $35 billion. Since the TVA sells enormous amounts of electric power but pays no taxes, the idea of selling the TVA to private utility companies has been repeatedly raised. It is estimated that private ownership of the TVA could bring in as much as $600 million annually in tax revenues. If this were to occur, management of dams for flood control and navigation could be given over to other federal agencies, such as the Army Corps of Engineers.

Donald J. Thompson

SEE ALSO: Dams and reservoirs; Hydroelectricity; *Tennessee Valley Authority v. Hill*; Snail darter; Tellico Dam; Watersheds and watershed management.

Tennessee Valley Authority v. Hill

DATE: 1978
CATEGORY: Animals and endangered species

Tennessee Valley Authority v. Hill *was a 1978 U.S. Supreme Court decision that forbade the completion of Tellico Dam across the Little Tennessee River because it would threaten the habitat of the snail darter, an endangered species of fish.*

Tellico Dam was part of a Tennessee Valley Authority (TVA) water resource and development project designed to control flooding, generate electric power, and promote industrial development in an economically depressed area. Opponents of the dam—a coalition of conservationists and local farm and landowners—initiated legal action to stop the project based on the presence of the snail darter, a small fish that had been placed on the federal endangered species list in October of 1975. The Little Tennessee River above Tellico Dam was considered its critical habitat. However, the U.S. District Court ruled that construction would be allowed to continue. Opponents immediately appealed, and the U.S. Court of Appeals reversed the lower court decision and enjoined construction. The TVA then appealed this decision to the U.S. Supreme Court, which upheld the U.S. Court of Appeals' ruling.

Chief Justice Warren Burger, writing for the Court, agreed with the Court of Appeals' opinion that the TVA would violate the Endangered Species Act (1973) if it completed the dam because the plain intent of Congress was to halt species extinction. The plain language of the statute, supported by its legislative history, revealed "a conscious decision to give endangered species priority over the primary missions of federal agencies." Since the statutory language and legislative history also revealed that Congress placed an incalculable value on endangered species, the Court would not engage in any "fine utilitarian calculations" and find that the loss of an almost completed dam at a cost of more than $100 million would outweigh the loss of the snail darter. The statute did provide hardship exemptions, but none applied to the Tellico Dam project, nor did continuing congressional appropriations for the dam constitute an implied repeal of the statute. Since the completion of the dam would destroy an endangered species, the Court held that the statute required an injunction forbidding its completion.

Justice Lewis Powell, joined by Justice Harry Blackmun, dissented. Condemning the Court's literalist interpretation of the Endangered Species Act, he argued that Congress could not have intended for the Court to give retroactive effect to the statute and disregard a congressional commitment for twelve years to complete the Tellico Dam project. He had no doubt that Congress would amend the statute so that the dam and its reservoir would serve their intended purposes instead of providing "a conversation piece for incredulous tourists."

Justice William Rehnquist, also dissenting, argued that the Endangered Species Act did not mandate the U.S. District Court to use its equitable powers and enjoin the TVA from completing the dam, nor did the District Court abuse its discretion in refusing to issue the injunction. In the face of conflicting evidence, the District Court had quite properly decided that the public harm from the failure to complete the dam outweighed the need to preserve the habitat of the snail darter.

Congress swiftly responded to the Court's decision by attaching an amendment to a 1979 energy bill that exempted the Tellico Dam from all federal laws, including the Endangered Species Act. The dam was completed by early 1980. The snail darter was not, as feared, extinguished by the impoundment of the Little Tennessee River because other colonies of the fish were discovered elsewhere.

William C. Green

See also: Dams and reservoirs; Endangered species; Endangered Species Act; Endangered species and animal protection policy; Snail darter; Tellico Dam; Tennessee Valley Authority.

Teton Dam collapse

Date: June 5, 1976
Category: Water and water pollution

On June 5, 1976, the Teton Dam, a large earth-fill dam in Idaho constructed by the U.S. Bureau of Reclamation, catastrophically failed. The released water flooded the town of Rexburg and killed eleven people.

The Teton Dam was located in a deep, narrow canyon on the Teton River, a tributary of the Snake River, in southeastern Idaho. It was 93 meters (305 feet) high, 975 meters (3,200 feet) long, and formed a reservoir that extended 27 kilometers (17 miles) up the canyon. Approximately 850,000 cubic meters (10 million cubic yards) of clay, silt, sand, and gravel were used to build the multilayered structure, a construction technique that the U.S. Bureau of Reclamation had previously used for approximately 250 other dams without a single failure. Extensive site investigations revealed that the fractured and porous bedrock of the area could be a problem. In an attempt to produce a barrier impermeable to water seepage, grout (a cement-based filler) was pumped under high pressure into drill holes on both sides and across the floor of the canyon.

Construction was complete, and the reservoir had been filled almost to capacity when two small springs were detected at 8:30 A.M. on June 5 in the lower wall of the canyon just below the dam. While attempts were being made to alleviate these flows, at 10:00 A.M. a large leak appeared in the dam itself, about one-quarter of the way up from the bottom and 4.6 meters (15 feet) from the canyon wall with the spring flow. This leak rapidly increased its discharge and began to erode material from the dam. Two 20-ton bulldozers were sent in to push boulders into the flow to stem it. By 11:00 A.M. the flow became so rapid that a whirlpool developed on the upstream side of the dam, and the bulldozers had to be abandoned. At 11:57 the dam was breached, and a tremendous wall of water roared down the canyon. The flow was so powerful that one of the abandoned bulldozers was carried 11 kilometers (7 miles) downstream. The town of Rexburg and 777 square kilometers (300 square miles) of farmland were flooded. Eleven people were killed, and damage estimates were as high as $1 billion.

An independent panel of experts concluded that the failure of the dam was caused by water traveling through fissures in the canyon wall, penetrating the grout curtain, and then moving through and eroding the core of the dam, until it failed. The Bureau of Reclamation was criticized for poor design of the grout curtain and dam core and for overreliance on past design practice without giving sufficient consideration to the porous rock at the Teton Dam site. The bureau was also criticized for not including any way to collect and safely discharge leakage, which should have been anticipated because of the presence of porous bedrock.

Gene D. Robinson

SEE ALSO: Dams and reservoirs; Flood control; Hydroelectricity; Watersheds and watershed management.

Thermal pollution

CATEGORY: Pollutants and toxins

Thermal pollution is an adverse environmental effect caused by heat, particularly waste heat from steam-electric power plants. Thermal effluent discharged into waterways from such plants can raise the water temperature, thereby causing damage to aquatic ecosystems.

The overwhelming majority of waste heat in industrialized countries comes not from factories but from steam-electric power plants such as coal-burning and nuclear power plants. Steam-electric power plants convert thermal energy from the combustion of fossil fuels or nuclear reactions into mechanical work and then into electrical energy. While the generators that convert the mechanical work into electrical energy are nearly 100 percent efficient, the rest of the plant is subject to maximum efficiencies imposed by the laws of thermodynamics and determined by the highest and lowest temperatures of the plant. Steam-electric power plants typically have efficiencies of about 40 percent or less, which means that 40 percent of the heat is converted into electrical energy, while the other 60 percent becomes unusable waste heat that must be removed.

As an example, consider a large electric power plant producing 1,000 megawatts (MW) of electricity. If its efficiency is 40 percent, the plant will have to produce 2,500 MW of heat (since 1,000 MW is 40 percent of 2,500 MW) to maintain this output. Waste heat will be produced at the rate of 2,500 MW – 1,000 MW = 1,500 MW. At a coal-burning plant, perhaps 200 MW of heat will be lost in and around the boilers and the rest of the plant, leaving 1,300 MW to be removed. Nuclear power plants usually run at lower maximum temperatures, so they have lower efficiencies and correspondingly produce more waste heat; in addition, less heat is lost around the plant itself, so more of the waste heat needs to be removed. Common methods of disposing of this heat include dumping it into rivers or lakes, or using it to evaporate water in cooling towers. Such disposal methods can adversely affect aquatic ecosystems or generate fog and ice.

There are three major methods of removing waste heat. The least expensive is referred to as "once-through" cooling, in which water from a stream, lake, or other body of water is used to cool the steam, after which the water returns to its source at a higher temperature. This method may result in temperature increases of several degrees in the body of water.

A second method is the use of artificial lakes or cooling ponds, which may be up to several square kilometers in size. The heated water from the power plant is discharged into one end of the pond, while water to be used for cooling is drawn from the bottom of the pond at the other end. The water in the pond cools naturally by evaporation; therefore, the pond's water source must be continuously replenished.

A third method is the use of cooling towers, either evaporative or nonevaporative. Evaporative towers, as their name suggests, cool water from the power plant by promoting evaporation. Some evaporative towers produce natural drafts, while others use fans to mechanically induce a draft. Natural draft towers may be more than 100 meters (328 feet) high, while mechanical draft towers are often much smaller. Nonevaporative cooling towers allow moving air to cool pipes containing the heated water from the power plants, but they are less popular because they are expensive to build and operate.

Since most cooling methods ultimately lead to the evaporation of substantial amounts of water, large power plants are usually located adjacent to rivers, which provide a source of water. The major ecological effects of thermal pollution occur in natural rivers and lakes and involve fish and other aquatic organisms. These organisms typically thrive when the temperature remains within a narrow range and may die if the water changes to lower or higher temperatures. For example, if a population of largemouth bass that are acclimated to a water temperature of 20 degrees Celsius (68 degrees Fahrenheit) are exposed to temperatures as low at 4.4 degrees Celsius (40 degrees Fahrenheit) or as high as 32 degrees Celsius (90 degrees Fahrenheit) for one or two days, about 50 percent will die. In addition, the sudden changes in temperature that are encountered by fish swimming into thermal effluent can produce thermal shock and almost instantaneous death if the change is sufficiently large.

A power plant discharges thermal effluent into a nearby waterway. Thermal pollution raises the temperature of the water into which it is emptied, causing damage to aquatic organisms. (Jim West)

All chemical reactions are increased by heat, so thermal pollution can lead to more rapid physiological processes in fish and other aquatic organisms. In certain circumstances this can cause increased growth rates and shorter life spans, leading to decreased populations and less biomass in the ecosystem; in other circumstances it may lead to increased populations and biomass, or may extend the growing system. Ecologists have established that a temperature change of a few degrees can have significant effects, both short term and long term, on aquatic ecosystems.

In addition, evaporative cooling methods inject large amounts of water vapor into the atmosphere. During humid weather this may lead to fog, which could be dangerous to motorists on nearby roads; during cold weather it may lead to icing of roads, trees, and buildings.

Laurent Hodges

SUGGESTED READINGS: Thermal pollution was an active area of research in the 1960's through the 1980's, when the major books on the subject appeared. These books tend to emphasize either the engineering aspects or the biological effects of thermal pollution. The classic book on the engineering aspects is Frank L. Parker's *Physical and Engineering Aspects of Thermal Pollution* (1970). A good reference on the biological effects is T. E. Langford's *Ecological Effects of Thermal Discharges* (1990). A general reference is Donald S. Miller's *Thermal Discharges* (1984).

SEE ALSO: Nuclear power; Power plants; Water pollution.

Three Gorges Dam

DATE: construction begun in 1993
CATEGORY: Preservation and wilderness issues

The proposed Three Gorges Dam in China threatened to alter a scenic rivercourse, produce far-reaching economic consequences and environmental changes, and displace 1.2 million people.

During the late twentieth century the People's Republic of China began pursuing a program of rapid industrialization in an effort to convert from a largely rural agrarian society to a developed economy. Among the proposed projects was the construction of the massive Three Gorges Dam, which was approved by the Peoples' Congress in 1992.

For many centuries, the Yangtze River has been a lifeline of transportation and trade through central China. Originating in the Tibetan highlands and the Himalaya Mountains in the west, it courses 6,300 kilometers (3,900 miles) eastward to Shanghai on the East China Sea. As it cuts through the Wu Mountains in central China before traversing the broad and flatter coastal plain, it forms a stretch of deep gorges with looming stone walls, narrows, and bends called the Three Gorges. This stretch is about 200 kilometers (120 miles) long and typically 100 to 300 meters (330 to 990 feet) wide.

Construction of the Three Gorges Dam, with an associated hydroelectric project and five-stage locks to raise and lower ships as they move up and down the river, began in 1993 just upstream from the city of Yichang, near the entry to the Three Gorges at a site called Sandouping. The plans called for the construction of a concrete dam over 2 kilometers (1.2 miles) long that would straddle an island in the river and rise 184 meters (600 feet) above the river bed. This would create a huge reservoir lake, partially filling and broadening the upstream river and its tributaries. It would extend as much as 600 kilometers (370 miles) upstream from the dam and submerge 150,000 acres of land. As such, the Three Gorges Dam, which was slated for completion in 2009, is the largest dam construction project ever attempted.

The projected benefits of the dam include controlling the periodic flooding in the region, some of which is devastating to large numbers of people and destructive to property. The electric power generated with the controlled fall of water through twenty-six turbines would provide electricity for central China and the economically booming coastal cities. In terms of megawatts of hydroelectric power, the dam was projected to

Xiling Gorge, the third of the Three Gorges, just upstream from the Three Gorges Dam site. The dam would create a reservoir that would partially submerge upstream gorges, damage the environment, and displace 1.2 million people. (Robert S. Carmichael)

produce 40 percent more energy than Itaipu Dam on the Brazil-Paraguay border (the world's largest dam after the Three Gorges Dam) and 160 percent more than Grand Coulee Dam in Washington State in the United States. This power would help replace the burning of China's high-sulfur coal to meet increasing demands. Coal combustion power plants (which have supplied three-quarters of China's energy needs) create large amounts air pollution, including particulate material and sulfur oxide gases that cause acid rain. They also emit large quantities of carbon dioxide, of concern for contributing to global warming. The dam's reservoir lake would provide water for irrigation of agricultural land and could have potential for recreational development. It would also make upriver passage for ships possible through the Three Gorges stretch of the river.

Many negative societal and environmental impacts have been associated with the Three Gorges project. Some of the prospective changes and consequences were widely discussed in China both inside and outside of government circles during the 1980's, in probably the first such environmentally driven public dispute in that country. The debate led to popular opposition to the government plans. The objections and concerns, from both domestic and international environmental organizations, caused a delay in approving and starting the dam project, as well as funding difficulties. The Three Gorges stretch of the Yangtze River is dramatically and spectacularly scenic, and is fabled in history, literature, and art. It has the same attraction for the people of China that the Grand Canyon has for Americans or the deep coastal fjords have for Norwegians. The reservoir created by the impoundment of water behind the dam would widen the waterway, partially fill the canyons, and degrade the dramatic scenery. Further, the reservoir would submerge the living space of over 1.2 million people who reside in nearly five hundred cities, towns, and villages; hundreds of early habitation sites, temples, and sculptures dating back many centuries; and the habitats of many species.

The dam also threatens the rich farmland and grazing areas along the downstream riverbanks. The reservoir would trap silt and other sediment normally carried by the river to the coastal plain to replenish soil in agricultural lands. Unless measures are taken periodically to flush or dredge the silt, it would progressively accumulate, fill in the reservoir, reduce its capacity to store water during flooding, and impede shipping and harbors; within several decades, the accumulated silt could negate many of the intended benefits of the dam. Further, there is concern that the still, trapped water of the reservoir would become a repository for the sewage, industrial pollution, and toxic waste from upstream. Furthermore, the prospect, however slim, of the dam, with its colossal pool of water, being damaged or weakened by one of the region's destructive earthquakes is a daunting one.

Robert S. Carmichael

Suggested readings: "China's Three Gorges: Before the Flood," *National Geographic* 192 (September, 1997), and "The Pulse of China," *Time* (June 29, 1998), both provide good, easy-to-read overviews of the dam and problems associated with its construction; on the Internet, see the dam's Web site (www.tgpdam.com).

See also: Dam and reservoirs; Flood control; Hydroelectricity; Sedimentation.

Three Mile Island nuclear accident

Date: March 28, 1979
Category: Nuclear power and radiation

On March 28, 1979, the Metropolitan Edison nuclear power plant at Three Mile Island on the Susquehanna River near Harrisburg, Pennsylvania, sustained what was at the time the most serious accident in the history of the U.S. nuclear power industry. The accident exposed serious weaknesses in plant operations and government oversight and prompted reform of the U.S. nuclear industry.

The Three Mile Island plant (TMI) was designed with two pressurized water reactors, units 1 and 2,

A Pennsylvania state police officer and security guards patrol the front gate of the Three Mile Island nuclear power plant after it was shut down following an accident in March of 1979. The near-disaster fueled public criticism of the nuclear energy industry. (AP/Wide World Photos)

which generated electric power by boiling water into steam that spun the blades of a turbine generator. The heat to convert water to steam was produced by fission of uranium in the reactors' cores. These were submerged in water and encapsulated in a containment building forty feet high, with walls of steel eight inches thick. Because the radioactive coolant water was under pressure, it could be superheated to 575 degrees Fahrenheit without boiling. When it reached that temperature, it was pumped to a steam generator, where, in a secondary system and under less pressure, it heated cooler water to steam, which spun turbine blades and propelled a generator. The steam then passed through a condenser, where it changed back to water and began the circuit back to the steam generator.

THE ACCIDENT

At 4 A.M. on Wednesday, March 28, a valve inexplicably closed, interrupting the water supply to the steam system. The main water pumps automatically shut down, which decreased the steam pressure and shut down the steam turbine a few seconds later. This interrupted the transfer of heat from the reactor cooling system, where the pressure began to rise. A pressurizer relief valve opened, which reduced some of the pressure but also allowed radioactive water and steam to drain into a tank designed for excess water. The valve should have shut off after thirteen seconds; however, it remained open for more than two hours, during which time the primary coolant water continued to drain.

Less than one minute later, emergency backup pumps automatically engaged to maintain the water supply in the secondary system. No water was actually added, however, because two valves that controlled the flow had been closed for routine maintenance two days earlier. (Nuclear Regulatory Commission rules required that a plant be shut down if these valves remained closed for more than seventy-two hours.) Instead, two minutes into the crisis, as the temperature continued to rise and steam pressure to decline, the emergency core coolant system began to operate, adding water to the reactor core.

There were no meters to measure the depth of water in the reactor core, but the technicians believed that there was sufficient water present. To prevent the pressure in the primary cooling system from rising too high, they turned off one emergency pump, and a few minutes later reduced the other one to half speed. This would have been proper procedure if the system had indeed been filled with water. In fact, the reactor core was not covered with water, and temperatures continued to rise.

At eight and one-half minutes into the crisis, technicians opened valves to fill the secondary system with water and draw heat away from the primary system. With the relief valve stuck open, though, the primary cooling water was still draining into the excess water tank, which overflowed and spilled its radioactive contents onto the containment building's floor. This activated suction pumps, which removed the water to a tank in the nearby auxiliary building. This tank too overflowed, and at 4:38 A.M., radioactive gases began to be released into the atmosphere.

By this time, the twelve-foot-tall fuel rods in the reactor, which should have been covered with water, had become exposed. With no cooling system in operation, temperatures had continued to rise, leading to a partial melting of reactor fuel; the zirconium shields around the rods reacted with the steam and released radioactive debris and hydrogen, which collected in the containment building. At 6:50 A.M., a general emergency was declared.

Early Wednesday afternoon, some of the hydrogen in the containment building exploded. Hydrogen continued to be created by the exposed fuel rods, giving rise to fears that the hydrogen bubble at the top of the reactor building could self-ignite and result in a meltdown. Controlled and uncontrolled radiation leaks from the plant continued through March 28 and 29. Lack of information, poor communication among the numerous agencies involved, and some degree of sensationalist news reporting fueled mounting public alarm. On Friday, March 30, Governor Richard Thornburgh ordered an evacuation of pregnant women and small children living within five miles of the facility. Administrators considered ordering a general evacuation but feared that it might set off an evacuation panic and result in more injuries than it would prevent. Finally, on Sunday, April 1, when President Jimmy Carter visited the facility, it was announced that the hydrogen bubble had shrunk and no longer posed a danger.

Effects

The Three Mile Island accident exposed weaknesses in U.S. nuclear power plant design, management, and operation, in U.S. emergency preparedness, and in the workings of the Nuclear Regulatory Commission (NRC). The matter was investigated by a presidential commission and congressional committees and internally in the nuclear industry. An immediate response to the event was the closing of seven reactors similar to those at TMI and a delay in restarting others that had been shut down for maintenance. The Nuclear Regulatory Commission, which was eventually completely restructured, placed a temporary moratorium on the licensing of all new nuclear reactors; several reactor projects were canceled outright. Other countries reassessed their nuclear industries; Japan closed one reactor and postponed restarting nine others. In addition, the event mobilized the campaign of opponents to nuclear power worldwide. Longer-term results included changes to the design and operation of all nuclear power plants in the United States.

The TMI accident prompted widespread reconsideration of nuclear power and a reassessment of its relatively low economic cost in light of its risks, which were tragically demonstrated by the 1986 nuclear disaster at Chernobyl. Perhaps the most lasting effect of the TMI crisis was on the general public's faith in industry and government representatives. The actions of Metropolitan Edison and government officials during the crisis made it clear that their first priority had been not the health and safety of the public but rather the safeguarding of their own and the plant's reputations. Projections about the potential results of a nuclear disaster and uncertainty about the long-term health effects of radiation contributed to growing public mistrust and the militancy of opponents of the nuclear industry.

John R. Tate

SUGGESTED READINGS: Philip L. Cantelon and Robert C. Williams's *Crisis Contained: The Department of Energy at Three Mile Island* (1982) evaluates of the Department of Energy's performance during the TMI emergency. Robert Del Tredici's *The People of Three Mile Island* (1980) contains interviews with local residents and others connected with the event. Raymond L. Goldsteen and John K. Schorr's *Demanding Democracy After Three Mile Island* (1991) surveys the changes experienced by the local population in the wake of the accident. Mark Stephens's *Three Mile Island* (1980) is an account of the incident by a staff member of the presidential commission.

SEE ALSO: Antinuclear movement; Chalk River nuclear reactor explosion; Chelyabinsk nuclear waste explosion; Chernobyl nuclear accident; Hanford Nuclear Reservation; Nuclear accidents; Nuclear power; Nuclear regulatory policy; Windscale radiation release.

Tidal energy

CATEGORY: Energy

The energy generated during the rise and fall of the tides may be cleanly and safely converted into electrical power. However, large-scale tidal-power installations may entail severe environmental consequences, including decimation of fisheries; destruction of the feeding grounds of migrating birds; damage to shellfish populations; interference with ship travel, port facilities, and recreational boating; and disruption of the tidal cycle over a wide area.

Tidal-power projects can be important sources of local electric generation because they produce energy that is free, clean, and renewable; neither air nor thermal pollution is involved, and exhaustible natural resources are not consumed. Only a limited number of places on the globe offer the potential for such power installations, however, because a vertical tidal rise of 5 meters (16.4 feet) or more is required. Installations must also be near major population centers in order to minimize transmission requirements, and a natural bay or river estuary is required to store a large amount of water with a minimum of expense for dam construction. The seawater impounded behind the dam at high tide produces a hydrostatic head so that electricity is generated as the water passes through the dam's turbines when sea level falls. If the turbines in the dam are reversible, power can be generated on both an incoming and an outgoing tide.

Existing tidal power plants are found on the Rance River near St. Malo, France (240 megawatts of power), on the Annapolis River in Nova Scotia, Canada (17.8 megawatts), in the People's Republic of China (3.2 megawatts), and in the former Soviet Union (0.4 megawatts). The installations in China and the former Soviet Union are small, and data relating to their environmental effects are not available. The Rance River plant has been in continuous operation since November, 1966, and is the world's largest tidal-power installation. It bridges the estuary with a dam nearly 0.8 kilometers (0.5 miles) long and provides power for 300,000 people.

Richard H. Charlier reports in *Sea Frontier* magazine that the environmental impact of the Rance River dam has been limited to a modification of fish species distributions, the disappearance of some sand banks, and the creation of high-speed currents near the sluices and the powerhouse. Tidal patterns have also changed, with the maximum average rise reduced from about 13.4 meters to 12.8 meters (44 feet to 42 feet) and a corresponding increase in the height of the mean low-tide level.

The environmental impact of the smaller Annapolis River plant in Nova Scotia, which became operational in 1984, is described by David Holt in *Maclean's* magazine. He reports that clam diggers claim the plant generates silt, which is destroying clam beds in the basin behind the dam, and that landowners upstream from the dam believe raised river levels are causing increased erosion and flooding. The Nova Scotia Power Corporation had already settled with one landowner whose house suffered a cracked foundation and shifted toward the river as a result of erosion.

Several tidal-power projects were proposed for the United States during the early and mid-twentieth century but were never built because of environmental concerns. A proposed tidal-power plant on the upper St. John River in Maine was halted, for example, because damming the river would have destroyed a unique stand of a rare wildflower. The flower was later found growing elsewhere. Objections cited for other projects included the possible effect on historic and archeological sites, or the presumed economic and social impacts on Native American communities such as the Passamaquoddy Indians.

Shortly after the dramatic jump in world oil prices during the 1970's, the Tidal Power Corporation, a venture owned by the Nova Scotia government, proposed building a major tidal-power project in the Bay of Fundy, which lies between Nova Scotia and New Brunswick in eastern Canada. This plant would have been the world's largest tidal-power installation, producing 4,560 megawatts of power—nearly twenty times the output of the Rance River plant and more than three times the output of Hoover Dam on the Colorado River in the United States. A major feature of the project was to be a dam 8.5 kilometers (5.3 miles) long across the Bay of Fundy, which has the largest tidal range in the world, averaging more than 15 meters (50 feet). The enormous scope of the project forced scientists to pay close attention to its anticipated environmental consequences, and these appeared to be so severe that the project was never begun.

Disrupted bird migrations were predicted after the dam's completion because of the submergence of the tidal mudflats behind the dam, where large numbers of semipalmated sandpipers and other shore birds annually gorge on mud shrimp before beginning their fall migrations to wintering grounds in South America and the Caribbean. Damage to fish stocks was also predicted because of repeated passage of the fish through the dam's turbines as the tides rose and fell. Particularly affected would have been the American shad, a member of the herring family, which migrates to the Bay of Fundy each year from as far away as the St. Johns River in Florida in order to fatten itself on mysid shrimp living on the tidal mudflats. Oceanographers also used computer modeling to show that dam construction would alter tidal patterns over

The La Rance tidal power plant—located at the mouth of the Rance River near St. Malo, France—is nearly 800 meters long and provides electricity for 300,000 people. (Électricité de France)

a broad area, resulting in tidal levels 10 percent higher and lower as far south as Cape Cod, Massachusetts, 400 kilometers (250 miles) away. They predicted that these tidal changes would flood coastal lands and threaten roads, bridges, waterfront homes, water wells, sewage systems, salt marsh areas, harbors, and docking areas along the entire coast.

Donald W. Lovejoy

SUGGESTED READINGS: A good technical discussion of the tides is provided in the Open University Course Team's *Waves, Tides, and Shallow Water Processes* (1989). For easy-to-read descriptions of the tides and tidal power, see Samuel Carter III's *Kingdom of the Tides* (1966), Edward P. Clancy's *The Tides: Pulse of the Earth* (1968), and Francis E. Wylie's *Tides and Pull of the Moon* (1979).

SEE ALSO: Alternative energy sources; Dams and reservoirs; Hydroelectricity.

Times Beach, Missouri, evacuation

DATE: 1983
CATEGORY: Human health and the environment

In 1983 the United States government purchased the town of Times Beach, Missouri, and relocated the entire population because of the threat of dioxin contamination. The events leading up to this action and the subsequent evaluation of it reflected the national concern with environmental safety issues and served to highlight the difficult decisions that government agencies must address.

During the 1960's a chemical plant in Verona, Missouri, produced Agent Orange, a chemical compound that was used during the Vietnam War. A by-product of this production was a highly toxic compound called dioxin, which was stored in large tanks. Syntex Agribusiness purchased the chemical plant in 1969 and leased part of the plant to Northeastern Pharmaceutical and Chemical Corporation (NEPACCO), which manufactured hexachlorophene. The production of hexachlorophene, a popular skin cleanser, also created the by-product dioxin, which was added to the same large tanks previously used. Independent Petrochemical Corporation (IPC), one of NEPACCO's suppliers, hired Russell Bliss to get rid of the compounds being stored in the tanks. He accomplished this by mixing the stored wastes with recycled oil, contracting with local towns, and spraying the oil on roads. Later, NEPACCO paid Bliss to dispose of additional wastes, which he did by spraying more roads. In 1971 Times Beach, Missouri, a small town in the southwestern part of the state, sprayed thousands of gallons of recycled oil onto its unpaved streets in an effort to control dust. The oil was contaminated with dioxin. However, in 1971 the toxicity of dioxin had not been publicized and was not widely known.

Soon after the spraying, problems began. Horses, dogs, cats, chickens, rodents, and birds perished. Within a few months, children became ill. By 1981 government officials began recommending that people vacate the area.

Since dioxin binds tightly to soil and degrades very slowly, high levels of dioxin remained even ten years after the spraying. Soil tests conducted at Times Beach in November, 1982, verified the presence of high levels of dioxin in the town. Dioxin levels in some parts of the Times Beach area reached 100 to 300 parts per billion (ppb). The United States Centers for Disease Control (CDC), at the time, believed that dioxin levels above one part per billion posed a potential risk to human health. The CDC recommended the temporary evacuation of Times Beach until more tests could be conducted. In February, 1983, using between $33 million and $40 million of Superfund money, the federal government purchased Times Beach and relocated the entire population, about 2,300 to 3,000 people.

Disagreements and debates continued to surround the events at Times Beach. In 1991 Vernon Houk, a top CDC official, stated that in light of new information obtained about dioxin, he believed that the CDC overreacted by evacuating the population of Times Beach, claiming that the danger was not great. Although dioxin had been shown to cause chloracne, a serious skin

A gate blocks access to the community of Times Beach, Missouri, which was evacuated in 1983 after tests revealed the presence of large amounts of dioxin in the town. (Jim West)

disease, and had been related to cancer and birth defects in humans, subsequent studies showed that exposure to dioxin was not a huge cancer threat unless the exposure was unusually high. In 1992 Edward Bresnick, chairman of the independent dioxin review panel of the Environmental Protection Agency (EPA), also reported that the government overresponded at Times Beach.

Reports downplaying the toxicity of dioxin surfaced stating that significant increases in disease had not been documented at Love Canal (another Superfund site located in the state of New York) or at Seveso, Italy, the site of an accidental release of dioxin-contaminated vapor in 1976. To further evaluate the toxicity of dioxin, the EPA began a series of studies in 1991. The hope of the EPA was that it could prove once and for all that dioxin contamination need not be a source of concern. In reality, they proved the opposite. The first drafts of the study pointed out more, rather than fewer, problems with dioxin. The study showed that dioxin was particularly damaging to animals exposed while in utero. Dioxin also affected behavior and learning ability and acted like a steroid hormone. The studies reported extensive effects on the immune system.

The citizens of Times Beach brought hundreds of lawsuits against Syntex, the company many thought was responsible for the dioxin contamination. None of these lawsuits succeeded. A jury in St. Louis in 1988 rejected the cases of eight plaintiffs, citing a lack of medical evidence to support the claims.

The Superfund cleanup of Times Beach continued to generate debate. In a controversial decision, federal and state environmental officials decided to incinerate the contaminated soil of twenty-seven sites in Missouri, including Times Beach. Environmentalists feared the dangers of smokestack emissions and accidents and vehemently opposed incineration of contaminated soil. Their concern was justified: On March 20, 1990, a power outage caused thousands of pounds of dioxin-contaminated pollutants to be discharged into the air. Although the risk caused by this pollution is not known, two environmental activist groups filed a federal lawsuit. A third organization, the Dioxin Incinerator Response Group, also opposed to incineration, proposed that storing dioxin-contaminated soil in capped drums was feasible and would be a safe alternative to burning.

In July, 1997, the U.S. Justice Department's Environmental and Natural Resources Division announced that the clean-up of the Times Beach, Missouri, Superfund site was complete and that the land was once again fit for human use. Begun in 1984, the cleanup cost an estimated $200 million. A 409-acre state park, named Route 66 State Park for the historical road that runs through it, was developed at the site.

Louise Magoon

Suggested readings: Times Beach Evacuation Web site contains a good summary of events (http://www.earthbase.org/home/timeline/1983/times beach/). A discussion of the Superfund law and sites may also be found on the Internet (http://www.usdoj.gov/opa/pr/1997/July97/28lenr.htm). Information on how government agencies can use scientific information can be found in the *Gale Environmental Almanac* (1993), by Russ Hoyle (1993).

See also: Agent Orange; Dioxin; Superfund.

Tobago oil spill

Date: July 19, 1979
Category: Water and water pollution

In 1979 two oil tankers collided off the coast of Tobago in the Caribbean Sea, causing a large oil spill and significant loss of human life.

On the evening of July 19, 1979, the weather off of the northern tip of the Caribbean island of Tobago was rainy with gusty winds. The Liberian-registered *Aegean Captain,* weighing 210,257 tons, was bound from the Netherlands Antilles to Singapore with a cargo of transhipped Arabian crude oil. The *Atlantic Empress* was also Liberian registered and operating under charter to Mobil Oil. This vessel was en route from the Persian Gulf to Beaumont, Texas, with a cargo of Arabian crude. At 325 meters (1,066 feet) long and 292,666 tons, it was a very large vessel.

The two tankers were both using radar for collision avoidance but indicated the pictures were "fuzzy" because of the rain. Each vessel was unaware of the other's presence until it was too late. They did not sight each other until they were approximately 183 meters (600 feet) apart. Because of their size, such vessels require miles to stop. The *Aegean Captain* was on an easterly heading and upon sighting the *Atlantic Empress* began a sharp left turn. The navigation rules allow for turns to the right only. The *Atlantic Empress* was northbound towards Texas, yet at the time of the collision it was on a southerly heading for unknown reasons. At 7:15 P.M. on July 19, the two vessels collided 29 kilometers (18 miles) north of the northern tip of the island of Tobago.

The bow of the *Aegean Captain* collided with the starboard (right) side of the *Atlantic Empress* and drove deep into the center of the other ship. The two vessels were locked together as fires broke out on both ships and oil began to spill into the sea. The *Atlantic Empress* sustained major casualties, as twenty-six of its crew were killed in the collision. Three crewmembers on the *Aegean Captain* were also killed. The captain of the *Aegean Captain* then backed his vessel away from the *Atlantic Empress* even though the *Atlantic Empress*'s captain asked him not to, fearing his ship would sink.

Salvage tugs stationed nearby responded almost immediately to the collision. The *Atlantic Empress,* on fire and leaking oil badly, was taken in tow. The tugs applied for permission to enter several local ports where fighting the fires would have been more efficient, but in all cases permission was denied. The vessel was towed out into the Atlantic Ocean, where it was racked by several explosions and sank. The *Aegean Captain,* on the other hand, was taken to local shipyards after the fires and oil leaks were stopped.

The fact that both vessels were carrying high-quality Arabian crude oil was actually an asset in this case. This crude oil is almost 25 percent gasoline. Consequently, in the collision a large proportion of the oil burned rather than fouling local beaches and fishing areas. A large percentage of the remaining oil evaporated, and dispersants were used to treat the rest before it came ashore. In total, 270,000 tons (2.14 million barrels) of oil were lost. Twenty-nine lives were lost, and damage to the two vessels and the environment came to $54 million in insurance claims.

Robert J. Stewart

See also: *Amoco Cadiz* oil spill; *Argo Merchant* oil spill; *Braer* oil spill; *Exxon Valdez* oil spill; Oil spills; *Sea Empress* oil spill; *Torrey Canyon* oil spill.

Torrey Canyon oil spill

Date: March 18, 1967
Category: Water and water pollution

On March 18, 1967, the tanker ship Torrey Canyon *grounded off the coast of England, spilling crude oil into the sea.*

The *Torrey Canyon* was built at Newport News shipbuilding yards in Virginia in 1959. It was later modified at Sasebo, Japan, in 1964. The ship was 297 meters (974 feet) long, 38 meters (125 feet) wide, and had a draft of 16 meters (52 feet) with eighteen cargo tanks capable of carrying 120,000 tons of oil. The vessel was owned by Union Oil Company of Los Angeles, California. It was chartered to British Petroleum and flew the Liberian flag while carrying predominately Italian officers and crew.

When the *Torrey Canyon* departed Mina ala Ahmadi in the Persian Gulf in early 1967, it was loaded with 119,193 tons of Kuwaiti crude oil. It was bound for Milford Haven in Wales. The last leg of the voyage was to go past the Scilly Isles and into the Bristol Channel. The Scilly Isles lie about 34 kilometers (21 miles) off Lands End along the Cornwall coast of England. The Seven Stones Shoal lies in between the Scilly Isles and Lands End. The *Torrey Canyon* normally passed outside of the Scilly Isles. On this voyage, however, the captain chose to pass between the islands and the shoal. The vessel ran aground on the Seven Stones Shoal on March 18.

Once the *Torrey Canyon* was aground, it began leaking oil. A salvage effort was quickly undertaken to try to refloat the vessel. At the same time, detergent was sprayed over the spilled oil to help disperse it. After one week of work, the

The wreckage of the Torrey Canyon *oil tanker burns off the coast of Cornwall, England, in March of 1967. The British Royal Air Force dropped explosives near the vessel in an attempt to burn the oil; although the bombs started several small fires, the effort failed.* (Archive Photos)

vessel was broken into pieces while salvagers were attempting to tow it off the shoal. By this time the oil had blanketed the sea within an 80 kilometer (50 mile) radius of the vessel; on some of the beaches where it had begun to wash ashore, the oil was 46 centimeters (18 inches) thick.

The salvagers were not equipped to deal with the volume of oil spilled. After twelve days of salvage efforts, air attacks on the stricken vessel began. The British Royal Air Force dropped explosive bombs and incendiary devices around the vessel in an attempt to burn the oil. Some small fires were started, but this technique failed. The oil continued to spread over the coasts of England and France. Both governments called in troops to help remove the oil from the beaches. Pump trucks and boats were used to literally pump oil off the beaches. Bulldozers and other heavy equipment were used to remove contaminated sand and rock. The largest single problem arose through the use of detergents to remove oil from rocky or hard-to-reach places in both England and France. Many of these detergents were toxic and killed everything in their path.

Many of the spill's effects were felt long after the oil was gone. An estimated 75,000 sea birds were killed, and an unknown number of other animals of all types also died. Some of these deaths were directly caused by the effects of the oil, while others were caused by the detergents and heavy equipment used to remove the oil.

Robert J. Stewart

SEE ALSO: *Amoco Cadiz* oil spill; *Argo Merchant* oil spill; *Braer* oil spill; *Exxon Valdez* oil spill; Oil spills; *Sea Empress* oil spill; Tobago oil spill.

Trans-Alaskan Pipeline

DATE: completed July, 1977
CATEGORY: Energy

The plan to construct an oil pipeline across Alaska presented many technical challenges that exemplified the potential conflict between supplying energy needs and protecting the environment.

In December, 1967, oil was first discovered by test-drilling at Prudhoe Bay on the North Slope of Alaska. It soon became evident that this was the largest petroleum field in the United States. However, transporting the oil from Prudhoe Bay to a port at Valdez, on the south coast of Alaska, would require the construction of a pipeline that would traverse 1,000 kilometers (620 miles) of federal land. The U.S. Geological Survey was assigned the task of conducting an environmental impact assessment.

After an exhaustive investigation, the federal agency recommended against construction on the trans-Alaskan route. Its objections were technical, geological, and ecological. The route crossed difficult terrain in an arctic region where local environmental damage could be severe and long lasting. The geological hazards included active earthquake fault zones in southern Alaska, mountains (the Brooks Range in the north and the Alaska Range in the south), thirty rivers—many of which flooded periodically—and unstable soil and permafrost (permanently frozen soil). Construction, subsequent pipeline operation, and potential accidental rupture or spillage would disturb the ground, water, fragile vegetation, and wildlife, including migratory routes of land species such as caribou.

In addition, transporting the oil southward from Valdez would require oil tankers to face the hazards of docking in the Arctic and the possibility of spills and pollution. The U.S. Geological Survey favored a longer inland pipeline route through Canada to refineries in Chicago, Illinois. However, others wanted to keep the construction and its economic benefits within Alaska and within U.S. territory. The report was overruled, and the U.S. Congress exempted the pipeline project from the law requiring a favorable environmental impact statement before work could begin.

Construction on the pipeline was started in April, 1974, by a consortium of eight major oil companies named the Alyeska Pipeline Company. It was projected to cost $900 million. By the time it was completed in July, 1977, costs had reached nearly $8 billion—the most expensive privately financed construction project in history. Part of the overrun was caused by redesign and construction techniques adopted to minimize environmental impact. The pipeline, 1.2 meters (4 feet) in diameter, extends 1,300 kilometers (800 miles) from Prudhoe Bay across Alaska to Valdez Arm, an inlet off Prince William Sound on the Pacific coast. The terminus of the pipeline is across the inlet from the town of Valdez. The first supertanker was filled with oil at Valdez in August, 1977—almost ten years after the North Slope discovery was made. The pipeline can deliver oil at 1.6 million barrels of oil per day. The oil revenues for Alaska from state taxes have enabled the state to abolish its personal income tax and distribute a substantial annual cash dividend to all its state residents.

The arctic ecology, with its fragile plant and animal life and slow recovery, is particularly susceptible to damage from a pipeline break or crude oil leak. Efforts were made during construction and subsequent pipeline monitoring to minimize any spillage. Despite these attempts, in June, 1981, a valve ruptured and spilled five thousand barrels of oil onto the soil. However, the greatest environmental and ecological damage occurred in March, 1989, when the *Exxon Valdez* supertanker, loaded with over 1.2 million barrels of crude oil, ran aground in Prince William Sound. Over 240,000 barrels (10 million U.S. gallons) of oil spilled into the water. This was one of the worst environmental disasters in the history of the United States, not only for the extent of the contamination and its impact on wildlife and fishing but also because of the remote location and arctic climate, which made cleanup and reclamation so difficult.

Robert S. Carmichael

SEE ALSO: Environmental impact statements and assessments; *Exxon Valdez* oil spill; Fossil fuels.

Workers install the final section of the Trans-Alaskan Pipeline in 1976. The completed pipeline can transport 1.6 million barrels of oil per day. (Library of Congress)

Turtle excluder devices

CATEGORY: Animals and endangered species

When fitted properly with a turtle excluder device (TED), a shrimp boat net will only catch shrimp and small, nontarget fish of similar size. Larger marine life such as sea turtles, sharks, and large fish can escape through a hatch near the end of the net.

Each year fifty thousand sea turtles are accidentally trapped in drift nets on shrimp boats. An estimated eleven thousand of them die. The large majority of those turtles killed are loggerheads, with a small percentage being Kemp's ridleys. Since sea turtles must go to the surface to breathe approximately every fifty minutes, shrimp boats trawling for over one hour will drown sea turtles that are caught in their nets. A shrimp boat without a TED generally dumps 12 pounds of dead and useless by-catch overboard for each pound of harvested shrimp. TEDs are required on commercial shrimp boats in the United States by the National Marine Fisheries Service (NMFS), which is a division of the United States Department of Commerce. TEDs are used during sea turtle nesting months (April to September) along the southeastern Atlantic coast and the Gulf of Mexico.

Many commercial fishers protest the use of TEDs because they claim a 20 percent loss in their catches because the shrimp are dumped through the escape hatches. Fishermen frequently refer to TEDs are "trawler eliminator devices." The situation is comparable to the previous conflict between tuna fishermen and dolphin rights advocates. Citing 97 percent efficiency, the government has maintained strict regulations about the mandatory use of TEDS. The NMFS has indicated that dolphin fishermen have acclimated to dolphin-free nets and therefore requires shrimpers to modify their nets to prevent a high mortality rate of sea turtles in trawlers. Environmentalists claim that shrimp boats using TEDs are more efficient and economical since they reduce fuel costs by excluding the drag and excess weight of large sea turtles and other marine animals and fish. TEDs additionally prevent shrimp from being crushed or damaged by the weight of sea turtles pushing against them in the nets.

The NMFS required all shrimpers to start using TEDs in May, 1988. In 1989 the United States government passed legislation that banned the importation of shrimp from any country that did not use equipment on their shrimp boats to prevent turtle drownings. The Center for Marine Conservation (CMC), based in Washington, D.C., is one of the world's largest nonprofit agencies whose main activity is protecting marine animals, plants, habitats, and resources. The CMC was one of the major supporters of the ban on imported shrimp from countries that did not use TEDs. Earth Island Institute, based in San Francisco, also led the way in convincing the federal government to enforce this ban by asking politicians to voluntarily make saving sea turtles an international effort. The countries most widely affected by U.S. TED laws are Belize, Colombia, Costa Rica, French Guiana, Guatemala, Honduras, Mexico, Nicaragua, Panama, Trinidad and Tobago, and Venezuela. All of these countries have agreed to comply with the TED laws and have also promoted shrimp farming as an alternative to ocean trawling. However, the growing shrimp agriculture industry is sacrificing forest and fresh water supplies while depositing hazardous wastes and chemicals into ecosystems. Shrimpers in the United States are protesting the import of these inexpensive shrimp.

Dale F. Burnside with Welland D. Burnside

SEE ALSO: Commercial fishing; Dolphin-safe tuna; Drift nets and gill nets; Marine Mammal Protection Act.

U

Union of Concerned Scientists

DATE: established 1969
CATEGORY: Nuclear power and radiation

The Union of Concerned Scientists (UCS) is a nongovernmental organization created in 1969 by a group of faculty members and students at the Massachusetts Institute of Technology (MIT). The organization was established to oppose perceived misuse of science and technology, especially related to defense research at universities.

When the UCS was established in 1969, members called for the application of scientific research to pressing environmental and social problems rather than to military programs. Months of planning and preparation by eminent scientists at MIT led to a voluntary research stoppage and day of education and discussion on March 4, 1969. News of the planned meetings at MIT spread to other universities, and meetings and protests took place on campuses elsewhere that same day.

The organization has since expanded its interests beyond protesting military research and has grown to more than 75,000 citizens and scientists around the United States. Members combine scientific research with legislative advocacy, grassroots organizing, and media coverage to work toward a safer and healthier environment. A core group of researchers collaborates with others to provide credible scientific information for use in citizen advocacy and expert testimony. A full-time staff of forty-four people in three offices nationwide helps coordinate and lobby for diverse projects on such issues as nuclear power safety, renewable energy sources, arms control, global environmental trends, international family planning programs, agricultural biotechnology safety, and sustainable transportation policies.

UCS projects and reports such as "Renewables Are Ready" and "Powering the Midwest" have demonstrated the efficacy and promoted the use of renewable energy sources in several midwestern states. UCS advocacy has worked toward advancing prospects for renewable energy under electric utility deregulation by increasing requirements for dollars spent on research for renewable energy sources by utilities and establishing a trust fund to support this research and development. The UCS organized a Science Summit on Climate Change, which clarified the growing consensus among scientists about global warming. A World Scientists's Call for Action at the 1997 Kyoto Climate Summit, signed by 1,572 senior scientists, including 105 Nobel laureates, went to seventy-five world leaders, and helped shape the Kyoto Accords. A Sound Science Initiative provides information to 1,500 participating scientists on biodiversity loss, climate change, population growth, and ozone depletion through action alerts, information updates, and a Web site.

UCS staff members assist scientists in developing and publishing articles, letters to the editor, and editorials covering environmental issues in major newspapers throughout the country. The organization has worked on international family planning issues and has helped convince the U.S. Congress to restore funding for these programs. UCS research and advocacy has supported efforts to reduce reliance on single-passenger cars and provide incentives for cleaner vehicles. They have also participated in successful legislative efforts to advance clean car technologies in California and New England. The book *The Ecological Risks of Engineered Crops* (1996), by Margaret Mellon and Jane Rissler, explores issues about genetically engineered food and has significantly influenced U.S. Environmental Protection Agency (EPA) policy in this area. The organization has also waged campaigns to pressure

Congress and the president of the United States to adopt the Chemical Weapons Convention, the Comprehensive Test Band Treaty, and a START III Treaty.

Anne Statham

SEE ALSO: Antinuclear movement.

United Nations Environment Programme

DATE: established December, 1972
CATEGORY: Ecology and ecosystems

The United Nations Environment Programme (UNEP) was created by the United Nations (U.N.) General Assembly in Stockholm, Sweden, in December of 1972 to coordinate worldwide environmental activities.

The creation of UNEP resulted from the wave of concern that arose during the 1960's about all forms of pollution. UNEP's most widely known undertaking is Earthwatch, a general term used to describe the agency's efforts to facilitate the exchange of information on environmental issues. The Earthwatch program includes the Global Environment Monitoring System, the International Referral System for Sources of Environmental Information, the International Register of Potentially Toxic Chemicals, and the Global Resource Information Database.

UNEP was created after delegates from 113 countries attending the 1972 United Nations Environmental Conference in Stockholm, Sweden, were convinced that the quality of human life was threatened by expanding pollution. The delegates received reports from regional committees situated in every corner of the globe. In addition, there was an overall report, written by environmentalists Barbara Ward and René Dubos, based on the observations of prominent scientists and cultural experts in Europe, North America, South America, and Asia. The report was later published as *Only One Earth* (1972). The assessments, while not representing circumstances as threatening to human existence, did emphasize the need for immediate remedial action to preserve an acceptable quality of life. The Ward-Dubos report raised issues relating to water and air pollution (especially acid rain), dwindling rain forests, and complex problems arising from urban drift.

It was clear that the world needed an agency that would coordinate and monitor environmental information from existing organizations. Such an agency could identify trouble spots and suggest corrective measures before major environmental disasters occurred. Third World delegates, fearing economic consequences, were not enthusiastic about tackling environmental issues. Nonetheless, the majority of delegates voted to recommend to the U.N. General Assembly that it create the United Nations Environment Programme. In December, 1972, the General Assembly approved the recommendation and established UNEP's operations in Nairobi, Kenya. It was the first major U.N. undertaking to be located in an African country. Maurice Strong, the North American businessman who had served as secretary general of the Stockholm conference, was appointed UNEP's first director.

It was nearly three years before UNEP began to implement regional and global monitoring. In 1975 UNEP created the Global Environment Monitoring System (GEMS). GEMS was charged with collecting, collating, and dispensing information on the environment received from hundreds of U.N.-associated agencies around the world, including the International Referral System for Sources of Environmental Information (INFOTERRA) and the International Register of Potentially Toxic Chemicals (IRPTC). GEMS also encouraged governments to expand their monitoring activities and advised UNEP (through its Environmental Fund) on which international conservation strategies deserved financial support. UNEP subsequently placed hundreds of monitoring stations around the world to record information on water quality in rivers, lakes, and oceans. Air-quality measuring systems were also constructed. By 1981 data on air and water quality came to GEMS from most large urban areas, including twenty cities in developing countries. GEMS also assisted the accumulation of information relating to food con-

Klaus Topfer (left), head of the United Nations Environment Programme, meets with Indonesian president Suharto in 1998 to discuss ways to combat fires that threaten Indonesia's rain forests. (Reuters/Alui An/Archive Photos)

taminants and toxic substances affecting marine life.

UNEP's overall effectiveness did not prove to be all that its sponsors had hoped. Problems existed with its Nairobi location. Frequent power failures and the lack of technical expertise caused serious disruptions. It also proved difficult to convince developing countries to provide information or take seriously the need for conservation of resources. The leaders of Third World countries showed little interest in environmental issues and frequently failed to cooperate with GEMS or any other UNEP agency. In general, Southern Hemisphere countries were slow to collect environmental information and slower still to file reports with UNEP. Most of the information UNEP generated came from developed countries.

In 1990 GEMS, with the approval of UNEP, began the Global Resource Information Database (GRID), which received support from the National Aeronautics and Space Administration (NASA) for the development of necessary software. In addition, International Business Machines (IBM) Corporation contributed $6.5 million in computer equipment. As required by UNEP, GRID control was situated in Nairobi, but its most important computer centers were in Switzerland and the United States. The primary purpose of GRID was to provide sophisticated technological support for the responsibilities imposed by UNEP on GEMS. While GRID, through GEMS, greatly increased the efficiency of UNEP's operations, it could not improve the quantity or quality of information received from underdeveloped countries.

The degree of UNEP's effectiveness has stirred much debate. Its defenders point to the fact that its very existence testifies to a global awareness of the need for environmental action. Its detractors note that UNEP has made little impact in much of the world, particularly in less-developed countries where the protection of natural resources is most needed.

Ronald K. Huch

SUGGESTED READINGS: Anne M. Blackburn, editor, *Pieces of the Global Puzzle* (1986), includes essays by Maurice Strong and other world leaders in banking, industry, and conservation. *Changing the Global Environment* (1989), edited by Daniel B. Botkin et al., contains twenty-seven essays ranging over economic, ethical, and technological questions. All reinforce the need to develop a global vision for the environment. René Dubos, *The Wooing of the Earth* (1980), focuses on human management of the environment. *Only One Earth: The Care and Maintenance of a Small Planet* (1972), by René Dubos and Barbara Jackson Ward, is the report that inspired the environmental initiatives that came out of the conference in Stockholm. Branislav Gosovic, *The Quest for World Environmental Cooperation: The Case of the U.N. Global Environment Monitoring System* (1992), is the major work on the evolution of GEMS. Written by an insider, it provides an insightful account of GEMS operations from 1975 to 1992.

SEE ALSO: Dubos, René; Earth Summit; Global Environment Facility; United Nations Environmental Conference; United Nations Population Conference.

United Nations Environmental Conference

DATE: June, 1972
CATEGORY: Ecology and ecosystems

The United Nations Environmental Conference was held in Stockholm, Sweden, in June of 1972. Also known as the United Nations Conference on the Human Environment and the Stockholm Conference, it was the first intergovernmental conference dedicated to discussing environmental issues.

In 1967 Sweden began taking steps toward calling an international conference to discuss global environmental issues. The United Nations (U.N.) Environmental Conference was convened by the United Nations General Assembly in Stockholm in June, 1972, and was the first international conference dedicated to environmental issues. It was, in fact, the first international conference ever held to focus on a single issue. The Stockholm Conference marked the beginning of coordinated global action to manage environmental issues. Delegates from 114 nations attended. However, the Soviet bloc nations did not participate. The United States played a lead role, as it was itself passing, for the first time, environmental laws with regulatory power to enforce them. In addition, 250 nongovernmental organizations (NGOs) and 1,500 journalists attended the conference.

The conference had been preceded by lengthy preparatory meetings at which the participating nations reached consensus on many issues, leaving a lesser number to be dealt with at the conference. Most of the proposals written in advance of the conference emphasized the necessity for the universal involvement of every nation, and the principle of universality was implied in the conference proceedings. A series of regional meetings were held in Asia, Africa, and Latin America so that the conference participants would have a broad view of environmental problems and concerns. Among the considerations was the need to provide countries with scientific, technological, and planning information so that they could effectively address their national environmental problems.

The nations participating in the conference submitted papers describing how they were managing their own national environments. During the conference, differences of opinion between Northern Hemisphere countries and Southern Hemisphere countries surfaced. Since then, these differences have frequently arisen when the global aspects of the environment are being discussed in international forums. Developing countries in the Southern Hemisphere declared that environmental protection was an issue of the wealthy, developed countries, not of the poor. They stated that development was their most important concern, that it outweighed environmental concerns, and that the developed countries were using those issues to retard the poor countries' economic development. The developed Northern countries, in return, attempted to minimize the effect of many of the demands of the developing nations.

One economic issue that developing nations emphasized at Stockholm was the transfer of technology. They insisted that the advanced market economies should provide developing economies with environmental technology free of cost. A major part of this dispute was the demand by the developing states for access to biotechnology that used species from the tropical forests, since such forests were mainly found in thirteen developing states. The final statement of the conference minimized this demand. Another problem occurred over the developing countries' fear that the proposed global environmental monitoring satellite system would be a means by which the developed countries, especially the United States, would spy on them and their resources.

Two major documents resulted from the U.N. Environmental Conference: the Stockholm Declaration and the Stockholm Action Plan for the Environment. The declaration was the first document in which the states agreed on the need for global action to protect the world's environmental assets. It consists of twenty-six broad principles for managing the global environment. The preamble to the declaration expresses the urgency with which states must act, the magnitude of the task before them, and the complexity of the job of preserving the environment in which the world's population lives. According to the United Nations, more than 90 percent of the population of the world was represented at the Stockholm conference.

The points on which the delegates had agreed were put into the Stockholm Action Plan for the Environment, which contains 109 recommendations for global cooperation on environmental issues and 150 proposals for action. The intent of the plan is to define and mobilize states in a "common effort for the preservation and improvement of the human environment." The major recommendations include a call for four U.N. conferences to be convened on topics of environmental education, human settlements, protection of the world's cultural and natural heritage, and international trade in endangered species. Also recommended was the establishment of an international fund to be used as seed capital for housing improvement, coordination of programs for integrated pest control, an International Registry of Data on Chemicals in the Environment, an International Referral Service, and a global atmospheric monitoring system.

The United Nations system was given the mandate to provide leadership in carrying out the recommendations. In December, 1972, the United Nations Environment Programme (UNEP) was created to focus on and coordinate the environmentally related activities in which the U.N. system would be involved. One commitment made by the delegates in Stockholm was to regulate the trade in waste products. However, it was not until 1984-1985 that UNEP was able to get agreement for such regulation. Twenty years later, the 1992 United Nations Earth Summit was held as a follow-up to Stockholm to assess progress since 1972 and to identify and legislate environmental issues important in the changed political, economic, and environmental situation of the 1990's.

Colleen M. Driscoll

SUGGESTED READINGS: For a record of the proceedings and results of the U.N. Environmental Conference, see *Report of the United Nations Conference on the Human Environment* (1973). Lawrence Juda discusses the period of time leading up to the conference in "International Environmental Concern: Perspectives on and Implications for Developing States," in David W. Orr and Marvin S. Soroos's *The Global Predicament: Ecological Perspectives on World Order* (1979). *The Limits to Growth* (1972), by Donella H. Meadows et al., published by the Club of Rome, helped convince the states of the need to coordinate action in preserving the environment. Garrett Hardin's "The Tragedy of the Commons," *Science* 162 (December 13, 1968), added urgency to the need for nations to work together to manage the common resources. The article was later expanded into a book.

SEE ALSO: Biodiversity; Earth Summit; United Nations Environment Programme.

United Nations Population Conference

DATE: 1984
CATEGORY: Population issues

The United Nations Population Conference brought together representatives from 147 countries to strengthen and update the 1974 World Population Plan of Action. Population growth has global consequences, and the United Nations conferences have highlighted shared concerns about population issues and helped coordinate efforts to understand how population growth influences quality of life.

Although several population conferences were convened during the early part of the twentieth century, few nations participated and little was accomplished. This changed in 1974 when official government delegates considered demographic issues in Bucharest, Romania, under the auspices of the United Nations. Experts attended preliminary meetings focusing on future trends, economic and social development, resources and the environment, and family units, and formulated a draft for the World Population Plan of Action.

The United States advocated government-sponsored family planning services, which was a new concept. However, not all delegates were convinced that population growth was a problem. The economic gap between rich and poor nations had widened during the 1960's, and developing nations wanted more financial assistance from developed nations. Oil prices were being set by the Organization of Petroleum Exporting Countries (OPEC), which, along with a bloc of African, Asian, and Latin American countries known as the Group of 77, wanted to reform international economic exchange. Thus, conflict ensued at the Bucharest meetings over the relative importance of population versus development planning. The People's Republic of China's delegation argued that population growth was not a problem under socialism, and the leader of the Indian delegation, Karim Singh, coined the slogan "development is the best contraceptive." Despite these arguments, consensus was eventually achieved. The revised World Population Plan of Action stressed the need to conserve and share the world's resources, reduce population growth, protect women's rights, and support responsible parenthood.

Between 1974 and 1984 the world's population increased by 20 percent, and 90 percent of that growth occurred in developing areas. This population explosion resulted in a different climate at the United Nations Population Conference in Mexico City in 1984. Many governments of developing nations had reversed their positions on population growth, as exemplified by China's aggressive one-child policy. Furthermore, many of these nations were deeply in debt. Market-oriented investment strategies were being pursued by the most successful East Asian and Latin American developing nations with the approval of wealthier countries such as the United States, now under more conservative political leadership. Reducing growth rates seemed useful with or without economic reform.

The biggest surprise in Mexico City was the change in position by the U.S. delegation led by James Buckley, an ex-senator and president of Radio Free Europe, who described population growth as a "neutral factor." The U.S. delegates refused to support population programs that practiced coercion or provided abortions and also argued that fertility would decline in settings with free market economies. However, the impact of the United States' change in position was less than expected. Except for the Vatican, there was unanimous consent that the primary elements of the World Population Plan of Action should be reaffirmed and extended. The new recommendations included strategies to integrate development and population planning, and proposals for local and international efforts to eliminate hunger, illiteracy, unemployment, poor health and nutrition, and the low status of women. There were also recommendations about population distribution and movement, as well as condemnation of unlawful settlements, which reflected increased concerns about population growth's impact on global stability.

Soon after the 1984 conference, the United States ceased its financial support for the United Nations Fund for Population Activities (UNFPA) and the International Planned Parenthood Federation (partially reversed only after President Bill Clinton took office in 1993). By 1990, however, most developing nations had family planning programs. The United Nations-sponsored International Conference on Population and Development (ICPD) in Cairo, Egypt, in 1994 echoed the two prior gatherings, but this time the approximately twenty thousand attendees—who represented more than 180 governments, the United Nations, nongovernmental organizations (NGOs), and the media—recognized the importance of reducing population growth. The United States returned to its former position of supporting population growth control, but there were still contentious discussions before consensus was finally achieved on the last day.

The Vatican actively challenged the delegates by opposing abortion, cohabitation without marriage, and sexual deviation, although many felt the Vatican's implicit, more significant goals were to limit female empowerment and access to birth control. Women's groups lobbied against the Vatican and representatives from some Muslim countries; in spite of the conflict, agreement was achieved on the worth of smaller families and slower growth rates. There were also many firsts. The Vatican approved the final Program of Action, NGOs played a prominent role at the conference and helped draft the Program of Action, environmental issues were placed into a population context, and the reproductive health and rights of individual women were accentuated at the expense of demographic goals or a serious consideration of the impact of consumption patterns on quality of life. An estimated $17 billion per year was suggested in order to implement the Program of Action.

In a follow-up debate in the United Nations General Assembly, developing nations in the Southern Hemisphere pointed out that they would need assistance from developed nations to implement the program. Leaders of Southern Hemisphere nations were concerned that increased economic demands might reduce the resources available for existing programs.

Joan C. Stevenson

Suggested readings: A superb background and history of the first two conferences and details about the preparations for Cairo are provided in Stanley P. Johnson's *World Population—Turning the Tide: Three Decades of Progress* (1994). More detail on United Nations efforts to control population can be found in Stanley P. Johnson, *World Population and the United Nations: Challenge and Response* (1987), and Richard Symonds and Marchael Carder, *The U.N. and the Population Question: 1945-1970* (1973). The United Nations published the *ICPD Programme of Action* (1994). Excellent reviews of the conferences are always provided by the inexpensive publications of the Population Council, including "The Cairo Conference on Population and Development: A New Paradigm?" *Population and Development Review* (June, 1995), by C. Alison McIntosh and Jason L. Finkle.

See also: Population-control and one-child policies; Population-control movement; Population growth; Sustainable development; World Fertility Survey; Zero Population Growth.

Urban parks

Category: The urban environment

Urban parks are rarely mentioned when environmental issues are discussed. Instead, emphasis has centered on addressing environmental degra-

Central Park in Manhattan, New York City, provides open space and a relaxing environment for urban dwellers. (Archive Photos/Joan Slatkin)

dation or wilderness preservation. However, urban parks reflect efforts to reform cities and environmental consciousness by reintegrating nature into the urban landscape.

The appearance of the earliest urban parks in the mid- to late nineteenth century arose from a broader urban reform movement. Reformers responded to rapid industrialization, urbanization, and land development by calling on cities to create and develop urban parks. This reflected a new environmental ideology that sought to reconcile nature and the city and change the way urban residents related to the natural world. From 1860 to 1900, cities spent millions of dollars to build parks, including New York's Central Park, Boston's Emerald Necklace, San Francisco's Golden Gate Park, and Chicago's Lakefront.

Frederick Law Olmsted, Sr., one of the most influential landscape architects of the nineteenth century and leader of the urban park movement in the United States, participated in nearly every major urban park project in the nation. Olmsted believed that urban parks functioned as the "lungs of the city." He argued that parks improved public health by increasing access to fresh air, trees, and sunshine—natural amenities that the gritty, industrial cities lacked. Parks also promoted spiritual restoration by providing an antithesis to the anxieties of the human-made, unnatural urban landscape. Finally, parks could prevent urban vice or riots by providing outlets for the urban underclass. Olmsted's park design legacy remains a cornerstone in urban planning. During the twentieth century, the design and purpose of urban parks shifted from the Olmstedian emphasis on the park as pleasure ground to an emphasis on the park as a multiuse recreation facility. Urban parks were designed to include baseball fields, golf courses, tennis courts, ice-skating rinks, playgrounds, and even zoos.

Much of the impetus for the environmental movement of the late 1960's and early 1970's began in cities. Air pollution from automobiles and industrial manufacturing, water pollution from industrial dumping and sewage disposal, and toxic waste disposal were among the most pressing environmental issues. Urban residents began to effectively mobilize around these issues and demanded change.

Although much of the resulting legislation addressed pollution problems in the cities, many citizens and environmental activists were also concerned that accelerating urban expansion and suburbanization were destroying the last remaining open spaces surrounding many cities. Nearly one hundred years after the first urban park movement, a new urban park movement emerged. It was tied to two developing political coalitions: the national environmental movement and the antigrowth movement. In some cities, local groups lamented the insufficiency of parks and pressured civic leaders to protect existing open space, to increase the size of existing parks and open spaces, and to control urban development. Environmentalists argued that open spaces should be protected as unique and important ecosystems. In many cities, outlying open spaces became integrated into regional and city park systems.

Urban parks were not just local issues. The subject of parks and recreation also became national political issues. In the United States, for example, a congressional report released during the 1970's stated that there were not enough outdoor recreation opportunities in U.S. cities. The National Park Service was charged with acquiring and maintaining national recreation areas (seashores, lakeshores, and recreation areas) located in or near urban areas. Among the new national urban parks that were designated were Cape Cod National Seashore (Massachusetts), Gateway National Recreation Area (New York and New Jersey), Golden Gate National Recreation Area (California), Cuyahoga Valley National Recreation Area (Ohio), and Santa Monica Mountains National Recreation Area (California).

Urban parks represent one way to reconnect to nature and to "green" the city. Greening the city takes many forms: creating urban farms, planting trees along street meridians, cleaning up neighborhood parks, and rehabilitating derelict or abandoned sites into small parks or gardens. These projects are often carried out by neighborhood groups, nonprofit organizations, or local schools. These modest, small-scale efforts represent the fundamental cultural rediscovery of nature within the urban environment and can provide powerful learning experiences.

Urban parks are also integrated into local economic development projects. During the 1980's and 1990's many cities redeveloped their waterfronts and lakefronts. Many redevelopment projects interspersed open spaces, small parks, picnic tables, and water fountains among office buildings, hotels, and residences. These projects reconnected urban residents to the natural environment while simultaneously revitalizing the local economy. The integration of parks and nature in such developments indicates the degree to which environmental sensitivity has become important in urban design.

Urban parks continue to play an important role in city life. Many are mixed-use places that offer a range of activities from solitude and reflection in a quiet, shady grove to social interaction in outdoor cafés and aquariums. Such parks are symbolic of efforts to redefine broader relationships between humans and the environment through the reintegration of nature in the most human of all creations, the city. The environmental movement has become increasingly concerned with changing people's perceptions and behaviors toward the natural environment. Urban parks, however large or small, whether self-contained or integrated into a mixed-use site, testify to the interdependence, not opposition, of nature and people.

Lisa M. Benton

SUGGESTED READINGS: Galen Cranz, *The Politics of Park Design: A History of Urban Parks in America* (1982), discusses the numerous policy decisions that must be made during the design of urban parks in the United States. Peter Schmidt, *Back to Nature: The Arcadian Myth in Urban America* (1990), explores the social and psychological implications of the urban design movement. David Gordon, editor, *Green Cities* (1990), is a collection of essays on various aspects of green spaces in urban areas.

SEE ALSO: Greenbelts; Olmsted, Frederick Law, Sr.; Open space; Urban planning.

Urban planning

CATEGORY: The urban environment

Urban planning ensures that private development decisions result in public benefits as well as private profits and that the expansion of government services is efficient and effective. It is increasingly important as a means to protect and improve the natural environment.

Urban planning involves the application of long-range, comprehensive, rational decisions by governmental agencies to the growth and development of urban centers and surrounding suburbs. Such planning ensures the timely availability of essential infrastructure and government services, including roads, schools, and utilities; a satisfactory quality of life for all residents; the compatibility of adjacent land uses, including buffers between incompatible uses; the rational allocation of uses on the land and the highest and best use for each parcel of land; the prevention of waste; the maintenance of public aesthetics, including open spaces; the promotion of social diversity and diversity in both private and public services and facilities, including housing and commerce; the orderly expansion of existing uses into new physical locations; the preservation of historically and culturally important buildings; and the protection of the natural environment.

Urban planning began in antiquity with the planned cities of Greece and pre-Columbian America. Hippodamus, a Greek lawyer who lived in the fifth century B.C.E., designed the first street layout on a gridiron pattern. The Greeks limited the size of their cities to the carrying capacity of the surrounding farmland. Once the existing site was maximized, the *paleopolis*, or old town, helped found a second *neopolis*, or new town, on a new site. Urban planning continued in Europe through Roman and medieval times into the Renaissance and Enlightenment periods. Many cities in the New World were planned cities, including Jamestown, Philadelphia, and Washington, D.C.

The technocrats of the nineteenth century developed the first planned urban renewal proj-

Many urban communities in the United States have implemented development plans to ensure that expansion provides maximum benefits for residents and causes minimal environmental impact. (Douglas Long)

ects. For example, beginning in 1853, Napoleon III tore down nearly one-half of the buildings in Paris, France, in history's largest public works and urban renewal project in an effort to improve the street system. In the early twentieth century, the first zoning regulations to separate land uses were enacted in Rotterdam, Frankfurt, and New York City. New York City enacted the first regional plan in 1929, regulating housing, highways, building height, and open spaces.

Before the twentieth century, the planners' primary emphasis was on design aesthetics and the delivery of services to meet human needs, including sanitation, transportation, housing, and basic utilities. In the early twentieth century, social scientists such as Earnest Burgess and Homer Hoyt began to emphasize the expansion and succession of existing land uses as the basis for planning. They were concerned with the orderly growth and development of the urban area, the outward expansion of existing land uses into new plots of land, the orderly succession of one land use after another on a single plot of land, and the correction of urban social problems resulting from urban growth.

In the mid-twentieth century, Lewis Mumford and others in the "new communities" movement believed that piecemeal, incremental, successive patterns of growth should give way to totally planned, functionally integrated communities. Also in the mid-twentieth century, Ian McHarg, a European planner, began a new emphasis on the physical environment in urban planning. Using a series of geological, geographic, and current land-use overlays, he mapped those plots of land best suited for each of several types of land use.

In 1973 the Council on Environmental Quality recognized a growing consensus that control over land use is probably the most important single factor in improving the quality of the environment. Effective planning reduces environmental problems stemming from land-use patterns that impose conflicting or unsustainable demands on the environment. One suggestion is to compare alternative uses of the land according to the relative demands each use places on the environment and to implement the one that minimizes human impact.

Many social scientists argue that effective urban planning requires a shift in power from the private to the public sector. Public support for urban planning in many European countries allows government agencies to effectively counterbalance private sector power in urban development. Nations with centralized authority, including many developing nations and nations with socialist politics, also tend to effectively limit private power. In the United States, private sector freedom is equated with liberty. Government intrusions and regulations are kept to a minimum. The American private sector is allowed maximum leeway in a free market system to pursue their interests and to develop and dispose of their property as they see fit.

In addition, the pluralist nature of electoral politics and government decision making in the United States provides the private sector with multiple opportunities to influence public decisions through political appointments, campaign finance, and lobbying. Governmental jurisdictions tend to be fragmented into multiple governmental units (cities, counties, and special districts), each with a unique set of functions, fiscal resources, expertise, and constituencies. These units of government often lack the time, resources, or interest to develop long-range, comprehensive, rational urban planning. Free market advocates argue against increased government urban planning activities and claim that the private self-interest of economically active individuals ensures that land uses are rationally allocated across the landscape and that urban development does not destroy the environmental life-support system upon which the survival of the urban area depends.

Gordon Neal Diem

SUGGESTED READINGS: Bernard H. Ross and Myron A. Levine, *Urban Politics: Power in Metropolitan America* (1996), presents an argument for shifting decision-making power from the private to the public sector. *Environmental Quality* (1973), written and published by the Council on Environmental Quality, presents an environmental argument for land-use planning. Ian McHarg, *Design with Nature* (1969), advocates planning in accordance with the limitations and possibilities presented by the natural landscape.

SEE ALSO: Greenbelts; Land-use policy; Mumford, Lewis; Open space; Planned communities; Road systems and freeways; Sustainable development; Urban parks; Urbanization and urban sprawl.

Urbanization and urban sprawl

CATEGORY: The urban environment

Urbanization is the concentration of human populations in dense settlements where social, economic, and government services can efficiently and effectively meet human needs. When these settlements become overcrowded, some individuals, businesses, and industries migrate to the fringes of the urban area, where population concentrations are less dense but urban services remain available. Such migration is called urban sprawl. The concentration and eventual diffusion of urbanization impacts existing land uses, the physical environment, the survival of species of plants and animals, and the aesthetics of the landscape.

Urban growth and development is usually controlled by two forces. First, municipal government authorities regulate growth through urban planning, zoning, and a variety of land-use ordinances. Sprawl is uncontrolled and unregulated development outside the administrative boundaries of the zoning and land-use authority of municipalities and outside the conscious and deliberate direction of those authorities. Second, social and economic forces combine to encourage outward growth of the urban area in one of three consistent patterns, described as concentric-circle, sector, or multinuclear patterns. Sprawl often involves some break in this historic pattern. Uncontrolled growth accompanying sprawl often results in a mix of incompatible land uses placed adjacent to one another on the same developed physical area.

PROBLEMS WITH URBAN SPRAWL

Sprawl results in discontinuous leap-frog or checkerboard patterns of development and strip development along transportation corridors, with skipped areas remaining undeveloped. This creates inefficiencies in providing urban services to the sprawl area. Sprawl also results in less-than-maximum utilization of existing developed land in the urban center. Instead of converting existing developed land to new uses, developers establish new developments on less-expensive land in suburban areas adjacent to the existing urban area.

Urban sprawl generally increases the total amount of land impacted by human activity, the amount of natural areas converted to recreational uses, the amount of wasteland, and the residential, industrial, institutional, and infrastructure uses of land. Wasteland is land disturbed by humans to such an extent that natural uses cannot be restored and future development on the land is restricted. Wastelands include soil-borrow pits, quarries, debris landfills, and construction material storage sites. Urban sprawl generally decreases agricultural uses and timber uses of land and decreases the amount of natural barren or rocky lands, river and stream floodplain areas, and native timber- and grasslands.

Sprawl is encouraged by a variety of social and economic forces. First, it is often a consequence of increasing heterogeneity of the urban population and is a means to limit or escape from interactions between dissimilar social, economic, and ethnic groups. Second, sprawl allows individuals and businesses to escape from the negative effects caused by concentrations of undesirable land uses—such as industrial, commercial, and low-income residential zones—and relocate to suburban areas with more pristine and affordable real estate. Third, sprawl is encouraged by landowners in suburban areas seeking to maximize profits from investments in land. Finally, sprawl permits individuals and businesses to escape taxes and regulations imposed in core urban areas.

Some researchers contend that the sprawl problem is overstated. In the United States, for example, less than 5 percent of all land is developed, and about three-fourths of the population lives on 3.5 percent of the land area. Three-fourths of the states retain more than 90 percent of their land area in rural uses, with less than 10 percent of their land area impacted by urban development. The amount of rural, park, and wildlife areas protected from development is nearly twice as large as the area developed as urbanized areas.

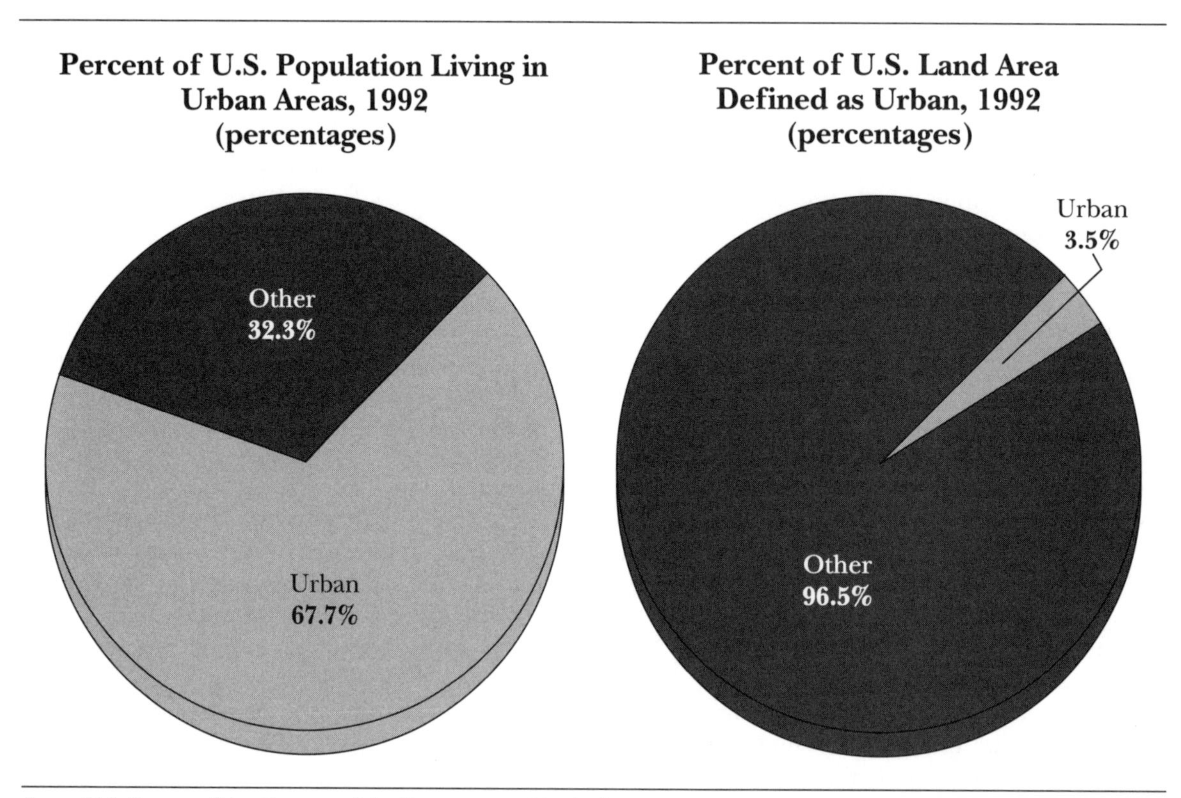

Source: U.S. Bureau of the Census and the Department of Agriculture.

Environmental Impacts of Sprawl

The first significant impact of urban sprawl on the environment is the abnormal greening of the physical area. In most cases, total greening is reduced through destruction of forests, grasslands, and floodplains to make way for development. This decreases the number and variety of species able to occupy the space and creates microclimate effects, such as regional warming. However, in some cases total greening is increased through irrigation and the introduction of cultivated lawns, orchards, and other plantings in areas that are naturally arid or barren. This results in the introduction of new species into the environment, increased pollen counts, and a variety of microclimate effects, such as increases in regional humidity. In either case, the presprawl natural ecology is dramatically changed, with resulting negative impacts on the survivability of plant and animal species displaced by or unable to adapt to the new environment.

In the course of sprawl development, existing natural ecosystems are destroyed, while presprawl plants and animals are killed, displaced, or replaced. Most presprawl wildlife retreats in the face of development. However, some species are able to adapt to the sprawl environment and find the mix of land uses, the dispersion of human activities, and the residue of human activity conducive to their survival. Among the wildlife that benefit are scavengers, such as pigeons, rats, raccoons, and opossums; vermin hunters, such as falcons, foxes, and coyotes; and foragers, such as deer, squirrels, and rabbits. Benefiting plant species include all those that homestead on disturbed soil or wasteland or that thrive in the open sunlight of most new suburban developments.

Sprawl reduces the amount of productive agriculture land and economically important forests. Productive agriculture land and forests are lost either by development of the land for residential, industrial, or institutional uses, or by the transformation of productive farms and forests into hobby farms and parklands in an effort to preserve green space without preserving the productive purpose of the agricultural lands or forestlands.

Finally, sprawl increases the total acres of land surface impervious to rainfall penetration, such as roadways, parking lots, and slab building foundations, thus increasing rainwater runoff and the possibility of flooding, soil erosion, and ecosystem destruction.

Impact on Air and Water Resources

Other consequences of sprawl include increases in noise, land area devoted to highways and roads, and public-utility impacts as land is cleared for underground water, sewer, and utility pipes, and for aboveground utility cables.

The impact of sprawl on air and water resources is both negative and positive. The increase in human population that accompanies sprawl increases the concentration of significant amounts of unnatural substances in the soil, water, and air and also produces abnormally high concentrations of natural substances at levels that may cause undesirable health effects, corrosion, and ecological change. However, studies also indicate that the population dispersal associated with sprawl actually reduces air pollution by dispersing both the mobile and stationary sources of pollutants. Increases in air pollution from automobiles associated with sprawl may be less than the air pollution produced by traffic gridlock, mass transit buses, and trains in more dense urban areas.

Subsurface water supplies and surface water courses are less impacted by sprawl than by denser patterns of development. Denser urban development increases the demands on water resources, runoff and the possibility of flooding, and the likelihood that water courses will be channeled and hardened by concrete and other construction materials. Denser urban development also makes it more difficult for subsurface water to replenish itself.

Many municipalities attempt to limit sprawl by refusing to extend essential services such as water lines, sewer lines, and road systems outside their municipal boundaries. Rural areas may attempt to limit sprawl through zoning restrictions on development, farmland protection ordinances, environmental impact regulations, and special development-impact fees levied on new development to recover the public costs associated with constructing roads, schools, and other facilities necessary to provide services to the newly developed areas.

Gordon Neal Diem

Suggested readings: Syed Muzamil Mujtaba, *Land Use and Environmental Change Due to Urban Sprawl: A Remote Sensing Approach* (1994), studies environmental impacts of sprawl in one region of India. Samuel Staley, *The Sprawling of America: In Defense of the Dynamic City* (1999), argues that the human and environmental impacts of sprawl are minimal and develops a sprawl index to measure the extent of sprawl. James Green, *Economic Ecology: Baselines for Urban Development* (1969), discusses the impact of sprawl on land values, the real estate market, and externalities, including environmental impacts. Jean Gottman and Robert Harper, *Metropolis on the Move: Geographers Look at Urban Sprawl* (1967), provides an overview of both human and environmental problems resulting from sprawl. The U.S. Department of Housing and Urban Development publishes *Metropolitan Development Patterns: What Difference Do They Make?* (1980), which summarizes research on a variety of sprawl issues.

See also: Air pollution; Automobile emissions; Open space; Population growth; Road systems and freeways; Urban planning.

V

Valdez Principles

Date: 1989
Category: Philosophy and ethics

The Valdez Principles are a set of guidelines for environmentally friendly corporate behavior developed in 1989 by the Coalition for Environmentally Responsible Economies (CERES), a coalition of fourteen environmental groups, and the Social Investment Forum, an association of more than three hundred investors, bankers, and brokers. Patterned after the antiapartheid Sullivan Principles designed to discourage investment in South Africa, the Valdez Principles were originally named to commemorate the 1989 Exxon Valdez *oil spill; later, they were redesignated the CERES Principles.*

At the time of the principles' enunciation, CERES members controlled some $150 billion in various funds. Members sought to use their financial leverage to promote environmentally sound practices, rewarding or punishing corporations by investing or withholding money on the basis of environmentally related behavior.

Companies that endorse the Valdez Principles pledge to adopt environmentally sustainable practices rather than simply comply with government-imposed regulation. Stockholders, investors, and pension-fund managers can use the principles to determine whether a company's behavior is environmentally responsible.

Companies that sign the principles pledge to protect the biosphere by monitoring and reducing the emissions of hazardous substances; to safeguard habitats affected by their operations and to protect open spaces and wilderness while preserving biodiversity; to work toward the sustainable use of renewable natural resources; to conserve nonrenewable natural resources through efficient use and careful planning; to reduce and, where possible, eliminate waste through source reduction and recycling; to handle and dispose of waste through safe and responsible methods; to conserve energy and improve the energy efficiency of their internal operations and of their goods and services; to use environmentally safe and sustainable energy sources; to minimize environmental, health, and safety risks to employees and communities; to reduce or eliminate the use, manufacture, or sale of products and services that cause environmental damage or health or safety hazards; to inform customers of the environmental impacts of products or services and to try to correct unsafe use; to correct conditions they have caused that endanger health, safety, or the environment; to inform in a timely manner everyone who might be affected by conditions caused by the company that might endanger health, safety, or the environment; to seek advice and counsel through dialogue with persons in communities near their facilities; to refrain from taking action against employees for reporting dangerous incidents or conditions; to inform managers about pertinent environmental issues and consider demonstrated environmental commitment in selecting managers; to conduct an annual self-evaluation of their progress in implementing the principles; and to complete an annual CERES report to be made available to the public.

The annual report and a list of signatories is available at the CERES Web site (http://www.ceres.org). Prominent signatories include Coca Cola, General Motors, International Telephone and Telegraph, Polaroid, and Sun Oil.

Allan Jenkins

See also: Coalition for Environmentally Responsible Economies; *Exxon Valdez* oil spill.

Vegetarianism

Category: Philosophy and ethics

Vegetarianism is the adoption of a meat-free diet. Many vegetarians believe that feeding, slaughtering, and disposing of livestock and their waste products is an inefficient use of the planet's limited resources. Others believe that eating lower on the food chain reduces their chances of ingesting environmental pollutants and other potentially harmful substances.

There are many forms of vegetarianism, and not all people who claim to be vegetarians eat the same diet. According to the North American Vegetarian Society, ten million Americans consider themselves vegetarians, though approximately two-thirds of them sometimes eat meat. Traditional vegetarians, also known as "lacto-ovo" vegetarians, eat no meat but do consume dairy products. Pure vegetarians, known as "vegans," consume no animal products at all, and many will not wear or otherwise use products of animal origin, such as leather and wool. There are also those who call themselves "pesco" vegetarians or "pollo" vegetarians—meaning they eat fish or chicken, respectively—but as a rule, anyone who eats meat is not really a vegetarian.

The late Indian nationalist and spiritual leader Mohandas Gandhi once said, "Eat simply, so that others may simply live." For many, adopting a vegetarian lifestyle is the embodiment of his message. According to Frances Moore Lappe, author of *Diet for a Small Planet* (1971), it takes 3.25 acres to support the average American meat-rich diet. Feeding a lacto-ovo vegetarian requires 0.5 acres, while a pure vegan diet requires only 0.17 acres. Put another way, if the amount of grain fed to cattle were converted into bread, it would be enough to provide every human on Earth with two loaves each day.

Water is also an important consideration. Producing 1 pound of beef requires up to one hundred times more water than producing 1 pound of wheat. According to a 1981 article in *Newsweek*, "the water that goes into a 1,000 pound steer would float a destroyer." In addition, groundwater contamination from animal waste is a growing problem. Livestock on U.S. feedlots produce twenty times as much excrement as the country's human population—more than one billion tons per year. Animal waste contains high levels of nitrogen, which can be beneficial to the soil, but will convert to ammonia and nitrates if not processed right away. If not treated properly, the runoff can leach into the groundwater supply, causing algae overgrowth and oxygen depletion in the nation's lakes and rivers and eventually reaching reservoirs and wells used for drinking water.

Animals not confined to the feedlot pose a different problem for the environment. Overgrazing and land use in the western United States has become a major battleground between ranchers and environmentalists. According to a report in *Fortune* magazine, the United States Bureau of Land Management spends $40 million maintaining federal grazing lands, yet collects only $18 million in fees. Cattle and sheep grazing on public land compete with wildlife for grass and water, and up to 1.5 million predators (such as wolves, bears, and coyotes) are killed each year to protect free-ranging livestock.

The impact in South and Central America is even more dramatic. Cattle ranching is one of the leading causes of the destruction of the rain forest, as ranchers employ slash-and-burn agriculture to clear grazing area for their livestock. Unfortunately, this newly created range lacks the topsoil necessary for sustained grazing and must be abandoned after only a few years.

"You are what you eat" is a well-worn axiom, but it contains a simple truth about the place of humans on the food chain. A diner sitting down to a plate of filet mignon is eating not only part of a steer but also everything that steer ingested during its life cycle. In the case of animals raised on feedlots, this goes well beyond grain and hay. Grain and corn destined to become cattle feed are sprayed with pesticides and larvacides, while the animals are injected with any combination of hormones, growth stimulants, tranquilizers, antibiotics, and appetite stimulants. While only minute traces of these products can be found in meat sold for public consumption, the long-term effects on humans are not known.

The scenario is no different for marine life.

Fish and mollusks literally "breathe" their environment, and contaminated water leads to contaminated seafood. This is especially true of shellfish, such as oysters and scallops, who feed by filtering the water. Predators who eat fish—for example, bears, eagles, and humans—consume the toxins accumulated by that fish as well as by all the smaller fish that fish has eaten. In addition, ground fish meal is used as both fertilizer and a component of some livestock feeds.

For many, choosing vegetarianism and eating lower on the food chain is their way of reducing the amount of toxins in their diets, but there is no perfect solution to the problem. Most fruit, vegetables, and grain products have also been chemically treated, especially those imported from countries where farming regulations are less strict. Even buying "organic" products does not guarantee safety because there are no industry standards for products carrying such a label.

P. S. Ramsey

SUGGESTED READINGS: One of the most influential books on the environmental impact of vegetarianism is *Diet for a Small Planet* (1971), by Frances Moore Lappe. A more disturbing picture of the effects of a traditional diet on animals and their environment can be found in John Robbins's *Diet for a New America* (1987). To learn more about the environmental effects of pesticides on the food chain, see *Silent Spring* (1962), by Rachel Carson. For those considering adopting some form of vegetarian diet, a good source of information is *The Gradual Vegetarian* (1985), by Lisa Tracy.

SEE ALSO: Animal rights; Animal rights movement; Singer, Peter.

Vernadsky, Vladimir Ivanovitch

BORN: March 12, 1863; St. Petersburg, Russia
DIED: January 6, 1945; Moscow, Soviet Union
CATEGORY: Philosophy and ethics

Vernadsky's 1926 book The Biosphere *inspired a new vision of humankind's role in shaping the earth's environment*

Vladimir Vernadsky was a Russian professor of mineralogy, crystallography, and biogeochemistry who developed the concepts of the biosphere and noösphere. A pioneer in the field of biogeochemistry, Vernadsky was given professional direction early in his life. An older cousin who was a retired army officer and an independent man of extensive reading remarked to Vernadsky that "the world is a living organism." Profoundly impressed, Vernadsky within a few years began his scholarly studies of the earth's physiology—the ways in which its matter and biota, including humankind, interact and affect one another and their common planetary environment.

Vernadsky was graduated from St. Petersburg University in 1885 and earned his Ph.D from the University of Moscow in 1897. He was professor of crystallography and mineralogy at Moscow University from 1890 until 1911. Following the 1917 Russian Revolution, he spent three years at the Sorbonne University in Paris, where he wrote extensively on the subjects of geochemistry and biochemistry, crystallography and mineralogy, geochemical activity, marine chemistry, the evolution of life, and futurology, displaying all the signs of a polymath. From 1926 until 1938, he directed the State Radium Institute in Leningrad; he was among the earliest scientists to recognize the tremendous importance of radioactivity as a source of thermal energy. He established the first Soviet national scientific academy, the Ukrainian Academy of Science, in 1928, serving simultaneously as its president and as the director of the Academy of Science's Leningrad biogeochemistry laboratory. Vernadsky founded the field of biogeochemistry, and it was the principal one in which he gradually gained international distinction.

A man of broad scholarly talents based on his mastery of several scientific specialities, Vernadsky became best known outside the Soviet Union for his publication of *La Biosphère* (*The Biosphere*, 1929), a study in which he elaborated upon his theory of the biosphere. "Biosphere" was a term Vernadsky borrowed from Eduard Suess (1831-1914), a Viennese professor of structural geology and eminent scholar who suggested the existence of an ancient supercontinent. To Suess—who had first used the word at the end of a monograph about the Alps—and Vernadsky, "biosphere" referred to the total mass of living organisms that processed and recycled the energy and nutrients available in the environment. This activity occurred inside a thin veneer of life that circled the globe.

Vernadsky was concerned that the importance of life in the entire structure of the earth's crust had been underestimated—when it was not ignored altogether—by his scientific colleagues. He elaborated an imaginative theory that, like the lithosphere, the atmosphere, the hydrosphere, and the sphere of fire—Earth's reliance upon the Sun—the biosphere formed another of the concentric circles enveloping the earth.

In the early 1940's, having fixed the word "biosphere" in the scientific lexicon, he added another word and therefore still another concept: that of the "noösphere." *Noös* is the Greek word for "mind"; Vernadsky believed that the "sphere of the mind" represented a new power altering the face of the earth. Defined precisely in the manner of science, the noösphere was neither a sphere, like the atmosphere or lithosphere, nor was it a physical phenomenon. Yet the noösphere had physical consequences, for the human mind, in Vernadsky's words, had become, for the first time, "a large-scale geological force" that was reshaping the planet.

As scholars familiar with Vernadsky's scientific achievements have noted, Vernadsky's imaginative conceptions of the biosphere and noösphere predated James E. Lovelock's inspired Gaia hypothesis and paralleled some fundamental ideas integral to it. Lovelock, who first expounded the Gaia hypothesis in 1972 and published further specifics during the 1980's, fully acknowledged Vernadsky's importance to his work.

Alexander Scott

SEE ALSO: Biosphere concept; Gaia hypothesis.

Waste management

CATEGORY: Waste and waste management

Waste management concerns the physical by-products of human activity that cannot be reintegrated into the ecological biomass cycle. These by-products include solid, liquid, and airborne substances that are potentially harmful to living organisms. As the human population grows and the use of manufactured materials expands, disposing of waste becomes more challenging.

According to the World Watch Institute, world production of manufactured materials (not counting recycled materials) increased nearly 2.5 times between the early 1960's and the late 1990's. In industrialized countries the increase is far greater: The United States, for example, has seen an eighteenfold increase in materials production since 1900. The average U.S. citizen throws away an estimated 2 to 8 pounds of garbage daily, and although studies demonstrate that the per-person production of waste remained approximately the same throughout the

Trash is dumped at an incineration facility in Detroit, Michigan. As with other waste management options, incineration has its advantages and disadvantages: Although it reduces the need for landfill space and may be used to generate electricity, it also releases pollution into the atmosphere. (Jim West)

twentieth century, the sharp rise in population and expanding industrial base meant greater total accumulations of waste. Furthermore, the types of waste changed.

Waste is commonly categorized as domestic, or solid, waste, and industrial, or liquid, waste, although the distinction is not absolute. Both may contain toxic substances, but the percentage of toxins in industrial waste is likely to be higher, and the types of waste are disposed in different ways. The smoke emitted from industrial processing of materials and vehicle exhaust are additional types of waste, although they are commonly thought of as pollution rather than waste.

Solid Waste

Solid waste is the familiar garbage that households and businesses in the United States have sent to the dump since garbage collection began late in the nineteenth century. The largest portion, more than 40 percent, consists of paper products, especially newspaper and containers. Yard waste, food debris, plastic containers and wrappings, bottles, metals, and appliances are also regularly thrown away. About 1 percent of this waste involves hazardous materials, typically insecticides, beauty aids, and cleaning products. Construction waste accounts for a large share—about 12 percent—of solid waste and may contribute a higher proportion of hazardous materials, such as solvents and paint.

Although most of these materials are solid, when dumped together they can soak up rainwater and then ooze chemical-laden liquids. This leachate may filter down into the groundwater and pollute nearby streams and wells. If it contains toxic elements, such as the lead or mercury from batteries, the leachate can be dangerous to health. The odor from rotting garbage may also foul the air, seldom enough to be harmful but still repellent to people living nearby. It can attract animal scavengers, which may become infected with diseases from the garbage and spread them to other animals or even humans, especially if feces are part of the waste.

In order to combat these effects, sanitary landfills place a plastic lining under the waste to contain leachate and cover each day's load of garbage under a thin layer of soil. Pipe systems also disperse methane gas produced by rotting organic materials. The landfills are therefore less dangerous to human health or the environment, but many old, abandoned sites were not so well engineered. They may continue to dribble harmful chemicals into groundwater for decades and emit methane, which is flammable. Numerous small, illegal dumps and litter compound the problem.

Measures to reduce the amount of waste deposited in landfills have partially succeeded. Recycling has drastically cut the total paper, metal, and glass waste in some U.S. states and industrialized countries. The use of garbage disposals and composting has caused the proportion of organic materials to decline. However, such reductions did not eliminate solid waste. By the end of the twentieth century, cities were finding it increasingly difficult to find room for new landfill sites, even when the space was urgently needed. Stringent regulations about the geological composition of landfills reduced the number of usable sites, while objections from citizen action committees, known as "not in my back yard" (NIMBY) groups, also eliminated sites near populated areas.

Facilities used to incinerate waste, which sometimes powered electrical generators with the resulting heat energy, also faced objections because burning could release health-threatening materials, such as dioxins, into the air. Moreover, a significant proportion of waste, such as appliances and concrete, cannot be eliminated by burning. Tires, too hazardous to burn, float to the surface in landfills, causing continuous problems for waste managers; they often end up stacking the tires in immense piles that, if accidentally ignited, can burn out of control and create large clouds of black fumes.

Industrial Waste

The effluent stream of by-products from factories, as well as chemical and petroleum refineries, is made up of water, solid filings and cuttings, liquid solvents and oil derivatives, and semisolid sludge. The solid components are usually no more hazardous than household wastes, although medical waste—particularly tainted blood and used "sharps," such as needles and scalpels—may pose the additional danger of spreading disease. However, liquids and semiliquids sometimes contain a high proportion of hazardous chemicals. Rain also leaches chemicals, such a cyanide and mercury, out of the smelted tailings from mines. Agricultural fertilizers and pesticides can enter groundwater or streams as well. Because these liquid wastes rapidly spread through waterways and groundwater, they are often collectively known as toxic waste.

Industry now uses all ninety-two naturally occurring elements on the periodic table, and the isotopes of some of these are radioactive. Nuclear weapons manufacturing in particular leaves radioactive debris, but medical procedures that use radioactive tracers and scientific instruments may also create radioactive wastes. This nuclear waste continues to emit radiation for thousands or hundred of thousands of years, and improperly stored radioactive materials have been associated with increased risk of disease for people, animals, and plants.

During the 1980's and 1990's federal and state regulations brought industrial waste management under rigorous control. Facilities known as secure landfills are designed to contain nonradioactive industrial wastes in tightly lined, self-contained areas. Incinerators reduce the waste to harmless ash while releasing few or no harmful particulates into the atmosphere. Separate repositories store nuclear wastes deep underground in leak-proof containers.

The public is seldom reassured by such measures, however. Leakage occasionally occurs from secure landfills. Near-zero toxic emissions from incineration means that some toxins do, in fact, escape into the atmosphere. In addition, nuclear repositories may not be catastrophe proof; for example, an earthquake could crack open containers, releasing radioactive material into groundwater supplies. Although waste managers insist that these dangers are minimal, the news media brings them to public attention, and NIMBYs regularly resist the opening of new secure landfills and radioactive waste repositories. State governments often object as well, as was the case when the Nevada legislature stalled the

U.S. Municipal Solid Waste, 1994

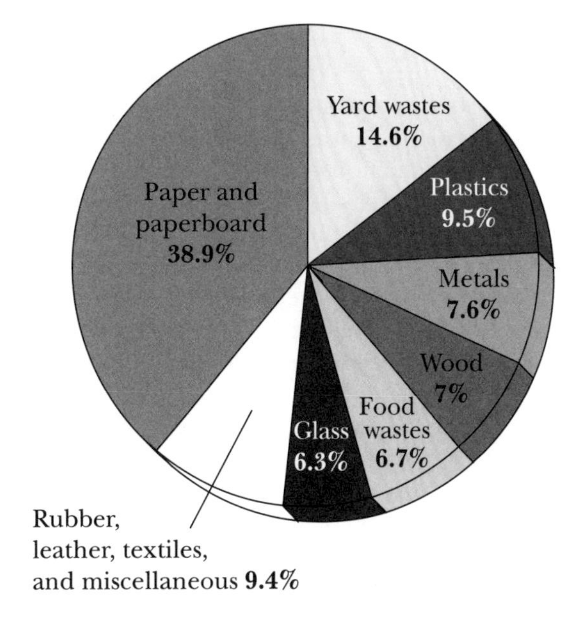

Source: U.S. Department of Commerce, *Statistical Abstract of the United States, 1996*, 1996. Primary source: Franklin Associates, for Environmental Protection Agency.

Note: Total U.S. municipal solid waste generated in 1993 was about 209 million tons, or 4.4 pounds per person per day. Not included in these figures are mining, agriculture, industrial, and construction wastes, junked automobiles and equipment, or sewage.

construction of a nuclear repository at Yucca Mountain. Many old facilities, built before strict government oversight, remain in use and could leak toxic materials into the environment undetected. The memory of deadly chemical leaks, such as that discovered at Love Canal in New York in 1976, and of released radioactive material, such as the plutonium that escaped the Hanford Nuclear Reservation in Washington State, makes the public wary of hazardous wastes.

As a result of citizen concern, most new hazardous waste disposal sites are now located far from population centers. This has created a new peril. The waste must be transported, primarily by trucks and trains, to a facility. Traffic accidents and train derailings en route can dump extremely dangerous chemicals straight into water or the atmosphere. Evacuations of residents near such accidents, while not common, increased during the 1990's. Even if people are rescued, however, plant and animal life is not safeguarded.

ENVIRONMENTAL CONSEQUENCES

Many critics of waste management insist that only source reduction—a drastic decrease in the use of raw materials—will make waste disposal safe. Accordingly, during the 1990's some countries, notably Denmark and Germany, sought to reduce virgin material use as much as 90 percent by intensifying recycling. In the United States, Superfund legislation sets aside federal funds to pay for cleanups of the most dangerous hazardous waste sites. Other industrial countries have similar projects. Still, only a fraction of sites receive attention, and until source reduction goals are met, household and industrial wastes will continue to swell landfills with environmentally hazardous substances. Illegal dumping of hazardous waste exacerbates the danger.

Scientists disagree about how severely wastes damage the environment, but there is agreement that repercussions are evident and likely to increase. Methane from dumps, smoke-stack emissions, and vehicle exhaust contain greenhouse gases, which are implicated in global warming. Nutrients released from sewers, as well as runoff from agriculture and mining, degrade the environment of rivers and streams, harming aquatic life and leaving the water unusable without special treatment. The waterborne wastes that reach the ocean, supplemented by ocean dumping of toxic materials, alter and sometimes destroy offshore ecosystems, as is the case for many coral reefs worldwide.

Roger Smith

SUGGESTED READINGS: *Garbage and Waste* (1997), edited by Charles P. Cozic, presents non-technical articles of opposing views on such topics as the alleged waste crisis, recycling, and public health hazards from waste. R. K. Gourley surveys the sources and effects of agricultural, industrial, and nuclear waste, as well as the challenges they pose to national governments, in *World of Waste: Dilemmas of Industrial Development* (1992). Timothy C. Jacobson's *Waste Management: An American Corporate Success Story* (1993) finds that waste-management professionals in the United States have brought about a change in cultural attitudes so that waste disposal is more environmentally safe. In *The Politics of Garbage: A Community Perspective on Solid Waste Policy Making* (1996), Larry S. Luton discusses a controversial incinerator project near Spokane, Washington, to illustrate the political and economic issues that can pit local government, citizens, and businesses against each other. *Rubbish: The Archeology of Garbage* (1992), by William Rathje and Cullen Murphy, supplies an overview on the waste produced by modern culture and how it will affect the future. Jennifer Seymour Whitaker finds that waste management is in crisis in *Salvaging the Land of Plenty: Garbage and the American Dream* (1994) and proposes that the only real solution is to generate less waste. Succinct, general discussions of waste management, pollution, and toxic wastes occupy chapters in *Eco-Facts and Eco-Fiction: Understanding the Environmental Debate* (1996), by William H. Baarschers, a retired chemistry professor.

SEE ALSO: Hazardous waste; Landfills; NIMBY; Nuclear and radioactive waste; Sewage treatment and disposal; Solid waste management policy; Superfund; Waste treatment.

Waste treatment

CATEGORY: Waste and waste management

Waste treatment is any practice designed to alter the physical, chemical, or biological composition of waste in order to concentrate or neutralize it. The degree of treatment required for any waste stream depends upon the characteristics of the waste, the maximum discharge limit, and the final disposal requirements imposed by regulatory agencies. Waste-treatment facilities utilize a number of processes to achieve the desired degree of treatment.

Rapid advances in technology and industrialization have resulted in the discharge of increased quantities of wastes into the environment. Adverse environmental effects develop if their concentrations exceed the natural capacity of air, water, and land systems to assimilate them. In the United States, where pollution control is exercised by the federal government, virtually all of the environmental control legislation has been written and passed since the end of World War II, when dramatic increases in urban density and industrialization occurred.

The principal wastes associated with industrial and municipal facilities can be categorized as organic and inorganic; solid (suspended and dissolved); acid and base; and hazardous. Organic wastes are oxygen demanding and lower the amount of dissolved oxygen in the receiving waters. Suspended solids settle and cause benthic deposits. Acid and base wastes destroy natural buffers, and acid rain has a potentially devastating effect upon aquatic life and forests. Hazardous wastes—which include explosive, flammable, volatile, radioactive, toxic, and pathological wastes—can cause serious damage to people or property. Storage, collection, transportation, treatment, and disposal of such wastes require special caution.

Waste-treatment processes can generally be divided into three categories: physical, chemical, and biological. Physical treatment is used to concentrate wastes, reduce volume, and separate different components for further treatment or disposal. Chemical treatment is used to precipitate, detoxify, or destroy hazardous properties. Biological treatment utilizes microorganisms to stabilize organic wastes.

There are more than twenty types of physical treatment processes—also called unit operations—commonly used for handling wastes. Removal of wastes is achieved by physical forces. Common examples of physical treatment processes are sedimentation, centrifugation, flotation, evaporation, drying, distillation, stripping, carbon adsorption, ion exchange, membrane processes, freeze crystallization, and solidification. Sedimentation, centrifugation, and flotation processes are typically used to remove suspended solids. Evaporation, drying, and distillation are utilized to concentrate wastes. Stripping removes ammonia and volatile organic compounds. Carbon and resin adsorption is used for removal of organic solute from aqueous waste streams. Their applications include separation or removal of phenol, fat, color, pesticides, carcinogens, and chlorinated hydrocarbons. Membrane processes such as reverse osmosis, ultrafiltration, and electrodialysis utilize synthetic membranes and are used for concentration of industrial and hazardous wastes, as well as desalination of brackish water. In freeze crystallization, the waste stream is cooled for separation of pure water (in the form of ice crystals or solid ice) from the contaminants, which concentrate in liquid. Solidification is the transformation of hazardous waste into a nonhazardous solid product by fixation or encapsulation.

Chemical processes are used for the treatment of industrial wastes and are usually used in conjunction with other methods to achieve the end result. Common chemical treatment processes used for waste treatment include neutralization, oxidation, reduction, precipitation, hydrolysis, catalysis, chlorinolysis, electrolysis, photolysis, and incineration. Neutralization is adjustment of pH by either acids or bases. This process has a wide application in treatment of wastes from many industries. Oxidation is used mainly for detoxification of hazardous wastes. Chlorine, ozone, hydrogen peroxide, and potassium permanganate are excellent oxidizing agents often used in the presence of ultraviolet light. Reduction is achieved by use of a reducing agent such as sulfur dioxide. As an example, hexavalent chromium, a very toxic substance, is reduced to trivalent chromium, which is much less toxic, and then precipitated. Hydrolysis, catalysis, chlorinolysis, electrolysis, and photolysis are destructive processes used to break chemical bonds. Incineration is thermal destruction of hazardous organic wastes.

Biological processes involve biochemical reactions that take place in or around microorganisms. Generally, organic compounds are decomposed in suspended or attached growth reactors. The most common biological waste-treatment processes utilize a biological reactor for stabilization of organic matter, followed by solid-liquid separation. Wasted solids, referred to as sludge, are combined with other solids and treated separately. Anaerobic sludge digestion is used for stabilization of sludge and high-strength wastes. With proper control, biological processes are reliable and environmentally sound; chemicals are not added, and operational costs are relatively low.

Syed R. Qasim

SUGGESTED READINGS: *Hazardous Waste Management* (1994), by Michael D. LaGrega, Phillip L. Buckingham, and Jeffrey C. Evans, provides an overview of processes involved in treating hazardous waste. *Pollution Prevention* (1992), by Louis Theodore and Young C. McGuinn, discusses techniques for dealing with pollution problems. *Environmental Science and Engineering* (1996), by J. Glynn Henry and Gary W. Heinke, contains information in some of the technology used to treat waste products. *Environmental Science* (1989), by Charles E. Kupchella and Margaret C. Hyland, provides a good overview of the scientific techniques used to study and deal with various environmental problems.

SEE ALSO: Hazardous waste; Solid waste management policy; Waste management.

Water conservation

CATEGORY: Water and water pollution

Water conservation is a reduction in water use. Exponential growth in worldwide population has made the conservation of water an increasingly important consideration.

In 1998 world population approached six billion, with an annual growth rate of 1.6 percent. If this growth rate were to continue, world population would double to twelve billion by the year 2040. Since water is a basic human necessity, it follows that the amount of water needed for human existence must also roughly double during this time. While it is theoretically possible to double available water by 2040, the likelihood that this can or will happen is low—for several reasons. First, although there is enough water available globally, it is not uniformly distributed around the world when and where it is needed. Second, the costs involved in doubling available water would be astronomical. Third, even if all needed moneys were readily available, it would be difficult to build the necessary facilities and have them operational by the time they would be needed.

The greatest potential for water conservation in residential settings is in bathrooms. Traditional toilets use 19 to 26 liters (5 to 7 gallons) per flush. However, low-flow toilets 13 or fewer liters (3.5 gallons) per flush—have been installed in new homes and have replaced traditional toilets in older homes. Baths and showers also use large amounts of water. Low-flow shower heads have contributed to water conservation efforts. A great deal of water is also wasted in kitchens. Water can be conserved at sinks by simple means, such as collecting food scraps for compost piles rather than putting them through garbage disposals. Low-flow faucets are also available. Dishwasher water can be conserved by running dishwashers only for full loads; the same is true for clothes washers. Both of these washers might be redesigned to use less water. Another potential source of water conservation outside homes is in plant and lawn watering. Water can be conserved by decreasing lawn sizes and using plants that require less water.

Business and industrial settings use copious amounts of water. In some cases, water conservation measures may result from simple changes in the way things are done, such as substituting sweeping for washdown of plant floors. In most cases, however, water conservation results from process changes of some type. For example, by converting from a water-cooled ice maker to an air-cooled one, a restaurant may reduce water consumption by 70 percent.

The largest amounts of water, however, are used in agriculture for irrigation. Following a history of relatively cheap water, farmers have tended to waste great amounts of water through improper or excessive application. Water can and has been saved through computerized timing of water application. Otherwise, modifications of irrigation procedures have helped. For example, drip irrigation, which applies water slowly and uniformly at or below soil level adjacent to plants via mechanical water outlets, has produced water savings over traditional methods of simply spraying water onto the soil.

As with other commodities, pricing can affect water usage. Water-conserving rates exhibit increasing unit costs as volume used increases. One simple model determines a certain rate for the average amount of water a household might be expected to use, with a much higher rate for all water used over that average amount. Yet another water conservation measure is finding and repairing leaks, which can occur in water distribution systems, homes, and industrial plants. Documented leakages have amounted to losses of more than 50 percent of the water sent through a distribution system. Leaks can be found by performing water audits.

Water reuse may be thought of as another form of water conservation. Ample opportunities for water reuse exist during industrial processes. For example, wastewaters that were being treated on-site and discharged into receiving streams might be reused as cooling water. While it is possible to treat sewage for reuse by households, most people are not yet ready to accept such a conservation measure.

There are some negative aspects of water conservation. First, water conservation may lead to reduced revenues for water utilities, which may then be forced to increase rates. Second, conserving water over the long-term may reduce the "slack" in the system and make short-term drought savings difficult to achieve or reduce the amount of water available for water rationing. Third, water conservation programs can be expensive and require considerable amounts of up-front money, while actual water savings develop slowly over a period of time. Such programs are not always cost effective.

Jack B. Evett

Suggested readings: For an easy-to-read book on ways to conserve water, see *Drought Busters* (1991), by Slater and Orzechowski. For a book on water scarcity, see Sandra Postel's *Last Oasis: Facing Water Scarcity* (1992). For a more technical overview of water conservation, see William O. Maddaus's *Water Conservation* (1987). For complete coverage of all aspects of water conservation, see the American Water Works Association's proceedings for *CONSERV90* (1990), *CONSERV93* (1993), *CONSERV96* (1996), and *CONSERV99* (1999).

See also: Drinking water; Irrigation; Population growth; Renewable resources; Water-saving toilets; Water use.

Water pollution

Category: Water and water pollution

Water pollution is the deterioration of natural water by human activities such as mining, agriculture, and improper waste disposal. Such activities may raise the normal concentration of dissolved or suspended constituents in natural waters so that they become harmful to organisms or plants.

Before the Industrial Revolution in the nineteenth century, humans produced only minimal amounts of refined metals and organic materials. Production of various alloys of copper, tin, lead, and zinc by heating the mineral ores or by using natural copper metal was also minor. During the Industrial Revolution, however, cast iron began to be produced by heating iron ores at high temperature by burning charcoal, and other metals began to be produced by other methods. For example, nickel, aluminum, titanium, cobalt, platinum, chromium, niobium, and molybdenum were discovered during this time. The extensive use of such metals resulted in massive increases in exploration, mining, and production, as well as an increased use of energy, resulting in waste disposal problems and contamination of water supplies. In addition, growing urban populations produced concentrations of untreated human and animal wastes and associated disease-producing organisms in natural waters.

Inorganic Constituents

There are both natural and human sources of water contamination. Humans may increase the natural contamination by, for example, mining natural sources and disposing of the waste, which may leach out dangerous constituents. Animal or plant health may be affected by enrichment or deficiency of certain dissolved constituents. Rock or soils may have low concentrations of substances such as selenium, potassium, phosphorous, copper, cobalt, molybdenum, zinc, or iodine, which cause health problems in animals. Although humans may need to supplement their diet with these constituents for optimal health, substances such as selenium, radioactive elements, copper, and zinc can be concentrated enough in some drinking waters to be harmful.

Humans mine many elements and use them for manufacturing and other purposes faster than they weather out of natural rocks, producing a variety of atmospheric and water pollutants. For example, fertilizers have high concentrations of soluble nitrogen and phosphorous compounds, and animal wastes contain high nitrate. Thus, the use of fertilizer can produce high nitrate (a nitrogen-oxygen compound) and phosphate (a phosphorous-oxygen compound) in natural waters. Nitrate can combine with hemoglobin so that oxygen transport in the body is inhibited. This is a potentially serious threat to babies, as they can literally turn blue, become sick, and die if high nitrate waters are consumed over a long period of time. Water high in phosphorous can stimulate the growth of organisms such as algae. As the abundant algae die and drop to the bottom of a body of water, they may use up the dissolved oxygen in the water, which may, in turn, cause the fish to die.

The use of table salt (sodium chloride) to melt roads in the winter or storage of the salt can produce high dissolved sodium and chloride concentrations in natural waters. In the past, deep groundwater with high salt concentrations that was brought up with petroleum was placed in "evaporation pits" on the surface until the water evaporated. This would allow the salt to slowly leak into the water supply in the ground. Now petroleum companies are required to inject these salty waters back into the ground at the level from which they came.

Another major pollutant in water results from the acidity produced by acid mine drainage and acid rain. Acid mine drainage results from the chemical reaction of sulfide minerals such as pyrite (iron sulfide) with water and oxygen from the atmosphere to produce sulfuric acid and dissolved metals in water. Most acid mine drainage comes from small amounts of pyrite in coal mines or waste piles from coal miners. Some acid mine drainage results from metal mines and wastes such as those found in lead and zinc mines in southeastern Kansas. These acid waters can readily dissolve other metals, so acid mine waters may contain high concentrations of many poisonous metals.

Acid rain results from high concentrations of sulfur dioxide, carbon dioxide, and nitrogen oxide gases spewed out into the atmosphere by industry. These gases dissolve in the water in the atmosphere to produce the acidity. Thus, the worst acid rain occurs in the industrial areas of the eastern United States. Acid rain will produce acid lakes and streams in areas that have little natural capacity to neutralize the acid. This results in the destruction of organisms that cannot live in such acidic waters. In some areas that have abundant limestone (a rock composed of calcium carbonate), the acid rain will react with the limestone to neutralize the acidity, so there is little problem with acidity of the natural waters. Areas without rocks that can neutralize the acid continue to have problems with the acidity.

Organic Compounds

Organic compounds are compounds consisting of carbon in chemical combination with hydrogen, oxygen, sulfur, chlorine, or nitrogen. There are many thousands of organic compounds currently manufactured, so their classification is complex. However, they may be simplistically grouped as alkanes, benzene derivatives, chlorinated hydrocarbons, and pesticides.

Alkanes are straight chains of carbon atoms combined with hydrogen. Benzene derivatives consist of six-membered rings of carbon combined with other constituents chemically attached to the carbon atoms. Alkanes and benzene derivatives with six to ten carbon atoms are the organic compounds found in gasoline. They are also found in other fuels such as diesel fuels. Alkanes in gasoline are not very soluble in water, but benzene derivatives are. Alkanes are also

Raw sewage flows into the Ganges River in India. Improper waste disposal is one of the primary causes of water pollution. (AP/Wide World Photos)

more easily degraded by bacteria than are benzene derivatives. A major pollution problem in areas around gasoline stations has been leakage of the gasoline into the groundwater system.

Groundwater is the name for any water found under the earth's surface. The upper portions of the groundwater contain both air and water in the pore space between mineral grains; the lower portions of groundwater only contain water in the pore space. The surface between the upper and lower zone is called the water table. Wells are usually drilled into the water table so some of the more soluble hydrocarbons of gasoline that has leaked into the groundwater system can dissolve in water and move in the saturated water zone. The insoluble alkanes and benzene derivatives of gasoline can also move as a separate fluid plume above the water table since they are lighter than water. The maximum allowable concentrations of these benzene derivatives are much lower than for the individual elements discussed previously.

Benzene derivatives are also fairly volatile. If they move in a liquid plume under a building, they can move as a vapor into basements or sewers, where they then may produce explosions or ill health to the residents of the building. One such explosion that occurred in Guadalajara, Mexico, in 1992 killed several hundred people.

Chlorinated hydrocarbons contain the element chlorine in one or more parts of the compound. Many of them—such as dichloroethane, tetrachloroethane, and chloroform—are among the most common organic pollutants found in waste disposal sites in the United States. Many are carcinogens (cancer-producing substances), and they become increasingly toxic at higher concentrations.

Pesticides

Pesticides are complex organic compounds that kill organisms such as insects. Examples of pesticides are dichloro-diphenyl-trichloroethane (DDT), malathion, alachlor, atrazine, and chlordane. Pesticides may be carcinogens, and some may not decompose readily in the food chain. DDT, for example, has long been banned in the United States because of its harmful effects and its slow decomposition in nature.

Pesticides vary greatly in the time it takes them to naturally decompose and move to the water table. At one extreme, pesticides such as prometon can last a long time and quickly move to the water table, thus rapidly contaminating the groundwater. At the other extreme, methyl parathion decomposes more readily and moves more slowly to the water table, and is thus less likely to contaminate the groundwater. Pesticides most often found in groundwater are dibromochloropropane (DBCP), aldicarb, carbofuran, chlordane, alachlor, and atrazine. In addition, a pesticide applied during times of high rainfall can rapidly move to streams, where the stream water can soak into the ground and contaminate the groundwater supply.

Another problem is that even if the original pesticide has been naturally destroyed by bacteria, the degradation products from the decomposition of the pesticide can be even more harmful to humans than the original pesticide. Few studies of these kinds of problems have occurred, and decay products from pesticides are often only poorly understood. The degradation products of a few pesticides such as aldicarb, however, have been studied for a number of years. Such products may have entirely different movement and stability than the original pesticide.

Radioactivity and Heat Pollution

The use of radioactive materials has produced special waste disposal and water pollution problems, especially since the radiation cannot be detected by the senses and can be very damaging if ingested. At one extreme, the radioactive element plutonium has a half-life (the time it takes for one-half of the radioactivity to decay) of about twenty-four thousand years, and it concentrates in the bones of vertebrates. This means that plutonium that has leaked into groundwater must be removed for hundreds of thousands of years. At the other extreme, radioactive materials with short half-lives that do not concentrate in organisms may not be of much concern, since the radioactivity will decay before it can be ingested.

Radioactive wastes are divided into low-level, intermediate-level, and high-level wastes. High-level wastes may be more than one million times more radioactive than what is considered acceptable. Low-level wastes may contain radioactivity up to one thousand times more radioactive than what is considered acceptable. Intermediate-level wastes have radioactivity between these ranges. A wide variety of low-level radioactive wastes are produced by hospitals, the nuclear industry, and research laboratories. These wastes are often sealed in drums and buried under a thin layer of soil or diluted in water to acceptable levels of radioactivity and flushed into the sewer.

High levels of radioactive waste are produced by nuclear fuel generation in fairly small volumes and account for about 95 percent of the radioactive waste materials. Many of the high-level wastes have been stored for decades in double-walled, stainless steel tanks that are air conditioned because of the intense heat given off by the radioactivity. Some of these tanks have leaked, and radioactive fluids have moved into the local groundwater system.

Thermal pollution is the heating of natural waters caused by the activities of industry, the burning of fossil fuels, and nuclear power production. Dumping heated waters directly into a river can kill many heat-sensitive organisms. Hot waters are often stored in ponds and allowed to cool to river temperature before being allowed to flow back into the river.

Prevention and Remediation

Efforts to prevent pollution of natural waters are easier and cheaper in the long run than trying to clean the water supply or remediate the source of pollution. Preventive approaches include keeping contaminants contained so they cannot escape into the water systems and banning the production and use of dangerous substances such as certain pesticides. For example, dangerous material could be disposed of by placing the material in a dry climate with low population density in geologic materials that are impermeable to water flow. This would minimize the chances that moving water would carry hazardous constituents to groundwater. The U.S. government has drafted plans to fuse high-level radioactive wastes in silicate material, which will then be stored underground in volcanic rocks in the Nevada desert. However, since the costs of transporting waste materials to the desert are high, municipal wastes are usually disposed of locally. This means that people who live in areas that have high rainfall must be especially careful to keep the waste contained in geologic materials such as unfractured mudrocks that are impermeable to movement of water.

Since prevention and remediation are expensive, there is an ongoing conflict between environmentalists and industry. Industry wants to avoid paying for proper waste disposal, whereas environmentalists want to keep pollution to surface water and groundwater to a minimum. In order for the problem to be solved, a reasonable compromise between the two viewpoints must eventually be reached.

Robert L. Cullers

SUGGESTED READINGS: Examples of specific water-quality problems are given in William M. Alley's *Regional Ground-water Quality* (1993). The environmental problems posed by various mining and industrial processes are discussed in *Mineral Resources, Economics, and the Environment* (1994), by Stephen E. Kesler, and *Resources of the Earth: Origin, Use, and Environmental Impact* (1996), by James R. Craig et al. A specific example of water pollution by atrazine is provided by D. Duncan et al., "Atrazine Used as a Tracer of Induced Recharge," in *Ground Water Monitoring Review* 4 (1991). An example of how nitrate is distributed in a given area may be found in D. D. Adelman et al., *Overview of Nitrate in Nebraska's Ground Water* (1985), Transactions of the Nebraska Academy of Science volume 13. A discussion of how constituents move in groundwater and how pollutants might be remediated can be found in *Ground Water Contamination: Transport and Remediation* (1994), by Philip B. Bedient et al. Distribution of specific pesticides in the United States is studied in *Pesticides in Surface Waters* (1997), by Steven J. Larson et al. Specific examples of nitrate pollution are included in Larry W. Center's *Nitrates in Ground Water* (1997). Other examples of how different kinds of water pollution may be cleaned can be found in *Alternatives for Ground Water Cleanup* (1994), by the National Research Council.

SEE ALSO: Acid deposition and acid rain; Hazardous waste; Heavy metals and heavy metal poisoning; Landfills; Lead poisoning; Mercury and mercury poisoning; Thermal pollution.

Water pollution policy

CATEGORY: Water and water pollution

Water pollution policy is determined by laws and regulatory agencies that deal with society's interactions with waterborne contaminants, including infectious disease organisms. Included are policies that prevent the entry of these agents into or lower their levels in aquatic ecosystems. Two primary goals pervade such policies: protection of human health and protection of natural aquatic resources.

Water pollution can be defined as any physical, chemical, or biological change in water quality that adversely affects living organisms or makes water unsuitable for desired uses. There are natural sources of water contamination, such as oil seeps. However, water pollution is generally caused by human activities. Categories of water pollutants include infectious agents such as bacteria and viruses; organic chemicals such as pesticides, plastics, and oil; inorganic chemicals such as acids, salts, and metals; radioactive materials; sediments; plant nutrients such as nitrates and phosphates; oxygen-demanding wastes such as manure and plant residues; and heat. Each of these categories presents unique scientific and technical problems that must be addressed through specific policies.

During the last twenty-five years of the twentieth century, water pollution policy was very effective at increasing the quality of water in the United States. However, about 40 percent of the

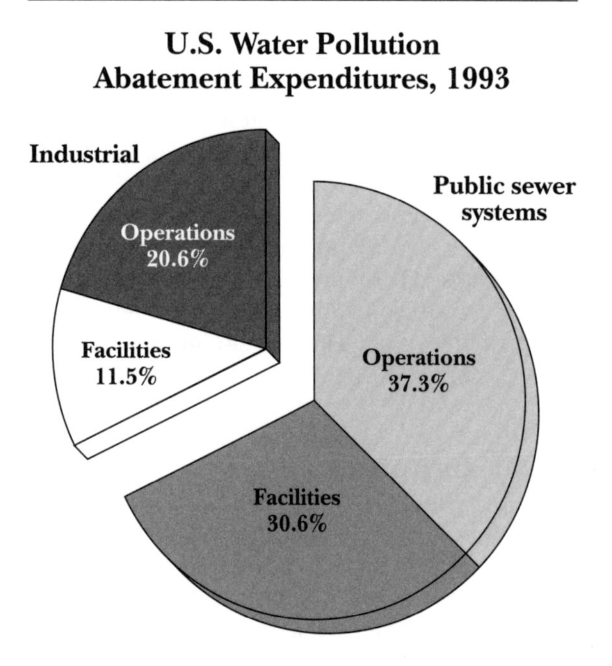

Source: U.S. Department of Commerce, *Statistical Abstract of the United States, 1996,* 1996.

Note: Total U.S. water pollution abatement expenditures for 1993 were between $32 and 33 billion.

bodies of water in the United States still do not meet water quality goals. The National Water Quality Inventory Report is the primary method for informing Congress and the public about water quality conditions. The report, required under section 305(b) of the Clean Water Act, is written every two years and characterizes the nation's water quality, identifies water quality problems of national significance, and describes various programs implemented to restore and protect water. During 1996 a total of 19 percent—almost 1.1 million kilometers (700,000 miles)—of the river miles in the United States were surveyed for water quality. Of the surveyed river miles, 64 percent had good water quality, while some form of pollution affected the other miles. Siltation was cited as the leading pollutant entering rivers, with agriculture cited as the leading source of the impairment. Despite the construction of new sewage treatment plants and improvements in older plants, municipal sewage effluent remains the second most common source of pollution in rivers.

In the same report, 40 percent of the lake acreage in the United States was surveyed. Of this, 61 percent of the acreage was reported to have good water quality. The remaining 39 percent was impaired to the greatest extent by nutrients (such as nitrogen and phosphorus) and metals. Agriculture and unspecified nonpoint sources were reported to be the leading sources of these pollutants.

U.S. Water Pollution Laws

Numerous laws have been passed in the United States that have some influence on water quality in the country. In response to increasing public concern about water pollution, Congress passed the Water Quality Act of 1948, the first federal legislation to directly deal with the issue. Its primary goal was to provide funding for the research and implementation of state water pollution control programs. Additional legislation and funding occurred through the Federal Water Pollution Control Act of 1956, which was drafted to combat water quality problems associated with increasing industrialization. The early emphasis of state control of water quality was extended in the Federal Water Pollution Control Act of 1965, which required states to adopt water quality standards and implementation plans. However, this act did not provide for sufficient enforcement mechanisms, and by 1972 only about one-half of the states had set water quality standards. Furthermore, many of the states did a poor job of enforcing the standards, particularly when applied to individual dischargers into their waters.

In 1972 Congress passed the Federal Water Pollution Control Act, also known as the Clean Water Act. The main objective of this act was to restore and maintain the chemical, physical, and biological integrity of the nation's waters. Some of the specific goals of the act were to eliminate all discharges of pollutants into the navigable waters by 1985; protect fish, shellfish, and wildlife, and provide for recreation by 1983; prohibit the discharge of toxic pollutants in toxic amounts; and provide financial assistance for the construction of publicly owned waste-treatment works. The act sought to accomplish these goals by combining state water quality standards with the technology-based approach of setting effluent limitations.

In 1987 the Clean Water Act was amended because Congress recognized that although significant progress had been made, significant water quality problems persisted. New provisions established a comprehensive program for controlling toxic pollutant discharges beyond that already provided in the act, added a program requiring states to develop and implement programs to control nonpoint sources of pollution, and authorized a total of $18 billion in aid for wastewater treatment assistance.

The goals of the Clean Water Act have not been entirely met, but there are many success stories. For example, in 1972 most experts estimated that only 30 to 40 percent of assessed waters met water quality goals, such as being safe for fishing and swimming. By the end of the 1990's, state monitoring data indicated that between 60 to 70 percent of assessed waters met state water quality goals. In addition, through Clean Water Act funding, the number of people in the United States served by sewage treatment plants increased from 85 million in 1972 to more than 173 million in 1998.

Other Important Policies

In order to protect the environment and human health in the United States, a number of other water quality bills have been passed by Congress. For example, some critics suggested that drinking water policies and enforcement of those policies were too lax, especially in rural water districts and small towns. Thus, the Safe Drinking Water Act of 1972 was passed. It regulates water quality in commercial and municipal drinking water systems by establishing minimum drinking water quality standards for every community. Among the contaminants regulated are bacteria; nitrates; metals such as arsenic, cadmium, chromium, lead, mercury, and silver; pesticides; radioactivity; and turbidity. The act also contains limited provisions for the protection of groundwater and aquifers.

In 1976 Congress passed the Toxic Substances Control Act, which categorizes toxic and hazardous substances, funds a research program, and regulates the use and disposal of poisonous chemicals. Before a new chemical can be manufactured in bulk, the manufacturer must submit a premanufacturing report to the Environmental Protection Agency (EPA) in which the environmental impacts are assessed, including those associated with disposal of the chemical.

In response to water quality problems associated with toxic waste dumps, Congress passed the Comprehensive Environmental Response, Compensation, and Liability Act (CERCLA) in 1980. This act is also known as the Superfund act because billions of dollars have now been spent on cleaning toxic waste dump sites. The act was amended in 1984 to include a section on a community's right to know about the presence of toxic materials in their area.

The London Dumping Convention of 1990 is an international treaty signed by the United States that calls for the cessation of the dumping of industrial wastes, tank washing effluents, and plastic trash into the world's oceans. The United States has passed legislation to support the provisions of the treaty. Another key international agreement is the 1972 Great Lakes Water Quality Agreement between Canada and the United States. This agreement and subsequent amendments in 1978 and 1987 affirmed the two countries' determination to restore and enhance water quality in the Great Lakes system, which includes the entire lake basin and the St. Lawrence River. The agreement focused on the eutrophication of the Great Lakes and the need to reduce loadings of phosphorous. Since then, the objectives have focused on the virtual elimination of persistent and toxic substances.

In 1995 the U.S. House of Representatives passed a bill that would have greatly weakened the Clean Water Act. However, the bill was abandoned because of a flood of public opposition and the threat of a presidential veto. On the other hand, supporters of the law would like to see stricter enforcement of existing standards. A specific issue that must be resolved is the problem of wet-weather flows, which have been identified as the largest remaining threat to water quality. Wet-weather flows include agricultural runoff, urban storm water, and sewer overflows. At issue is whether to detail wet-weather programs in the act or implement a flexible approach that would recognize the site-specific nature of intermittent wet-weather pollution. Another key issue is how to provide regulatory relief for industry, states, cities, and individual landowners. In addition, some people have suggested that the focus of water pollution policy should be shifted from developing and implementing water quality standards for effluents to changing industrial and other processes so that toxic substances are not produced in the first place.

Roy Darville

SUGGESTED READINGS: An excellent introduction to international water quality issues can be found in *Water Quality: Management of a Natural Resource* (1996), by James Perry and Elizabeth Vanderklein. An encouraging case study of international efforts to control water pollution is chronicled in Karl-Geert Malle, "Cleaning Up the River Rhine," *Scientific American* 274 (January, 1996). A thorough exposition of the problems faced in the clean-up of Chesapeake Bay is presented in Tom Horton, "Chesapeake Bay: Hanging in the Balance," *National Geographic* 183 (June, 1993). A review of how water pollution policy is enforced is found in *Enforcing the Law: The Case of the Clean Water Acts* (1996), by Susan Hunter and Richard W. Waterman. For a view of water pollution policy from something other than the federal level, see Evan Ringquist, *Environmental Protection at the State Level: Politics and Progress in Controlling Pollution* (1993).

SEE ALSO: Clean Water Act and amendments; Water pollution; Water quality; Watersheds and watershed management.

Water quality

CATEGORY: Water and water pollution

The quality of water is defined by the solutes and gases dissolved in it, as well as the matter suspended in it. When human activity causes water to become unfit for the use for which it had previously been suited, the water is considered to be polluted or contaminated.

Only a tiny fraction of the earth's abundant water supply is available as fresh water for consumption. Once the water becomes contaminated or polluted, it must be restored before it can be utilized for its original intended use. The United States government has passed laws to help ensure that natural water resources are protected from contamination and that water quality meets certain standards before it is consumed. The Clean Water Act of 1977 directed each state to establish water quality standards for bodies of surface water. The Safe Drinking Water Act of 1974 mandated the United States Environmental Protection Agency (EPA) to establish drinking water standards for all public water systems serving twenty-five or more people or having fifteen or more connections.

When contaminants enter a water supply, the quality of the water is often compromised. The contaminants affect the water in such a way as to alter one or more water-quality parameters. Several parameters are used to characterize a given body of water, which can be broadly classified into either physical or chemical water-quality parameters. Physical parameters include turbidity, color, temperature, taste, odor, and amount of suspended solids. Chemical parameters include pH and hardness, as well as amount of dissolved solids, fluoride, metals, organics, nutrients, pathogens, and dissolved oxygen.

Water quality is perceived differently by different people. For example, a public health official is concerned with the viral and bacterial safety of water used for bathing and drinking, fishers are concerned that the quality of the water provide the best habitat for fish, and aquatic scientists are concerned about the habitats off all aquatic organisms. The state of the water and the nature of the concerned party will often determine which water-quality parameters must be measured. For instance, a raw wastewater entering a wastewater treatment plant does not need to be tested for dissolved oxygen. However, parameters such as amount of organics and metals are often important. On the other hand, dissolved oxygen is extremely important for the health of a river or lake.

In an attempt to devise a standard system for comparing river water quality in various parts of

Concentrations of Main Constituents in Some Natural Waters
In Milligrams per Liter

Constituent	Estimated Average World River Water	Average Seawater	Shallow Groundwater	Deep Groundwater, Midland, Michigan
Calcium	13.0	410	48.0	93,500
Magnesium	3.4	1,350	3.6	12,100
Sodium	5.2	10,500	1.0	28,100
Potassium	1.3	390	1.2	11,700
Bicarbonate	52.0	142	152.0	low
Sulfate	8.3	2,700	3.0	17
Chloride	5.8	19,000	8.0	255,000

the country, the National Sanitation Foundation designed the water quality index (WQI). Developed in 1970, it is one of the most widely used water quality indexes. It can be used to compare the water quality of different rivers, monitor water quality changes in a particular river section, and compare the water quality of different sections within a river.

To determine the WQI, nine tests are performed. These include measures of dissolved oxygen, biochemical oxygen demand (five-day), pH, temperature, total solids, turbidity, nitrates, total phosphorus, and fecal coliform. The results from the tests are then given a numerical value. The sum of the nine values yields the overall water quality index. The values range from 0-100: 0-25 is very bad, 26-50 is bad, 51-70 is medium, 71-90 is good, and 91-100 is excellent.

A host of different pollutants can enter a water body and affect its quality. The principal water pollutants include disease-causing agents (pathogens), oxygen-demanding wastes, inorganic chemicals, organic chemicals, and sediment or suspended matter. The disease-causing agents include bacteria, viruses, protozoa, and parasitic worms. These pathogens enter the water from domestic sewage and animal wastes. They can cause a variety of diseases, including cholera, dysentery, giardia, hepatitis, and typhoid fever.

Oxygen-demanding wastes are organic wastes capable of being decomposed by aerobic (oxygen-requiring) bacteria. In the process of decomposing this waste, the bacteria consume oxygen. If aquatic plants and contact with air do not replenish the oxygen at a rate that is equal to or greater than the rate at which it is depleted, then the oxygen level will drop. This can be measured by a decrease in the dissolved oxygen content. If the level drops low enough, it will affect all aquatic organisms that depend on oxygen. The quantity of oxygen-demanding wastes can be determined by measuring the biochemical oxygen demand (BOD). This is the amount of oxygen needed by aerobic bacteria to decompose the organic materials over a five-day period at 20 degrees Celsius (68 degrees Fahrenheit).

Inorganic chemicals include toxic metals, such as mercury and lead, and plant nutrients, such as nitrates and phosphates. When nutrients enter the water, they can cause extensive algal growth. When the algae die and decay, oxygen is depleted, and aquatic organisms are killed. This process is known as eutrophication. The source of these nutrients is often agricultural runoff of fertilizers. Organic chemicals include pesticides and petroleum products. These threaten human health as well as aquatic life. They are often resistant to microbial decomposition and can persist within the environment for long periods of time. Sediment and suspended matter are insoluble particles of soil and other solids that become suspended in water. By weight, this is by far the largest water pollutant and is derived mostly when soil is eroded from land. It clouds the water, disrupts food chains, and often contains harmful substances.

John P. DiVincenzo

SUGGESTED READINGS: An excellent overview of water supply, development, pollution, and restoration can be found in "Water: The Power, Promise, and Turmoil of North America's Fresh Water" *National Geographic* (November, 1993). *Field Manual for Water Quality Monitoring* (1994), by Mark K. Mitchell and William B. Stapp, has a good discussion on the WQI and the measurement of the water quality parameters that go into its calculation. An easy-to-follow booklet published by the U.S. Geological survey, *A Primer on Water Quality,* talks about criteria for water quality and water quality management.

SEE ALSO: Clean Water Act and amendments; Drinking water; Groundwater and groundwater pollution; Water pollution; Water treatment.

Water rights

CATEGORY: Water and water pollution

Water rights safeguard the use, quality, availability, and enjoyment of water found underground and in streams, rivers, lakes, and ponds. These rights are endowed by ownership of land that borders a watercourse or contains water or groundwater. Specific forms of water rights vary considerably across North America. They are monitored by tradition, court decisions, state and federal laws, and various governmental bodies.

During the colonial period of the United States, most Americans considered water an abundant resource that needed scant government regulation. As the population of the nation grew, however, this view changed. A growing number of disputes over access to fresh water prompted lawmakers and courts to establish various doctrines of water rights to protect water as a precious and limited resource.

English common law provided the legal foundation in the eastern states. One of its chief provisions recognized riparian, or riverside, rights. These rights entitled landowners whose land abutted natural, free-flowing streams and rivers to use water for ordinary purposes such as bathing and drinking. They did not have the right to pollute, stem, or divert waters from their natural paths in a way that interfered with the water rights of other downstream riparians. With the arrival of the Industrial Revolution, riparian rights came under attack as factory owners tried to divert water from rivers for industrial purposes. Though many courts struck down such attempts, others rendered legal decisions that allowed riparian owners a "reasonable use" of water for industrial purposes, providing it did not excessively harm the rights of other riparians.

Even more radical changes took place in the American West during the late nineteenth century, when growing numbers of miners, farmers, and ranchers competed for scarce water sources with little regard for the riparian rights of the East. As a result, disputes over water rights multiplied. One of the more famous feuds took place along the Poudre River in Colorado when one group of settlers diverted water for irrigation purposes and deprived others downstream of sufficient water. The state legislature took up the matter in 1876 and established the Colorado doctrine, also known as the prior appropriation doctrine, which held that all water rights belonged to the first user to claim them.

Other western states quickly adopted versions of Colorado's "first in time, first in right" approach. Eventually, however, state courts modified the doctrine by applying a "reasonable use" standard test to water disputes. Under this provision, senior water users could be forced to reduce their water consumption if a challenger proved the senior user took more than a reasonable share of water. Unlike in the East, western water rights hold that water is a thing separate from the soil rather than an ingredient of it. This principle permits western landowners to remove water from their land and sell it as a commodity. In 1982 the United States Supreme Court ruled in *Sporhase v. Nebraska* that an individual even had the right to export water from one state to another.

Though water rights exist in many—and often conflicting—forms across the United States, they are all subject to state and federal authority. They are also being called upon to address a growing number of water disputes arising from a water shortages. While most feuds over water rights still take place in the semiarid west, conflicts also occur in fast-growing, water-rich states such as Florida and Georgia.

Water shortages are not unique to the United States. A 1997 United Nation study reported that one-fifth of the world's population lacked clean drinking water. In many developing nations, people face daily water rations and have no running potable water. The growing water scarcity is a result of human population growth, increasing agricultural demands, industrialization, urbanization, and a continuous degradation of a finite reserve of fresh water. The United Nations predicts that if present trends continue, two-thirds of the world's population will face critical water shortages by the year 2025.

Sovereign nations have the legal means and the authority to resolve disputes over water rights within their own borders. Such conflicts are harder to settle when several countries share the same river. Serious international quarrels arose among riparian nations along the Ganges, Niger, Mekong and several other river basins during the late twentieth century. Squabbles over access to the Jordan River in the parched Middle East have, on some occasions, provoked talk of war. International law offers only limited help in solving these problems. For example, nations have the right to exercise authority over any resources under their immediate control. On the other hand, a United Nations convention of 1972 and the Helsinki Rules of 1966 oblige countries to share water rights with other riparians, as long as they show regard for the needs of local populations and traditional water consumption practices. Generally, riparian nations rely on international treaties and agreements to establish mutual water rights. Without effective enforcement powers, however, they have few options other than persuasion, sanctions, or even military force to ensure treaty compliance.

Proposed solutions to the growing water shortage include improved water conservation and pollution reduction efforts, a community approach among riparian nations to establish equitable water rights, and privatization of government controlled water-delivery systems. Water experts at the World Bank and the United Nations have called upon nations to abandon the concept of water as an abundant and cheap resource to be subsidized by governments. Instead, they suggest, water should be viewed as an "economic good," such as oil or gold, that is subject to free market mechanisms. By doing, say the specialists, water's proper price would match its value and assure its own protection. Critics, however, contend that water is too precious to be turned over to commerce. Humans, they point out, have a right to a fair share of a scarce resource that bestows life.

John M. Dunn

SUGGESTED READINGS: "Comprehensive Assessment of the Freshwater Resources of the World: Report of the Secretary General, Economic and Social Council," *The United Nations* (April 7-25, 1997), gives a well-written and detailed report on the growing worldwide water shortages. Klaus Lanz, *The Greenpeace Book of Water* (1995), provides stunning photographs and a compelling exposition of global water supplies and ways to protect them. The seven-volume *Waters and Water Rights: A Treatise on the Law of Waters and Allied Problems: Eastern, Western Federal* (1967), edited by Robert Emmet Clark, is an authoritative and comprehensive discourse on water rights in the United States.

SEE ALSO: Prior appropriation doctrine; Riparian rights.

Water-saving toilets

CATEGORY: Water and water pollution

Water-saving toilets are modified to reduce the amount of water used to move human waste products from buildings to sewer lines and septic tanks.

It is estimated that an average family of four uses 340,000 liters (90,000 gallons) of water per year. Modest but conscious changes in water use and to plumbing fixtures in and around the house can save thousands of gallons each year. Such savings would reduce water and energy costs in individual homes. They would also help protect water resources for future generations.

Toilets are made to suit a wide range of budgets, decors, and functions. Thus, choosing the right toilet involves informed decision making. A good toilet has many important characteristics, such as trap size, bowl-water surface, toilet footprint, and tank lining. Trap size indicates the ability to flush well without clogging. Larger traps are less likely to clog. The bowl-water surface indicates the potential for a toilet to remain clean between regular scrubbings; toilets with larger bowl-water surface areas generally require fewer cleanings. The toilet footprint is the space that the toilet pedestal takes up on the floor; it indicates the area that will not need to be covered with bathroom tiles. Lined tanks do not "sweat" during hot weather. Sweating can be a problem if the bathroom is not adequately air conditioned.

For those people who are concerned about the effects of their lifestyles on the environment, choosing a toilet also involves the consideration of how much water the toilet uses. The toilet in a home of four is flushed at an average rate of thirty times per day. Flushing is generally considered to be the single largest source of water loss in the home because it accounts for about 38 percent of the water used each day.

The amount of water used by a toilet is called the flush rate, which is measured in gallons per flush (gpf) in the United States and liters per flush (lpf) in most other parts of the world. Typical toilets rate at 3.5 gpf or higher. However, designs have been introduced that rate at 1.6 gpf. Some of the best available rate at 1.5 gpf. Toilets that perform at these levels are called low-consumption, low-flush, ultra low-flush, or water-saving toilets. They come in a variety of engineering designs. The four designs that are the most common are gravity-tank, flushometer, pressurized-tank, and vacuum-assist toilets.

It is estimated that one could cut water consumption in the home by 25 percent or more by replacing an old model toilet with a low-consumption toilet. This reduction is both automatic and permanent. In order to achieve these savings on a larger scale, some communities in Canada and the United States now require that all new or replacement installations of two-piece tank-type and floor-mounted flushing toilets rate at no more than 1.6 gpf. Water management officials in these communities also provide incentives to owners of existing homes to install low-flush toilets.

Josué Njock Libii

SEE ALSO: Water conservation; Water use.

Water treatment

CATEGORY: Water and water pollution

An estimated twenty-five thousand people die each day from such waterborne diseases as typhoid fever, cholera, and dysentery. Consequently, safe drinking water is essential for everyone's health and welfare. To be safe for drinking, water must be free of disease-producing bacteria, undesirable tastes and odors, color, turbidity, and harmful chemicals. Since raw water is seldom pristine, most public potable water systems purify their water prior to distribution.

Many substances may occur naturally in raw water that are either harmful or unpalatable to people. Human discharge of many substances into the environment also contaminates water supplies. Contaminants, either natural or anthropogenic in origin, can be divided into three groups: organoleptic substances that pertain to the senses of vision, taste, and odor; inorganic and organic chemical substances, which could be toxic or aesthetically undesirable, or could interfere with water treatment processes; and harmful microorganisms, which usually result from human and animal wastes.

The organoleptic parameters must be reduced to very low levels for drinking water to be acceptable for public use. Color, turbidity, and particulate matter represent visual problems.

A water treatment plant in El Segundo, California. (McCrea Adams)

Color results from organic matter that leaches from soil or decaying vegetation. Turbidity results from suspended clay or organic matter that imparts a muddy and therefore undesirable appearance to the water. Particulate matter floating in the water is not only aesthetically undesirable but may also provide food for certain organisms. Decomposed organic material and volatile chemicals result in unpleasant tastes and odors in water.

Iron, manganese, and aluminum are metals that are commonly found in water. Other metals such as lead, copper, cadmium, and silver are occasionally present, as are the nonmetals nitrate, fluoride, and phenols. These chemicals have both natural and anthropogenic origins. Chemically synthesized compounds such as pesticides, herbicides, and polychlorinated biphenyls (PCBs) are particularly dangerous as they can enter the food chain and accumulate in animal tissue.

Although most bacteria are harmless and indeed essential to life, some varieties (pathogens) can cause illness. These waterborne diseases include cholera, typhoid, and bacillary dysentery, which are common in areas without properly treated water. Viruses are pathogenic organisms that are much smaller and much harder to control than bacteria. Common viral diseases include poliomyelitis and infectious hepatitis. Cryptosporidium and giardia are protozoan waterborne parasites that are found in surface waters. They cause a severe form of gastroenteritis that can be deadly in people having immune-suppressed systems, as with acquired immunodeficiency syndrome (AIDS).

The origin and characteristics of the raw water source govern the type of treatment necessary to provide safe drinking water. For example, groundwater may require only pH adjustment and minor disinfection if the source is relatively pristine. However, in heavily fertilized agricultural areas and locations where soluble iron and manganese are naturally present, ion exchange for nitrate removal and chemical treatment for iron and manganese removal may be needed.

Surface water generally requires many more types of treatment, such as screening, sedimentation, chemical treatment, clarification, filtration, and disinfection.

Bar screens to block fish and debris is a standard first step in treating raw surface water. The screens must be strong enough to prevent wood, game fish, and even shopping carts from getting

Maximum Allowable Concentrations of Toxins in Drinking Water

Substance	Maximum Concentration (milligrams per liter)	Notes
Arsenic	0.05	Very toxic; causes dermal cancers
Cadmium	0.005	Moderately toxic
Chromium	0.1	Necessary for life; toxic at high concentrations
Copper	1.3	Necessary for life; toxic at high concentrations
Fluoride	4.0	Used to prevent dental caries
Iron	0.05	Necessary for life; toxic at high concentrations
Lead	0.015	Decreases blood hemoglobin
Manganese	0.05	Necessary for life; toxic at high concentrations
Mercury	0.002	Concentrates in food chain
Selenium	0.05	Very toxic
Zinc	5.0	Necessary for life; toxic at high concentrations
Nitrate	10.0	Particularly toxic to infants
Benzene	0.005	Possible carcinogen
Xylene	10.0	Causes liver and kidney damage
Tetrachloroethane	0.005	Carcinogen; causes liver and kidney damage
Dichloroethane	0.005	Carcinogen; causes liver and kidney damage
DDT	0.00042	Carcinogen; causes liver damage and concentrates in food chain
Alachlor	0.002	Carcinogen; causes liver and kidney damage
Atrazine	0.003	Causes reductions in hemoglobin and body weight
Chlordane	0.002	Carcinogen; causes liver damage

into the treatment plant and damaging the machinery. The next step is usually a sedimentation basin where the larger suspended particles can settle out by gravity. This process may be accelerated by mixing chemicals with the water to form a flocculate precipitate, which helps settle the suspended particles. The chemical coagulation process removes natural color originating from peat, animal and vegetable debris, plankton, and other organic substances. Even after sedimentation, some of the finer particles in the water may still be in suspension and have to be removed by filtration. Sand filters provide an inexpensive and effective medium for the removal of fine solids in either raw water or partially treated water. Granular activated carbon (GAC) filters have been replacing conventional sand filters in recent years since they can remove a wide variety of undesirable organic compounds such as herbicides, pesticides, and chemical compounds that form naturally. They are also useful in the treatment of taste and odor. Indeed, many beverage manufacturers use GAC filters where water is a major component of the product, such as soda, beer, and fruit juice. Residential point-of-use kitchen filters for drinking water incorporate GAC filters as the major treatment technique.

The final treatment process is disinfection, since pathogenic bacteria can pass through both the sedimentation basin and filtration. Drinking water standards require the absence of the indicator organisms fecal *streptococci* and the coliform group of bacteria, specifically fecal *Escherichia coli*, in the distributed water. Disinfection, which is the killing of harmful bacteria, is usually accomplished by chlorination. Chlorine is a very effective biocide. However, one major disadvantage of chlorine is that it is very reactive and can produce compounds such as trihalomethanes that are potentially carcinogenic. Other compounds produced by chlorine have taste and odor problems. Ozone and ultraviolet light are also powerful disinfectants but do not have the residual properties of chlorine, which protects water from contamination as it travels through the distribution system.

Some of the water treatment plants built in the late twentieth century use a combination of ozonation for its effectiveness against cryptosporidium, giardia, and viruses; GAC filters for taste and odor control; and small amounts of chlorine as a residual biocide for the treated water in the distribution system.

Robert M. Hordon

SUGGESTED READINGS: Recent developments in sand filtration technology are covered in *Advances in Slow Sand and Alternative Biological Filtration* (1996), edited by Nigel Graham and Robin Collins. *A Practical Guide to Particle Counting for Drinking Water Treatment* (1999), by Michael Broadwell, is a reference guide to filtration performance. The American Water Works Association Research Foundation and Compagnie Generale des Eaux outline the design and operation of the ozone process within potable water plants in *Ozone in Water Treatment* (1991). Edwin E. Geldreich discusses the processes and issues related to water quality in transmission pipes and storage reservoirs in *Microbial Quality of Water Supply in Distribution Systems* (1996).

SEE ALSO: Chlorination; Clean Water Act and amendments; Desalination; Drinking water; Fluoridation; Water quality.

Water use

CATEGORY: Water and water pollution

Although water is the most abundant liquid on the earth and can be found almost everywhere, 99.35 percent of it is either too salty for human use or frozen. Available freshwater resources, like mineral resources, are unevenly distributed. Thus, water must be transported long distances to supply the needs of major metropolitan areas.

Raw water can come from either surface or ground sources. Surface sources include river systems, such as the Delaware River, which supplies Philadelphia, or stored water in reservoirs, such as the Quabbin Reservoir in Massachusetts, which is used for the Boston metropolitan area. Lakes also serve as natural reservoirs of surface water, as exemplified by Chicago and Milwaukee

U.S. Water Use Per Day

				Public Supply				
Year	Total (billions of gallons)	Per Capita (gallons)	Irrigation (billions of gallons)	Total (billions of gallons)	Per Capita (gallons)	Rural (billions of gallons)	Industrial and Misc. (billions of gallons)	Steam Electric Utilities (billions of gallons)
1940	**140**	1,027	71	10	75	3.1	29	23
1950	**180**	1,185	89	14	145	3.6	37	40
1960	**270**	1,500	110	21	151	3.6	38	100
1970	**370**	1,815	130	27	166	4.5	47	170
1980	**440**	1,953	150	34	183	5.6	45	210
1990	**408**	1,620	137	41	195	7.9	30	195

Source: U.S. Department of Commerce, *Statistical Abstract of the United States, 1996*, 1996.

Note: Data are for water "withdrawals"—water physically removed from a source. They do not include "consumptive" use of fresh water—water evaporated, transpired, or incorporated into products, plant, or animal tissue. Consumptive use in 1990, for example, totaled an additional 94 billion gallons.

on Lake Michigan. Groundwater sources range from a few wells serving a small community to a system of many wells serving a larger area, such as the Suffolk County Water Authority in Long Island, New York. The groundwater source varies from unconsolidated materials, such as the sandy deposits along the Atlantic and Gulf Coastal Plain and the stratified sands and gravels of glaciated areas, to consolidated rocks, such as sandstone and shale, where water is obtained from the fractures within the formation.

Water for public supply needs refers to water that is delivered to multiple users for domestic, commercial, industrial, and thermoelectric power purposes. The purveyor can be either public, such as New York City, or private, such as the Elizabethtown Water Company in New Jersey, which is investor owned. In the United States, public or private water systems are further defined by a minimum threshold of serving at least twenty-five people or having fifteen hookups. Public systems, which deliver potable water to a variety of users, must comply with federal and state safe drinking water standards.

Domestic water use is defined as water that is used for normal household purposes, such as drinking, food preparation, bathing, washing dishes and clothes, toilet flushing, lawn and garden watering, and home car washing. Households that obtain their water from on-site wells are not part of the public potable water system infrastructure. This self-supplied category is substantial. For example, the U.S. Geological Survey estimates that 42.8 million people, or 17 percent of the total population, were served by their own wells in 1990. Self-supplied domestic water systems are rarely metered, and minimal data exist as to the amount withdrawn.

Commercial water use includes water for motels, hotels, office buildings, commercial facilities such as shopping centers and fitness centers, institutions such as schools and prisons, and military bases. Industrial water use includes water that is necessary for processing, washing, and cooling in factories that make a variety of products. Industries that are major users of water include steel, chemical, paper, and petroleum refining. The thermoelectric category includes water used for electric power generation with fossil fuel, nuclear, or geothermal energy. Most of the water used by thermoelectric plants goes for condenser and reactor cooling. Only a small fraction of the water used in this category comes from public water systems.

The public use subcategory includes water used for fire fighting, street washing, municipal parks, and public swimming pools. Water lost in the collection and distribution system by leakage forms another subcategory. The remaining categories of water use include irrigation, livestock, and mining. The percentage distribution of diversions by public water systems by user category for the United States in 1990 was as follows: 57 percent domestic, 15 percent commercial, 14 percent public use and losses, 13 percent industrial, and 0.2 percent thermoelectric power.

Per capita consumption in gallons per day (gpcd) or liters per day (lpcd) is obtained by dividing the total amount of water diversions by the population served by public systems. The total diversions by public systems include the deliveries to domestic, commercial, industrial, and thermoelectric power users. Public use and losses are also included as part of the deliveries by public systems. Based on these assumptions, per capita consumption varied from a low of 413 lpcd (109 gpcd) for Rhode Island to a high of 1,302 lpcd (344 gpcd) for Nevada in 1990. The national average is 697 lpcd (184 gpcd). Generally, the per capita values are higher in the western states, presumably reflecting the lower precipitation and higher evapotranspiration.

Water use in most countries is a function of population served. Consequently, as the population increases, water consumption will also increase, which means that water purveyors continually need to expand their water-supply sources. The need for additional supplies of water has resulted in innumerable disputes over the years. In arid areas such as the Middle East, water is crucial for general use and irrigation. Thus, the decision by Turkey in the late 1990's to build large reservoirs in the headwaters of the Tigris and Euphrates Rivers may impact the downstream states of Syria and Iraq and deprive them of a portion of the flow, upon which they have come to depend. The allocation of the Jordan River among Israel, Jordan, Syria, and Lebanon in another politically sensitive arid area is intimately related to the possibility of sustained peace in the region. Egypt is totally dependent on the Nile River, which originates in Ethiopia and Lakes Albert and Victoria in central Africa. Any major diversion of the Nile water by the upstream states would have a severe impact on Egypt. Unless serious efforts are taken at local, state, national, and international levels to balance water use and availability, the prospects for increasing water-use conflicts will grow.

Robert M. Hordon

SUGGESTED READINGS: An invaluable collection of water-use data for the United States is the U.S. Geological Survey Circular 1081, entitled *Estimated Use of Water in the United States in 1990* (1993), by Wayne B. Solley, Robert B. Pierce, and Howard A. Perlman. An excellent review of water-use issues is contained in *Perspectives on Water: Uses and Abuses* (1988), edited by Donald H. Spiedel, Lon C. Ruedisili, and Allen F. Agnew. Water use and supply are concisely covered in *Water Resources Planning* (1996), by Andrew A. Dzurik. Global water quality trends that negatively impact freshwater supplies are discussed in *Imperiled Waters, Impoverished Future: The Decline of Freshwater Ecosystems* (1996), by Janet A. Abramovitz.

SEE ALSO: Drinking water; Los Angeles Aqueduct; Water conservation; Water quality; Water rights; Water-saving toilets; Watersheds and watershed management; Wells.

Watersheds and watershed management

CATEGORY: Water and water pollution

A watershed is an area bounded by drainage divides within which precipitation drains to a particular watercourse or body of water. Human activity can cause unanticipated changes in watersheds, affecting the hydrologic balance. Land use alters the balance between storage and dispersal of precipitation, in many cases increasing erosion, stream sedimentation, and flooding hazard.

Watersheds are defined at many scales: The Mississippi River watershed contains the Ohio River watershed, which in turn contains smaller watersheds. A fundamental part of the hydrologic cycle, the watershed collects and stores precipitation in soils, lakes, wetlands, or aquifers, and disperses water by evaporation, plant transpiration, surface runoff, springs, and baseflow to streams. Watersheds of different geographic regions have distinctive characteristics based on climate, topography, and soil type; therefore, the natural variability among watersheds is predictably large. In arid regions, precipitation occurs as intense, infrequent storms, with most of the water rapidly running off and eroding soil with little protective vegetation. Watersheds in humid areas are characterized by frequent, usually gentle rain that replenishes aquifers and sustains streams, springs, and wetlands.

Ecologically, the watershed provides habitat and nutrients for plants and animals, including humans. Land use can disrupt a watershed's ecology by disturbing habitat and nutrient cycling through soil loss and removal of native vegetation. The role of the watershed in environmental problems such as flooding, erosion, sedimentation, and ecological disruption has led to increased emphasis on the watershed as the basic unit for environmental management, rather than political units such as states or counties.

The 1954 Watershed Protection and Flood Prevention Act authorized the secretary of the U.S. Department of Agriculture (USDA) to manage watersheds in cooperation with states and local organizations, such as soil and water conservation districts. The driving idea behind the act is that floods are better controlled through management of runoff upstream in the watershed rather than through downstream engineering projects. The act required local interests to contribute up to 50 percent of the costs to ensure local support for watershed projects. In contrast, Army Corps of Engineers flood control projects originally were funded entirely by the federal government. The Watershed Protection and Flood Prevention Act is generally administered through the USDA Natural Resources Conservation Service, formerly the Soil Conservation Service.

Recognizing the need for basin-wide planning, the federal government created the Water Resources Council via the 1965 Water Resources Planning Act. This council created river basin planning commissions but fell into disfavor and lost funding because the river basins were too large for effective planning.

Characteristics of Selected Major Drainage Basins

River	Outflow	Length	Area	Average Annual Suspended Load
Amazon	180.0	6,300	5,800	360
Congo	39.0	4,700	3,700	—
Yangtze	22.0	5,800	1,900	500
Mississippi	18.0	6,000	3,300	296
Irawaddy	14.0	2,300	430	300
Brahmaputra	12.0	2,900	670	730
Ganges	12.0	2,500	960	1,450
Mekong	11.0	4,200	800	170
Nile	2.8	6,700	3,000	110
Colorado	0.2	2,300	640	140
Ching	0.06	320	57	410

Note: Rivers are ordered by outflow; outflow is multiplied by 1,000 cumecs (cubic meters of water per second); length is measured in kilometers; area is measured in square kilometers multiplied by 1,000; average annual suspended load is measured in millions of metric tons.

A major step in watershed management was taken with the 1972 Clean Water Act. With this act, land management began to include water quality control. Nonpoint sources of pollution were targeted, among them agriculture, forestry, mining, and waste disposal. Most states passed laws directing use of best management practices (BMPs) to protect or rehabilitate watershed functions. BMPs are widely accepted methods of preventing soil and water problems. The Clean Water Act provided for regulation of land use, initially through incentives. The 1985 Food Security Act provided incentives to control erosion on highly erodible croplands. The "swampbuster" provisions directed protection of existing wetlands and also provided incentives for wetland restoration. The 1986 amendments to the 1974 Safe Drinking Water Act encouraged public suppliers of drinking water to protect wellheads. The 1987 amendments to the Clean Water Act encouraged states to address non-point-source pollution.

State regulations on non-point-source pollution range from voluntary compliance with BMPs to strict enforcement of BMPs with fines for noncompliance. In general, however, they are becoming more detailed and comprehensive. The concept of the total maximum daily load (TMDL) permissible for non-point-source pollutants has been introduced, but determining appropriate loads is costly and difficult because it varies with land use and with watershed.

Historically, there has been a broadening of perspective on land and water management. Although management initially focused on a single farm or field, the watershed view is now widely accepted. The perspective is broadening with the concern about the greenhouse effect, in which atmosphere-biosphere-hydrosphere-terrasphere interactions are critical. The term "ecosystem management" may better reflect watershed management focus in the future.

Watersheds are managed for a spectrum of land uses, including water supply, settlement, grazing, crop production, forestry, and recreation. Management focuses on water, sediment, and wastes. Water management generally seeks to reduce runoff; exceptions are landfills and mine spoils in which infiltration is minimized. Sediment management seeks to prevent soil erosion or trap eroded sediment. Waste management seeks to better distribute the waste load and prevent it from reaching water. The appropriate strategy to achieve these management goals varies from problem to problem. For example, in forestry the strategy might include revegetating logged areas, diverting water from logging roads, and closing logging roads after use. The overall approach to management is to identify the problem and its source, and to select and implement BMPs. While the law requires that BMPs be considered, there is no definitive catalog of them. Many state agencies have written and assembled their own loose-leaf binders of BMPs for various land uses, which are available to the public. Public education and public participation in decision making have played an increasingly important role in sustainable watershed management.

Mary W. Stoertz

Suggested readings: A good general text emphasizing the movement and storage of water on and in natural watersheds is provided in Peter E. Black's *Watershed Hydrology* (1996). A critical look at U.S. policy and programs on watershed protection is found in *Entering the Watershed: A New Approach to Save America's River Ecosystems* (1993), by Bob Doppelt, Mary Scurlock, Chris Frissell, and James Karr. *Wildland Watershed Management* (1992), by Donald Satterlund and Paul Adams, presents fundamental theory and basic practices for those who make watershed management decisions. An insightful book with an international perspective is *Land, Water and Development: River Basins and their Sustainable Management* (1992), by Malcolm Newson.

See also: Erosion and erosion control; Flood control; Runoff: agricultural; Runoff: urban; Sedimentation; Wetlands.

Watson, Paul

Born: c. 1953; Canada
Category: Animals and endangered species

Watson founded the Sea Shepherd Conservation Society, one of the world's most aggressive environmental organizations.

A onetime officer in the Canadian Coast Guard, Paul Watson helped to found Greenpeace in 1971. By the mid-1970's, however, he found himself increasingly at odds with the group's philosophy. Initially a small band of dedicated activists who placed themselves at personal risk to interfere with such antienvironmental occurrences as nuclear tests, Greenpeace had grown into an international concern that directed much of its efforts toward fund-raising. Watson believed that the group's evolution was dissipating its effectiveness, and he was also chafing at the organization's commitment to passive protest in pursuit of its goals.

Although Watson himself believed that doing harm to living things was wrong, he believed that the use of nonharmful force against property could be justified to protect life. Moreover, he became increasingly convinced that passive protest was not working. In March, 1977, while protesting the slaughter of baby harp seals in Canada's Gulf of St. Lawrence, Watson seized a club from a seal hunter and threw it into the water. Such direct action conflicted with Greenpeace's expressed policy, and he was expelled from the group. When Greenpeace's board of directors charged Watson with vigilantism, he replied that, in the absence of an environmental police force to oppose environmental crimes, environmental vigilantes were bound to appear.

Paul Watson, the Canadian antiwhaling activist who founded the Sea Shepherd Conservation Society. (AP/Wide World Photos)

In the summer of 1977, Watson and several friends established an organization they at first called Earthforce. Headquartered in Vancouver, the group was dedicated to the use of direct action to protect the world's animals. Although Watson and his band of activists would achieve notoriety for defending marine life, their first mission was to travel to East Africa to document the killing of elephants for ivory. Earthforce presented its findings, including films of the illegal slaughter of elephants, to the U.S. government to help lobby for a ban on the importation of African ivory; for Watson, though, the preservation of marine mammals was a recurring goal. With the aid of a grant from the writer and activist Cleveland Amory, Watson purchased a retired fishing boat, which he christened the Sea Shepherd, and hired a crew.

On the Sea Shepherd's first voyage, Watson and the crew took the ship through four hundred miles of ice to the Gulf of St. Lawrence. There, they sprayed hundreds of baby seals with a red dye that made the animals' white pelts valueless to hunters. Soon, however, they were stopped and arrested by Canadian police. Undeterred, Watson and his allies would return to the gulf in 1982 and 1983, after which the gulf hunt was discontinued.

Publicity stemming from the group's sometimes violent encounters with sealers and from their brushes with the law had led to steady membership increases, and the scope and ambition of the society's activity expanded as well. In the spring and summer of early 1979, Watson and other Sea Shepherd members organized a virtual espionage ring that tracked the movements of the *Sierra,* a notorious "pirate" whaling ship that had killed an estimated four hundred whales a year since the 1960's. In July, 1979, the *Sea Shepherd* located the *Sierra* off the North African coast and followed the whaling ship to the harbor in Leixoes, Portugal. Outside the port, the larger and faster *Sea Shepherd,* crewed by only Watson and two assistants, twice rammed the *Sierra,* rupturing the whaler's side but causing no injuries to the crew. The Portuguese navy took the *Sea Shepherd* into custody, but the *Sierra* had incurred enormous damage and had to be towed to Lisbon for repairs. After Portuguese authorities threatened to turn over the *Sea Shepherd* to the *Sierra*'s owners as compensation, Watson and two friends sneaked into Lisbon harbor and scuttled the ship by opening its pipes.

Watson and the Sea Shepherd Conservation Society would repeat these and similar scenes many times in ensuing years, actively confronting whalers, sealers, and drift-net fishermen and often ending up in court. By the 1990's, the organization had grown to include thousands members and was operating a fleet of antiwhaling vessels. Opponents of Watson's methods would continue to portray him and his followers as fanatics and to accuse them of terrorist tactics; Watson, however, would reiterate with pride that the society's actions had never harmed a human being. In 1982, Watson published an account of the early history of the Sea Shepherd Conservation Society, *Sea Shepherd: My Fight for Whales and Seals* (1982), and in 1994 released an autobiography, *Ocean Warrior: My Battle to End the Illegal Slaughter of Marine Life on the High Seas.*

Robert McClenaghan

SEE ALSO: Endangered species; Greenpeace; Sea Shepherd Conservation society; Whaling.

Watt, James

BORN: January 31, 1938; Lusk, Wyoming
CATEGORY: Preservation and wilderness issues

Lawyer and bureaucrat James Watt was labeled a major antienvironmentalist during President Ronald Reagan's first term in office. From the time of his appointment as secretary of the interior in 1981 to his resignation in 1983, Watt used his office to weaken environmental policies that fell under his domain of authority.

James Watt's ancestors were nineteenth century homesteaders in Wyoming, laying claim to a large tract of land for ranching. Watt was both a rancher and a successful lawyer. As a child, he became familiar with the harsh, barren land of the family ranch, pumping water for the cattle, repairing fences, and performing other difficult chores. He recalled later in life that his early experiences trained him to challenge a hostile environment.

In 1962 Watt obtained a degree in law from the University of Wyoming; shortly thereafter, he moved to Washington, D.C., to assume a position as legislative assistant and counsel to Senator Milward Simpson of Wyoming. From 1966 to 1969, Watt worked with the U.S. Chamber of Commerce's Washington, D.C., office in natural resource and environmental pollution policy. He lobbied for prodevelopment business interests in such areas as public lands, mining, energy, and water resource development. During the administrations of Presidents Richard M. Nixon and Gerald Ford, Watt served in the Department of Interior as an assistant secretary responsible for water and energy resources, and as director of the Bureau of Outdoor Recreation.

During the 1970's Watt became closely associated with the Sagebrush Rebellion, a movement of western ranchers and entrepreneurs who opposed numerous federal regulations, which, in their view, inhibited profitable exploitation of natural resources. In 1977 Watt assumed the presidency of the Mountain States Legal Foundation, founded by brewing magnate Joseph Coors, to provide assistance for individuals who challenged government restrictions on strip mining, oil and gas exploration, mineral extraction, and grazing lands.

Watt was a logical choice to become secretary of the interior in the Reagan administration, which was committed to economic expansion and resource development with minimal government intrusion. From the beginning of his tenure, Watt worked to cut the department's budget and eliminate agencies and programs. He promoted measures to ease restrictions on oil and gas exploration on federal lands and in offshore waters, open more federal land for grazing and timber cutting, facilitate the construction of dams and reservoirs to improve irrigation, and restrict expansion of the national park system. Watt's actions were consistent with his ideology of economic growth with minimal government interference, and he felt called by his religious convictions to "follow the Scriptures which call upon us to occupy the land until Jesus returns."

Environmental groups strongly opposed Watt's appointment as secretary of the interior, arguing that the mission of the department—to manage federal lands in the public interest—ran contrary to his ideology. During his tenure, Watt refused to meet with environmental group leaders, suggesting that they were subversive and were weakening the United States. Watt finally left office in 1983 after making derogatory remarks about Senate members who had been appointed to a coal advisory panel. At the time of Watt's resignation, the Senate was working on a resolution calling for his dismissal.

Ruth Bamberger

SEE ALSO: Antienvironmentalism; Grazing; Sagebrush Rebellion; Wise-use movement.

Weather modification

CATEGORY: Weather and climate

Human activities can cause intentional or accidental changes in local weather situations. Many intentional weather modification experiments have focused on creating conditions favorable for agriculture.

Inadvertent weather modification, including fog formation and increases or decreases in precipitation downwind from large industrial sites, creates problems in some locales. Scientific attempts to modify weather deliberately have been pursued since World War II. The most popular techniques involve cloud seeding, the injection of cloud-nucleating particles into likely clouds to alter the physics and chemistry of condensation. Proponents of this technique claim that it may enhance precipitation amounts by 5 to 20 percent. However, some scientists believe that deliberate efforts to enhance precipitation often yield questionable results, even in favorable situations. In 1977 the United Nations passed a resolution prohibiting the use of weather modification for hostile purposes because of the threat to civilians. The United States signed the resolution but has continued defense research on operational weather modification in battlefield situations, as summarized in the U.S. Air Force position paper "Weather as a Force Multiplier: Owning the Weather in 2025."

Studies have field-tested various methods of weather modification; results have varied widely. Weather modification has been attempted in many countries around the world, by government agencies, agricultural cooperatives, private companies, and research consortiums. In agricultural areas farmers are convinced that hail suppression and precipitation augmentation have been achieved by weather modification. In some of these same locales, meteorologists have been unable to determine if weather modification has produced any change from what would have occurred without intervention. Attempts to duplicate weather modification efforts that have apparently been successful in one locale have often been met with questionable results. Meteorologists occasionally disagree among themselves as to whether a specific attempt at weather modification has succeeded. Reexamination of data from American studies undertaken in the past has led many scientists to conclude that the efficacy of cloud seeding has been overstated.

It should be understood that it is impossible to change the climate of an entire region for a desired outcome through weather modification. It is also impossible to end a drought by seeding clouds. Cloud seeding for agricultural purposes assumes that some enhancement of regional rainfall amounts over the course of the growing season will increase crop yields. Weather modification for hail suppression assumes that reduction in regional crop losses over the growing season is an attainable goal.

Selected Scientific Attempts to Modify Weather

Location	Weather Situation	Mode of Deployment	Results
New Mexico	no special selection of conditions	ground-based silver iodide generator	possible rain enhancement
Atlantic Ocean near Georgia	hurricane	airborne seeding	inconclusive
Colorado Rockies	no special selection of conditions	ground-based silver iodide generators	doubtful enhancement
Atlantic Ocean (Project Stormfury)	hurricanes	airborne seeding	inconclusive
Vietnam	military	airborne seeding	rain enhancement
Switzerland	thunderstorm (hail suppression)	rocket seeding	inconclusive
Bulgaria	thunderstorm (hail suppression)	rocket seeding using lead iodide	claimed 50 to 60 percent reduction in crop losses
Soviet Union	thunderstorm (hail suppression)	rocket seeding	claimed 50 to 95 percent reduction in crop losses

Inadvertent Weather Modification

Pulp and paper mills produce huge quantities of large-and giant-diameter cloud condensation nuclei (CCN); downwind of these mills, precipitation appears to be enhanced about 30 percent above what was observed prior to construction of the mills. It is also thought that the heat and moisture emitted by these mills may play an active role in precipitation enhancement. One specific study of a kraft paper mill near Nelspruit in the eastern Transvaal region of South Africa has indicated that storms modified by the mill emissions lasted longer, grew taller, and rained harder than other nearby storms occurring on the same day. Radar measurements supported the theory that hygroscopic particulates released by this mill accelerated or amplified growth of unusually large-diameter rain drops.

An egregious example of inadvertent weather modification is the formation of ice fog over Arctic cities in Siberia, Alaska, and Canada. During winter, cities such as Irkutsk, Russia, and Fairbanks, Alaska, experience drastic reductions in visibility as particles released by combustion act as nuclei for the formation of minute ice crystals. No techniques are available to modify ice fogs.

During an investigation of the meteorological effects of urban St. Louis, Missouri, conducted during the 1970's, it was found that urban summer precipitation was enhanced by 25 percent relative to the surrounding area. Most of the increased precipitation occurred in the late afternoon and evening as a result of convective activity. The frequency of summer thunderstorms was enhanced by 45 percent, and the frequency of summer hailstorms was higher by 31 percent over the city and adjacent eastern and northeastern suburbs. During the late 1960's, studies demonstrated that widespread burning of sugar cane fields in tropical areas released large numbers of cloud condensation nuclei. Downwind, rainfall decreases of about 25 percent were noted.

Cloud Seeding

For millennia, people attempted to influence the weather by using prayers and incantations. Sometimes rain followed, and sometimes no rain fell for extended periods. Scientists began attempting various techniques to modify weather during World War II. In 1946 Vincent Schaefer of the General Electric Research Laboratory observed that dry ice put into a freezer with supercooled water droplets caused ice crystals to form. On November 13, 1946, Schaefer demonstrated that dry ice pellets dropped from an aircraft into stratus clouds caused liquid water droplets to change to ice crystals and fall as snow. Bernard Vonnegut, a coworker, determined that silver iodide (AgI) particles also caused ice crystals to form. Project Cirrus involved apparently successful scientific attempts to seed clouds with ground-based AgI generators in New Mexico. These researchers then tried seeding a hurricane on October 10, 1947. The hurricane changed direction, making landfall in Georgia, resulting in a number of lawsuits against General Electric.

Early cloud-seeding experiments were empirical. AgI was dropped from aircraft, shot into clouds by rockets, or dispersed from ground-based generators. Researchers could selectively seed a pattern such as an "L" into a supercooled stratus cloud and see a visible "L" appear, thus "proving" that they could achieve results. When any rain occurred near a seeded area, it was attributed to the intervention. The apparent success of cloud seeding using AgI caused the technique to be modified and adopted in France, Canada, Argentina, Israel, and the Soviet Union. Wine-growing regions such as the south of France and Mendoza, Argentina, installed ground-based AgI generators. The Soviet Union opted for rocket-borne AgI, which was launched in agricultural areas during thunderstorms in an effort to suppress hail.

In 1962 the U.S. Navy and Weather Bureau began an ambitious cooperative plan to modify hurricanes called Project Stormfury. Only a few hurricanes were seeded in attempts to reduce the intensity of the storms. Proponents of Stormfury suggested that seeding of Hurricane Debbie in 1969 caused a reduction of 30 percent in wind speed on one day. The following day, no seeding was done, followed by another seeding attempt. The second seeding was thought to have caused a 15 percent reduction in wind speeds. Proponents believed that 10 to 15 percent reductions in wind speeds might result in a 20 to 60 percent reduction in storm damage if similar results could be achieved by seeding other hurricanes. Stormfury was terminated in the late 1970's, with no definitive results.

During winters between 1960 and 1970, the Climax I and Climax II randomized cloud seeding studies were conducted in the Colorado Rockies. Although it was initially thought that precipitation enhancements on the order of 10 percent may have resulted, more recent examination of the results appears to indicate that cloud seeding had no statistically discernible effect on precipitation. During the Vietnam War, the U.S. military attempted to increase precipitation along the Ho Chi Minh Trail in an effort to impede enemy forces. In the United States during the 1970's, some entrepreneurs deployed ground-based AgI generators in selected agricultural regions, billing farmers for their services. Aircraft delivery of AgI became increasingly popular. By the late 1990's, a number of private companies were delivering airborne cloud seeding services in various areas worldwide.

Cloud physicists have explored why cloud seeding might be effective. The evidence suggests that seeding increases the size of droplets or ice crystals, allowing them to fall as precipitation. Two concepts have emerged: a static mode theory, which assumes that natural clouds were deficient in ice nuclei, and a dynamic mode theory, which assumes that enhancement of vertical movement in clouds increases precipitation. The static mode assumes that a "window of opportunity" exists for seeding cold continental clouds during which clouds must be within a particular temperature range and contain a certain amount of supercooled water.

FOG DISSIPATION AND HAIL SUPPRESSION

During World War II, when improvements in visibility were crucial for military operations, efforts were made to dissipate fog. Fog can be dissipated by reducing the number of droplets, decreasing the radius of droplets, or both. Decreasing droplet radius by a factor of three through evaporation can provide a ninefold increase in visibility. Possible methods of fog removal include using dry ice pellets or hygroscopic materials, heating the air, and mixing the foggy air with drier air. Airports that are plagued by supercooled fog in winter, such as Denver and Salt Lake City, can dissipate the fog by dropping dry ice pellets. Dry ice causes some liquid water droplets to freeze and grow, evaporating the remaining liquid droplets and allowing the larger frozen ice crystals to fall. One way of clearing fog at military airports when there is a shallow radiation fog close to the ground is to use helicopters to provide mixing. Jet engines can be used as heaters, an expensive technique that has been used operationally in France.

Farmers and vintners worldwide fear damaging hailstorms that can devastate crops. There are three approaches to suppressing hail damage: converting all liquid water droplets to snow to prevent hail formation, seeding to promote growth of many small hailstones instead of larger damaging hail, and introducing large condensation nuclei to reduce the average hailstone size. Most weather modification proponents believe that seeding with lead iodide or AgI to cause many small hailstones to form can substantially reduce hailstone size. Because small hailstones are less damaging than large ones, this technique could potentially lessen (but not eliminate) crop losses. It has been claimed that rocket-borne lead iodide seeding in Bulgaria reduced crop losses from hail by 50 to 60 percent. Similar seeding operations in the former Soviet Union were said to have reduced crop damage by 50 to 95 percent. A randomized study in North Dakota over four summers claimed that seeding helped reduce hail severity.

Anita Baker-Blocker

SUGGESTED READINGS: *Human Impacts on Weather and Climate* (1995), by William R. Cotton and Roger A. Pielke, gives a comprehensive overview of weather modification. *Weather and Climate Modification* (1975), edited by Wilmot N. Hess, provides a look at the "glory days" when scientists and governments enthusiastically embraced weather modification. Graeme K. Mather, "Coalescence Enhancement in Large Multicell Storms Caused by the Emissions from a Kraft Paper Mill," *Journal of Applied Meteorology* 91 (1991), is a detailed study of the effects of Kraft paper mills on precipitation. Difficulties in assessing the effects of cloud seeding are exemplified by A. L. Rangno and P. V. Hobbs, "Further Analyses of the Climax Cloud Seeding Experiments," *Journal of Applied Meteorology* 93 (1993). Richard G. Semonin and Stanley A. Changnon, "METROMEX: Lessons for Precipitation Enhancement in the Midwest," *Journal of Weather Modification* 7 (1975), presents information gleaned from a study of urban effects on local weather. Military weather modification information is available in Tanzy J. House, James B. Near, William B. Shields, Ronald J. Celentano, Ann E. Mercer, and James E. Pugh, "Weather as a Force Multiplier: Owning the Weather in 2025," published by the U.S. Air Force in 1996.

SEE ALSO: Cloud seeding.

Wells

CATEGORY: Water and water pollution

A typical well is a hole bored into the ground to extract or inject fluid.

The earliest wells were excavated by hand. Wells are now driven by various mechanical methods depending on the depth and type of rocks through which it is bored. The borehole is cased with plastic pipe in shallow wells and steel pipe in deep wells to prevent caving of the walls. The casing is perforated at the depth from which production occurs. This section, called a screen, allows an exchange of fluid between the casing and surrounding rock. The space between the screen and the surrounding rock is filled with gravel to allow fluid to flow freely between the well and the aquifer. The space between the casing and the exposed rock in the upper part of the well is tightly sealed with impermeable grout to prevent contaminants from entering the bore from the surface.

Wells were originally used only to extract underground water. Today they serve many purposes, including extraction, injection, and monitoring of fluid below the surface. Water wells and oil wells are examples of producing wells that extract fluids from the subsurface. A typical water well draws water from either the surface aquifer or from deeper, confined aquifers. Oil or gas wells produce from deep rock strata. Wells are also used to remove contaminated groundwater and dewater saturated zones in which construction or other activity extends below the water table.

Injection wells are used to introduce fluids into the subsurface. They are used as a way to store water that could be lost by high evaporation rates or runoff, and oil is occasionally stored by pumping it into subsurface, impermeable salt caverns. Water or gas is pumped into oil-bearing strata to displace the oil and increase production. Hazardous waste is injected into deep levels of the crust as a means of disposal. Monitoring wells are used to determine variations in depth to the water table or monitor the migration of hazardous fluids. Monitoring wells are required around sites of potential groundwater contamination.

Groundwater is a major source of water for domestic, agricultural, and industrial use. It exists in the subsurface, filling pores and cracks in consolidated rocks and loose, unconsolidated sand, gravel, clays, or mixtures of these materials. The surface aquifer is the saturated zone that receives water by percolation down from the surface. This is the zone most susceptible to contamination by toxic substances from industrial and municipal waste, feed lots, septic tanks, crop fertilizers, pesticides, and herbicides. Confined aquifers are less susceptible to contamination because they are sealed from surface percolation by overlying impermeable beds. Confined aquifers can be contaminated when they are exposed to direct recharge or by boreholes that reach them from the surface. Disposal wells with corroded casings may serve as conduits for hazardous waste into subsurface water supplies. Improperly grouted wells may allow surface contamination to infiltrate water supplies by seepage along the outside of the casing.

René A. De Hon

SEE ALSO: Aquifer and aquifer restoration; Groundwater and groundwater pollution.

Welsh mining disaster

DATE: October 21, 1966
CATEGORY: Waste and waste management

On the morning of October 21, 1966, a sea of saturated coal sludge slipped from Merthyr Mountain in Aberfan, Wales. As a result, several homes were damaged and many people were killed. Of the 144 deaths, 116 were children buried in the waste.

In 1870 workers at Merthyr Vale, a coal mine near the village of Aberfan, Wales, began dumping the remains of processed coal onto nearby Merthyr Mountain, forming a coal tip. When Aberfan was struck with heavy rainfall in October of 1966, the black sludge loosened from its base and slid down the side of the mountain, crushing eighteen homes, the Hafod-Tanglwys-Uchaf farm, and the Pantglas junior high school. In addition to the rain, a hidden underground spring contributed to the saturation of the inert materials that sent the avalanche of mud into the village.

On October 22, the day after the disaster, the National Coal Board (NCB) acknowledged their accountability for the collapse of the coal tip and instantly sought procedures to eliminate similar tragedies. To aid the NCB, a Tribunal of Inquiry, headed by Lord Justice Edmund Davies, was appointed by the Welsh secretary of state to investigate the causes of the slide. The first measure taken to ensure Aberfan's safety was securing the coal tip on Merthyr Mountain. An inspection of 477 of the 1,753 coal tips used by the colliery was the next step in searching for evidence of instability that required prompt action. At the same time, information concerning the state of each tip was gathered to assist in the creation of regulations and correct supervision of them. The research resulted in a more explicit set of individual obligations and responsibilities, including detailed methods of management. One of the outcomes of the investigation was an observation of the wide diversity of characteristics that distinguished each coal tip. Consequently, the NCB decided to turn its attention to keeping a vigilant eye for changes occurring in any of them.

Despite allegations made by Aberfan's local newspaper that the slide was similar to previous slides, the NCB determined that the avalanche was a result of unusual environmental factors. The newspaper, however, did point out that the years leading up to the Aberfan disaster displayed a neglect of coal tip safety. The NCB, the Mines Inspectorate, the legislature, and other countries had disregarded the importance of maintaining definitive laws for colliery tips. To rectify the problem, specifically at Aberfan, drains and boreholes were constructed for improved drainage at the tips. In addition, the spring that caused most of the saturation was enclosed to prevent further water emissions. Yet the people of Aberfan opted for a complete removal of the coal tip, which was funded by the NCB, local agencies, and the Welsh government.

Several new practices arose from the disaster at Aberfan that involved Wales as a whole. The NCB, for instance, started a program to research how colliery tips are influenced by their environmental surroundings. Similar policies and programs were created for further understanding of coal tips and their regulation.

Carolynn A. Kimberly

SEE ALSO: Solid waste management.

Wetlands

CATEGORY: Preservation and wilderness issues

Wetlands are transitional areas between terrestrial and aquatic ecosystems that exhibit some of the characteristics of each. Wetlands are considered by many to be one of the most important ecosystems on Earth because of their high biodiversity and productivity and because of the large number of economic benefits that they provide.

Wetlands are often distinguished by having three major components: water, hydrophytic vegetation, and hydric soils. Wetlands must have water present for at least part of the year, though the depth and duration of flooding may vary considerably. Some wetlands may have water-saturated soil, while others are characterized by permanent flooding. Wetlands, at least periodically, support a predominance of hydrophytic vegetation adapted for life in saturated soil conditions. Wetlands are also characterized by having undrained, or hydric, soils. These are soils in which anaerobic conditions have developed because of long periods of saturation, flooding, or ponding during the growing season.

Defining Wetlands

Defining wetlands would seem to be a simple matter until a single description is applied to diverse wetland types over a large geographic scale with diverse climatic conditions. Therefore, no single, formal definition for wetlands exists. On the contrary, dozens of definitions have been written for specific reasons by specific interest groups and various federal agencies. The problem is of critical consequence to the regulated community, which consists of people who are subject to restrictions and limitations placed on them by various federal, state, or local laws. Inconsistent definitions place a severe burden on the private landowner who does not have adequate technical or legal knowledge.

In the United States, a regulatory definition has been developed so that the Army Corps of Engineers and the Environmental Protection Agency (EPA) can administer the permitting of dredging and filling of wetlands as prescribed in section 404 of the Clean Water Act. The term "wetland" is defined as

> those areas that are inundated or saturated by surface or ground water at a frequency and duration sufficient to support, and that under normal circumstances do support, a prevalence of vegetation typically adapted for life in saturated soil conditions. Wetlands generally include swamps, marshes, bogs, and similar areas.

Thus, jurisdictional wetlands—those that are subject to section 404 permitting—must possess all three key characteristics: hydrology, hydrophytic vegetation, and hydric soils. However, because many wetlands are not permanently wet and because water may not be seen during a single site visit, positive hydrology indicators must be found, which must be supported by wetland vegetation and soils. These strict requirements for wetland identification have caused many wetlands to fall into uncertain categories. For example, some wetlands can have the appropriate hydrology but fail to develop the appropriate wetland soils and vegetation; often two of the three characteristics can be confirmed but not the third. These situations continue to cause confusion among governmental agencies and private landowners.

Another significant issue concerning wetlands is that they form ecotones between upland and aquatic ecosystems. Thus, even if an area is identified as a wetland, it may be extremely difficult to determine its exact boundary because of the gradual, perhaps imperceptible, change in soil and vegetation characteristics. The problem of identifying areas as wetlands and defining the boundaries of those wetlands is known as wetland delineation. The ability to perform delineations requires advanced training, especially in the areas of soils and botany.

Function and Value

Not all wetlands perform the same set of functions or perform functions at the same rate or efficiency. Often, the size and location of a wetland in the watershed determines its functions. Wetlands are considered to have extremely high biodiversity and rates of productivity. They provide food, shelter, and water for various invertebrates and vertebrates, many of which may be endangered or threatened. For example, in 1991 the U.S. Fish and Wildlife Service listed 595 plant and animal species as endangered or threatened. Of this number, wetlands provide essential habitat for 40 percent of the endangered and 60 percent of the threatened species. Another function associated with wetlands is the maintenance of the quantity and quality of both surface water and groundwater. Many wetlands serve to recharge aquifers. Wetlands also accumulate sediments, nutrients, and many forms of water pollutants from its watershed. By removing these materials, wetlands serve to clean the water.

Wetlands are sometimes referred to as nature's sponges because of their ability to ameliorate the effects of stormwater runoff and reduce floodwater damage. Stormwater enters wetlands and spreads out over large areas and then is slowly released by the wetland. Increased property damage from flooding has been shown to occur following destruction of wetlands. The U.S. Army Corps of Engineers has estimated that if the Charles River wetlands were destroyed near Boston, Massachusetts, flood damage would increase by as much as $17 million annually. Other wetland functions are prevention of saltwater intrusion to groundwater and surface water supplies, protection against coastal erosion from storms, and regional and global climate stabilization.

A wetland value is anything that has worth or is beneficial to the environment or to people. Wetland products, such as timber, fiber, food, and fish, have commercial value and are easily measured. Some values of wetlands, as in other ecosystem types, are not easily quantified. Wetlands provide recreation and tourism, sociocultural significance, research, and education. However, wetland values can go far beyond monetary value. It is impossible to place value on the fact that wetlands provide habitat for a high percentage of endangered and threatened species around the world.

The value placed on wetland functions is, in many cases, the most important factor in determining whether the wetland is preserved or converted to some other use. Also, as society's needs and perceptions change, the value assigned to wetland functions may also change over time. Wetland values are often in conflict because of the large number of functions they can perform. For example, if the water level in a wetland is raised, waterfowl production may increase while timber production may decrease. Managers of wetlands may make decisions that are popular with one user group but unpopular with other user groups.

Loss and Degradation

Wetland loss is defined as a decrease in wetland area caused by the conversion of wetland to nonwetland areas. Wetland degradation is defined as the impairment of one or more wetland functions because of human activity. In most cases, wetland loss is difficult or impossible to reverse because of the complexity of wetland structure and function. Wetland creation—the formation of wetlands in formerly nonwetland areas—is becoming an increasingly common strategy to combat wetland loss. Wetland degradation is more easily reversed through a variety of applied science and conservation tools.

Wetlands are found on every continent except Antarctica. The exact size of global wetland areas is difficult to assess because of differences in wetland definitions and the lack of documentation in many of the world's countries. Wetlands have been estimated to cover about 8.5 million hectares (6 percent) of the land area of the earth, with the largest wetland areas found in tropical, subtropical, and the boreal regions. During the twentieth century, the world may have lost 50 percent of the wetlands that existed prior to that time.

Through the National Wetlands Inventory, the U.S. Fish and Wildlife Service possesses the most accurate and comprehensive information on wetland losses of any country in the world. The inventory estimated that the area of wetlands in the United States decreased from about 391 million acres during the 1780's to about 274 million acres in the mid-1980's, which represents a 30 percent loss of wetland area. However, the lower forty-eight states have lost 53 percent of their wetlands. The states with the highest wetland losses are California (91 percent), Ohio (90 percent), Iowa (89 percent), Indiana (87 percent), and Missouri (87 percent). However, the rate of wetland loss in the United States between 1985 and 1995 (117,000 acres per year) was

Activists protest plans to convert a wetland area into a boat launch ramp. Wetlands are managed for a variety of purposes; a management policy that pleases one group of people may anger another. (Jim West)

lower than the rate between 1975 and 1985 (290,000 acres per year).

The underlying causes of wetland loss and degradation are numerous. These include poverty and economic inequality; population pressures from growth, immigration, and mass tourism; social and political conflicts; high demand for wetland resources such as timber; drainage; diking and damming; air and water pollution; introduction of exotic species; natural events such as hurricanes; and economic policies. In the United States, approximately 80 percent of wetland loss has been a result of agricultural practices.

Conservation and Protection

Wetlands in the United States are protected through regulation, economic programs, and acquisitions. At the federal level, a confusing mix of programs and legislation simultaneously encourages and discourages wetland conservation. As early as 1903, President Theodore Roosevelt recognized that wetland loss had become significant. By executive order, he established Pelican Island in Florida as the nation's first wildlife refuge. The federal government also protects wetlands through several laws, including the Clean Water Act of 1972, which created a plan to control the discharge of dredged or fill materials into wetlands and other waters of the United States. The Army Corps of Engineers and the EPA share responsibility for implementing the program. The "swampbuster" program is part of the Food Security Act of 1985 and 1990. It seeks to remove federal incentives for the agricultural conversion of wetlands to nonwetlands. In conjunction with this act, the Wetland Reserve Program was created, which provides financial incentives to farmers to restore and protect wetlands through the use of long-term easements.

The North American Waterfowl Management Program was another milestone in the conservation of important wetland habitat. This plan was signed between Canada and the United States in 1986 to restore the declining waterfowl populations through habitat acquisition, development of economic incentives to change land-use practices, and improvement of water management. At the global level, the most significant wetland conservation work was done by the Convention on Wetlands of International Importance in 1971. This global treaty provides the framework for the international protection and wise use of wetlands.

U.S. presidents became active in wetland protection during the 1970's. President Jimmy Carter signed two executive orders that provide guidance for wetland and floodplain management and protection of these areas by federal agencies. President George Bush extended these efforts to recommend that the United States establish a national goal of no net loss of wetlands. This policy has become a major force for wetland conservation in the United States.

Despite all of this activity, several difficulties remain: There is no specific wetland law or policy; wetlands, because of the complexity of their values and functions, continue to be managed for a variety of purposes; and "no net loss of wetlands" applies to the loss of wetland acreage only, not to wetland functions and values. President Bill Clinton took a compromise position in wetland protection by reaffirming the "no net loss" policy and supporting the Wetland Reserve Program; however, he created section 404 exemptions for 53 million acres of previously converted wetlands and for small plots of land owned by families who want to build single-family houses.

Roy Darville

Suggested readings: An excellent text covering all aspects of wetlands is Mitsch and Gosselink, *Wetlands* (1993). For an overview of the world's wetlands, see *Wetlands* (1991), by Finlayson and Moser, and *Wetlands in Danger* (1993), by Patrick Dugan. See *Discovering the Unknown Landscape* (1997), by Ann Vileisis, for an excellent, detailed account of the history of wetlands in the United States. William Niering's *Wetlands* (1985) is an excellent resource for the identification of wetland plants and animals. It contains more than six hundred color plates and detailed descriptions of wetland species. Dennison and Berry, *Wetlands: Guide to Science, Law, and Technology* (1993), provides a practical and technical guide to many wetland issues.

SEE ALSO: Conservation; Convention on Wetlands of International Importance; Ecosystems; Flood control.

Whaling

CATEGORY: Animals and endangered species

Up until the mid-1980's, when the International Whaling Commission enacted a moratorium on whaling, commercial exploitation was responsible for a serious decline in the total whale population. This led to concerns over biodiversity and the possibility of the extinction of some species of whales.

Humans have hunted whales for thousands of years. Whales have provided meat, oil (used for lighting and as an industrial lubricant), and ambergris (used in the manufacture of perfume). Blubber was converted into soap, baleen was fashioned into objects such as women's corsets, and whale bones were ground into fertilizer. The abundant uses that humans made of whales quickly led to overhunting of certain species. Bowheads were increasingly scarce by the late seventeenth century, and the Atlantic gray whale was hunted to extinction sometime during the early eighteenth century. One hundred years later, the U.S. whaling industry went into decline as the sperm whale became increasingly difficult to find.

The advent of new technologies during the late nineteenth and twentieth centuries hastened the decline of whale populations on a global scale. As steam and diesel engines replaced sails, ships gained the speed necessary to pursue faster species of whale. Harpoons fired from cannons were tipped with grenades that exploded upon contact, immediately killing the whale and ending the era of epic struggles often depicted in art and literature. With the exception of right whales, all whales sink when dead, so engineers crafted inflation devices to float the carcasses. Factory ships introduced during the 1920's allowed fleets to process the whales at sea. Each of these developments increased the efficiency of whalers, leading to larger catches and placing more stress on whale populations.

Whaling virtually ceased during World War II, but at war's end several nations aggressively competed for the remaining whales. Between 1940 and 1980, an estimated 1.5 million whales were killed, nearly double the approximately 794,000 killed between 1900 and 1940. The peak year was 1961, when whalers captured more than 66,000 whales. Several species, including the world's largest mammal, the blue whale, neared extinction. While early expressions of concern over the fate of whales focused on the need to conserve remaining stocks for future commercial exploitation, new attitudes toward the environment prompted some people to call for an end to commercial whaling. The International Whaling Commission enacted the International Whaling Ban in 1986. By the mid-1990's, however, several nations, including Japan and Norway, argued that populations of some whale species had increased enough to permit a renewal of whaling.

Problems in assessing the number of whales, many of which travel over wide areas, made conflicts over the resumption of whaling difficult to resolve. Nonetheless, rough estimates indicated that most species remained in grave danger of extinction as the twentieth century drew to a close. Right whales, long the favorite of whalers because they floated after they were killed, had been reduced from a prehunting population of approximately 100,000 to between two thousand and three thousand survivors in 1990. Nearly 250,000 blue whales had once graced the world's seas, but in 1990 fewer than 12,000 remained. Other species experienced similar declines.

Ironically, the devastation benefitted at least one species of whale. Since the minke whale faced less competition for food from other whales whose populations were shrinking, the number of minkes actually doubled, reaching an estimated 880,000 by 1990. It was this growth that prompted Japan and Norway to request a return to minke whaling, in part because they claimed that the minke was responsible for the decline of the mackerel and other fish upon which their fishing industries depended. In addition to the minke whale, wildlife that benefitted from the disappearance of whales included seals and birds that eat krill. The reduction in the number of baleen whales consuming krill led to an increase in food supply for many animals, which resulted in population growth.

Many scientists fear that some whale species will never recover and will eventually become extinct. The pressing issue regarding the survival of most species of whale is that of biodiversity. With populations so small, species such as the blue whale might lack the genetic diversity to respond to environmental changes. Present populations might persist and even grow but may be unable to survive any significant alterations in the environment. The concern over biodiversity is linked to the growing awareness that the whales' habitat, the world's oceans, is becoming an inhospitable and even dangerous place. The impact of global warming, sound and chemical pollution, and the decline in fish populations on whales remains unknown, but evidence indicates that whales face habitat loss, just as many land animals do. Whales washed ashore on Canada's St. Lawrence River were so loaded with chemicals that they were declared to be toxic waste.

Given the pressures that pollution places on whales, many observers have argued against any renewal of whaling. Calls for an end to the moratorium on the hunting of minkes indicates that many people still regard whales as a commercial resource to be exploited. However, as the environmental movement gained support in countries such as the United States during the last decades of the twentieth century, whales became a powerful symbol of humankind's obligation to preserve wildlife. The debate between these two groups over whaling has revealed that international cooperation on environmental issues remains problematic, especially when it concerns the care of the oceans and their inhabitants. Moreover, the debate indicates that although whales live far from most humans, human management has become necessary to ensure their continued survival.

Thomas Clarkin

SUGGESTED READINGS: Peter J. Stoett's *The International Politics of Whaling* (1997) examines ethical issues and political conflicts surrounding whaling. Robert McNally's *So Remorseless: A Havoc of Dolphins, Whales, and Men* (1981) places whaling in the larger context of the deaths of marine mammals. Richard Ellis's *Men and Whales* (1991) offers a comprehensive history of whaling with numerous illustrations and photographs. David G. Campbell's *The Crystal Desert: Summers in Antarctica* (1992) includes discussions of Antarctic whaling and whales.

SEE ALSO: Endangered species; Extinction and species loss; International Whaling Ban; International Whaling Commission; Marine Mammal Protection Act.

Whooping crane

CATEGORY: Animals and endangered species

The endangered, migratory whooping crane is one of the rarest birds in North America. Named for its loud, bugling call, this regal crane has become an important symbol for wildlife conservation.

Standing at a height of 1.5 meters (5 feet) or more with a wingspan of up to 2 meters (7 feet), whooping cranes are the tallest birds in North America. They are white with black-tipped wings, black beaks, and a bare patch of red skin on the tops of their black-banded heads. Marshes are vital to the survival of whooping cranes. They nest in the tall marsh grasses and eat blue crabs and other food found in wetland areas. Unlike other cranes, whoopers are slow to mature. After four years, they choose a mate for life. Adult females produce only two eggs in a clutch. Often only one of these two chicks survives. For this reason, the whooping crane population has never been abundant.

In the nineteenth century the number of whooping cranes was estimated at 1,400. Two flocks were known to migrate to Canada from Louisiana and Texas. Their numbers rapidly dwindled as their wetland habitats were destroyed. The magnificent look of adult whoopers also made them targets for sport hunting. Every death of a whooping crane also ended the repro-

A whooping crane (center) follows an Idaho rancher who used his ultralight plane to lead cranes along a new, safer migratory route. The species became endangered during the late nineteenth century after much of its wetland habitat was destroyed by development. (AP/Wide World Photos)

ductive life of the surviving mate. By 1900 only about one hundred whooping cranes were left. The Migratory Bird Treaty, signed by the United States and Canada in 1918, made it illegal to hunt whooping cranes. Despite this, the number of whooping cranes dwindled to only fifteen by the late 1930's because of the continuing loss of wetlands.

In 1937 the wintering grounds of the Texas flock were set aside as the Arkansas National Wildlife Refuge. By 1944 this flock had grown from fifteen to twenty-one in number, but the last of six whoopers from the Louisiana flock died in 1948. Conservationists were delighted, however, by a discovery in 1954. Whooping cranes were found breeding in a secluded section of Wood Buffalo National Park in Alberta, Canada. To help ensure their survival, eggs were taken from these nests to be raised in captivity. Later these birds were returned to the flock, helping build the population. Another flock that does not migrate was established at Florida's Kissimmee Prairie Sanctuary using captively bred whoopers.

Conservation efforts have increased the whooping crane population to around 300. Of this number, only 157 are part of the wild flock in the Arkansas National Wildlife Refuge that migrates between Canada and the United States. The proximity of the refuge to shipping routes with barges carrying chemicals and fertilizers has raised fears among conservationists that an accidental spill could wipe out the whole flock. The next step is to establish a second migratory flock in the wild. Experiments using ultralight planes have been carried out in the hope that they can guide the whooping cranes in learning the migration route.

Lisa A. Wroble

SEE ALSO: Captive breeding; Endangered species; Endangered Species Act; Wetlands; Wildlife management; Wildlife refuges.

Wilderness Act

DATE: passed 1964
CATEGORY: Preservation and wilderness issues

The Wilderness Act, passed by the United States Congress in 1964, is designed to preserve land in its most natural condition.

The 1964 Wilderness Act established the National Wilderness Preservation System (NWPS), gave the U.S. Congress authority to designate wilderness areas, and directed the secretaries of the interior and agriculture to review lands for possible wilderness designation. The act initially set aside fifty-four areas—a total of 9 million acres of federal Forest Service land—for wilderness classification. In 1968 Congress began adding additional wilderness areas. As of 1999, there were 631 wilderness areas in forty-four states totaling nearly 104 million acres. Wilderness areas are located within national forests, wildlife refuges, and parks, and are managed by a host of agencies, including the Forest Service, the Bureau of Land Management (BLM), the National Park Service (NPS), and the Fish and Wildlife Service.

The land area of the United States totals 2.3 billion acres. By 1999 133 million acres had been designated or recommended for wilderness designation. Approximately one-half of this total—68 million acres—is in the state of Alaska and accounts for 28 percent of land in the state. Alaska is also the state in which the largest amount of land has actually been set aside: The Alaska National Interest Lands Conservation Act (ANILCA) of 1980 more than tripled the NWPS by establishing thirty-five new wilderness areas totaling more than 56 million acres. Excluding Alaska, less than 4 percent of land in the United States is classified as wilderness or has been recommended for such designation.

The act defines wilderness as federal land "where the earth and its community of life are untrammeled by man, where man himself is a visitor who does not remain." Although there are numerous exceptions, the act generally prohibits commercial activities, motorized and mechanical access, permanent roads, and human-made structures and facilities within wilderness areas. An area may be determined to be suitable for wilderness designation if it is

> an area of undeveloped land retaining its primeval character and influence, without permanent improvements or human habitation, which is protected and managed so as to preserve its natural conditions and which (1) generally appears to have been affected primarily by the forces of nature, with the impact of man's works substantially unnoticeable; (2) has outstanding opportunities for solitude or primitive and unconfined type of recreation; (3) has at least five thousand acres of land or is of sufficient size as to make practicable its preservation in use and in an unimpaired condition; and (4) may also contain ecological, geological, or other features of scientific, educational, scenic, or historic value.

Congress has the authority to designate areas as wilderness and uses its power to do so under the act. The intent is to make designations permanent and add new lands as Congress sees fit.

Although protected to preserve its natural conditions, a number of nonmotorized activities such as horseback riding, hiking, camping, fishing, and hunting are allowed in wilderness areas. Preexisting and valid extractive uses are also allowed to continue until the permits granted for such activities expire, are abandoned, or are purchased by the government. Preexisting grazing is also allowed to continue as long as it is consistent with sound resource management practices. In addition, the Wilderness Act honors all federal-state relationships with regard to state water laws and state fish and wildlife responsibilities. Activities that are generally not allowed in these areas include mining, timber harvesting, water development, mountain biking, and use of any motorized equipment such as snowmobiles and all-terrain vehicles. However, allowances in the act can be seen as a compromise between preservationists and those resource interests concerned with grazing, mining, timber harvesting, and water development.

Given the definition of wilderness and the amount of federally owned land in the western United States, congressionally classified wilderness is a particularly western phenomenon. While states such as Alaska, Arizona, California,

Idaho, and Washington each have more than 4 million acres of wilderness within their borders, such nonwestern states as Connecticut, Delaware, Iowa, Maryland, and Rhode Island have none.

While Congress makes the final decision of the suitability for additional wilderness areas, land management agencies such as the Forest Service and the BLM are often responsible for making official recommendations. For example, the Federal Land Policy and Management Act of 1976 directed the BLM to review land it administers for possible wilderness designation, while the Forest Service has gone through three Roadless Area Review and Evaluation (RARE) plans to determine suitable wilderness areas. Such plans have often been criticized for not designating enough acreage and refusing to designate land with great economic potential, thus leaving "rocks and ice" for wilderness classification and those lands with economic value under multiple-use management.

Martin A. Nie

Suggested readings: An extensive review and analysis of the Wilderness Act, wilderness areas, and their history and management is provided in John C. Hendee, George H. Stankey, and Robert C. Lucas, *Wilderness Management* (1990). Michael Frome's *Battle for the Wilderness* (1974) focuses on the struggle to pass the act while at the same time examining the American conservation movement. The conservation movement and its foundation are also explored in Roderick Nash's *Wilderness and the American Mind* (1967). Ross W. Gorte's Congressional Research Report *Wilderness: Overview and Statistics* (1994) provides an excellent summary of the act and statistics pertaining to it.

See also: Alaska National Interest Lands Conservation Act; Gila Wilderness Area; Nature preservation policy; Nature reserves; Preservation; Wilderness areas.

Wilderness areas

Category: Preservation and wilderness issues

Wilderness areas are natural, undeveloped areas in the United States that are protected by law under the Wilderness Act of 1964.

The Painted Desert Wilderness Area in Arizona. The Wilderness Act of 1964 mandated the designation of wilderness areas in which the impact of humans would be minimized. (Jim West)

Preserving areas of unspoiled nature is a relatively new idea and one that made sense only when the seemingly inexhaustible wilderness of North America was, in fact, nearly exhausted. In 1924, at the urging of Forest Service employee and influential conservationist Aldo Leopold, 574,000 acres of the Gila National Forest in New Mexico were set aside as the first federally protected wilderness, the Gila Primitive Area. As the system of primitive areas grew, environmentalists became concerned about inconsistent management and by the fact that the areas were protected only by agency policy and not by law. Environmentalists began lobbying for federal legislation that would designate and protect wilderness areas throughout the United States. However, the concept of wilderness was strongly opposed by many of those who made their living from natural resources such as timber, mining, and grazing. They saw these lands, which often had great economic value, as being "locked up" for the pleasure of a few.

On September 3, 1964, after eight years of debate and compromise, President Lyndon Johnson signed the Wilderness Act, creating the National Wilderness Preservation System (NWPS), which consisted of fifty-four areas totalling 9 million acres. According to the act,

> a wilderness, in contrast with those areas where man and his own works dominate the landscape, is hereby recognized as an area where the earth and its community of life are untrammeled by man, where man himself is a visitor who does not remain.

In addition, the act defined the mechanism for adding more areas to the system in the future. To be considered, an area must be at least 5,000 acres, "or of manageable size." This was a far cry from the early days of wilderness advocacy, when the minimum size was thought to be 500,000 acres or, as Aldo Leopold put it, "large enough to absorb a two-week pack trip." Designated wildernesses become part of the NWPS. All roads, structures, or other installations are prohibited in designated wilderness, as is the use of motorized equipment or any mechanical transport. These areas of wild nature have been, and continue to be, the focus of intense controversy regarding their designation and management.

A significant addition to the system came in 1975 with the passage of the Eastern Wilderness Act. The lack of pure, untouched wilderness in the eastern states led to the loosening of the strict standards of the original act to allow the inclusion of ecologically significant areas that showed more impact from human activities than would have originally been permitted. In this way, sixteen areas totalling 207,000 acres, from 22,000-acre Bradwell Bay in Florida to the 14,000-acre Lye Brook Wilderness in Vermont, were added to the system. As of the late 1990's, the wilderness system encompassed more than 650 areas, ranging in size from the 5-acre Oregon Islands Wilderness to the 9 million acres of the Wrangell-St. Elias Wilderness in Alaska, for a total of more than 100 million acres. Of this total, more than 60 percent is in the state of Alaska.

While wilderness opponents believe that this must certainly be enough, preservationists argue that less than 5 percent of the landscape of the United States is protected in its natural state. Some large areas continue to be fought over, such as the fragile Arctic coastal plain of Alaska, home of vast caribou herds and underlain by large oil deposits. Idaho, which boasts more designated wilderness than any of the other lower forty-eight states, still has millions of acres of undeveloped roadless land that many believe should be protected. Wilderness advocates also point out that many wilderness areas, as well as national parks and other protected lands, have illogical political boundaries, unrecognized by grizzly bears and other important wildlife species. They argue that areas between and adjacent to designated wilderness should often be protected as well, to create units based on natural, ecological boundaries.

After wilderness is designated, the focus shifts to maintaining the desired qualities of the area, leading to the paradox of "wilderness management." Though recreation is only one of the stated uses of wilderness—the others being scenic, scientific, educational, conservation, and historic—agency efforts and budgets are primarily based on managing the often vast numbers of human visitors. One of the stated purposes of wilderness is "primitive and unconfined recreation" but another is the protection of the resource itself. At what point do camping and trail restrictions, quotas, and permits impinge on this freedom?

Another issue is that of wildfire suppression. It is now understood that fire is an important component of most ecosystems, yet a century of fire suppression has left unnatural fuel conditions in many areas. Should managers allow wildfires—which would likely be larger and more destructive than natural, periodic fires of the past—to burn? Other major controversies center on the reintroduction of wildlife species such as wolf and grizzly bear to wilderness, the disposition of long-standing mining and drilling claims, and airplane flights over, or even into, remote wilderness areas.

Joseph W. Hinton

SUGGESTED READINGS: A Forest Service publication, *Wilderness Management* (1978), by John C. Hendee et al., is a textbook for wilderness managers but provides excellent introductory chapters on wilderness history and issues. Michael Frome's *Battle for the Wilderness* (1974) discusses American cultural attitudes toward wilderness and then describes the people and events leading to the Wilderness Act. *The Politics of Wilderness Preservation* (1982), by Craig W. Allin, focuses on the legislative path to the Wilderness Act, while Lloyd C. Irland's *Wilderness Economics and Policy* (1979) is an in-depth study of the economic issues at the heart of the wilderness controversy.

SEE ALSO: Ecosystems; Gila Wilderness Area; Leopold, Aldo; Marshall, Robert; Wilderness Act.

Wilderness Society

DATE: established 1935

CATEGORY: Preservation and wilderness issues

Organized in 1935, the Wilderness Society is a nonprofit organization that focuses on the protection and preservation of wilderness and wildlife in the United States.

In January, 1935, a group of eight dedicated conservationists organized the Wilderness Society in Washington, D.C. Among the participants were Robert S. Yard, publicist for the National Park Service (NPS); Benton MacKaye, the "father of the Appalachian Trail"; Robert Marshall, head of recreation and lands for the Forest Service; and Aldo Leopold, a wildlife ecologist at the University of Wisconsin. Leopold believed that the new society would form a cornerstone to help preserve America's vanishing wilderness.

After much dedicated work and pressure by the Wilderness Society, the Wilderness Act was finally signed into law by President Lyndon B. Johnson on September 3, 1964. This act formed the National Wilderness Preservation System (NWPS), which enabled the United States Congress to set aside selected areas in the national forests, national parks, national wildlife refuges, and other federal lands as units to be kept permanently unchanged by humans. There would be no roads, structures, vehicles, or any significant impacts of any kind in these selected areas. The Wilderness Act designated approximately 9 million acres as wilderness.

Since 1964, the Wilderness Society has been instrumental in the passage of bills that have contributed a total of 104 million acres to the NWPS. In particular, the Alaska National Interest Lands Conservation Act (ANILCA) of 1980 designated 56 million acres of pristine beauty, and the California Desert Protection Act of 1994 protected 8 million acres of desert lands. After a ten-year effort, the Wilderness Society was finally instrumental in the enactment of the National Wildlife Refuge Management Act of 1997, which provides for stronger protection of wildlife in all national wildlife refuges.

Despite the successes of the Wilderness Society, public lands continued to be compromised and degraded by air and water pollution, excessive development, road building, logging, cattle grazing, mining, and recreation in the 1980's and 1990's. By the late 1990's, the Wilderness Society had focused on the preservation of a number of high-priority areas, with the overall goal of creating a nationwide network of wild lands. Key campaigns were launched in Montana, California, Idaho, Nevada, Utah, Colorado,

Texas, Vermont, the Pacific Northwest, and the southern Appalachians.

In 1997, with backing from the society, Congress blocked a plan to create a massive gold mine just outside Yellowstone National Park by appropriating money from the Land and Water Conservation Fund to purchase the mining claims. Similarly, the society helped create public pressure that led Du Pont to defer plans for a titanium mine on the border of the Okefenokee National Wildlife Refuge in Georgia. Working closely with local and national groups during 1996 and 1997, the society convinced the federal government to withdraw a proposed logging plan for nine national forests in the Sierra Nevada. Additionally, in response to a Wilderness Society lawsuit, a federal judge blocked logging in four national forests in Texas, pointing out that it would be detrimental to wildlife habitats.

Alvin K. Benson

SEE ALSO: Conservation; Marshall, Robert; Nature preservation policy; Preservation; Wetlands; Wilderness Act; Wilderness areas.

Wildlife management

CATEGORY: Animals and endangered species

Wildlife management involves the control of wild species and the maintenance and improvement of their habitats. Such practices require coordination among many interest groups, including conservation officers, biologists, ecologists, fishers, hunters, park employees, wildlife enthusiasts, environmentalists, and state and federal governments.

Wildlife has been compromised by human activity throughout history. Animals have been hunted for food and skins, trees have been cut down for shelter, and plants have been collected for food, medicine, and clothing. As weapons improved, more animals were hunted. As humans learned to cultivate their own plants and herd animals, they required more space, so areas of wild habitat were cleared. With diminishing habitats, wildlife numbers also diminished. Many species became extinct without notice.

The rulers of ancient civilizations set aside areas as game reserves for hunting and protected certain forests for religious reasons. Medieval and European kings continued the practice of creating game reserves for sport hunting by the privileged class. These reserves created artificially high populations of sport species, such as grouse, mallards, and pheasants, while reducing the number of predator species, such as weasels, otters, wildcats, and badgers.

In the United States, however, the amount of land and wild animals seemed limitless. As land was cleared to make way for westward expansion, forests were lost, marshes were drained, coastlines were changed, and breeding grounds were destroyed. While some species failed to breed or produced fewer numbers, some animals, such as wolves, coyotes, and grizzly bears, came into constant conflict with humans. Early settlers were so intent on battling nature to ensure their survival that many never noticed the demise of numerous species.

Hunters in the United States lobbied for the protection of wildlife as early as 1888. During the early twentieth century, wildlife and habitats began to be conserved for their beauty and scientific value. Most of these efforts occurred through either conservation clubs or the establishment of reserves and parks through the Forest Reserve Act of 1903. It was not until the conservation movement of the 1960's and 1970's that wildlife management focused on ecology and the balance of nature.

The balance of nature is important for the environmental health of the earth. Natural cycles generally ensure that animals, plants, birds, fish, and insects are supported by the habitat in which they live. If one element, such as a predator, is removed or disrupted, overbreeding may result. Wildlife management assists in maintaining the balance between the number of animals and the amount of food and shelter available. In order to keep a proper balance, wildlife managers study animals, fish, and waterfowl. They tag the animals and track migration patterns to gain helpful information. They also catalog the plants, insects, soil types, and variety of species

Kenya Wildlife Service (KWS) director David Western stands next to the carcass of an elephant whose ivory has been poached in Kenya in February of 1998. Poaching is a major problem faced by wildlife management personnel throughout the world. (AP/Wide World Photos)

in different areas to learn how habitats manage to support their wildlife.

Overbreeding may lead to disease or lack of food. Hunting and fishing regulations help keep populations under control. To keep the number of certain species from becoming too low, legal hunting and fishing seasons are established. During years when wildlife populations are high, seasons may be extended. During years when wildlife populations are low, limits are set on the number each hunter or fisher can take. Natural predators may be reintroduced to an area as another method for maintaining populations.

Wildlife management also involves enforcing wildlife laws. State and federal law enforcement and conservation officers make sure that people enjoy protected habitat and park areas without littering, picking protected plants, or destroying the habitat in any other way. They also enforce laws prohibiting poaching—killing species out of season or taking more than the legal limit—and selling or harming endangered, threatened, or rare species.

Fines collected from wildlife enforcement and fees for hunting and fishing licenses help support a variety of wildlife programs, including the protection of breeding grounds. Many animals, birds, and waterfowl need specific habitats in order to breed. Wildlife management includes intervention to ensure that these breeding grounds remain undisturbed. Maintaining the plants, soil, and insects within habitats is also part of wildlife management. Maintenance of plants and insects may involve preventing or cleaning up pollution.

When populations are low, moratoriums on hunting may be announced until they repopulate. Most often, a species needs help to repopulate. Wildlife managers may create protected habitat areas or restore breeding grounds, such as marshes. Restoration is costly, however.

For species such as the whooping crane and

the black-footed ferret, captive breeding proved successful. Captive breeding involves breeding species in a controlled environment, then releasing them into the wild or into a protected habitat. A hunting moratorium and nesting boxes helped wood ducks recover from endangered status. Artificial nesting boxes are placed in areas where predators and humans will not harm the eggs. This method was also used with snow geese. The program was so successful that their numbers almost tripled in two decades, forcing the U.S. Fish and Wildlife Service to investigate ways to reduce flock populations before disease set in. The snow geese recovery program is a prime example of how challenging wildlife management can be.

Lisa A. Wroble

SUGGESTED READINGS: A thorough history of wildlife in America, from human encroachment to loss of species, is available in *Wildlife in America* (1987), by Peter Matthiesson. The book includes a list of endangered or threatened wildlife, a chronology of wildlife legislation, and an extensive bibliography. *Averting Extinction: Reconstructing Endangered Species Recovery* (1997), by Tim W. Clark, describes the problems involved in wildlife conservation. Jean F. Blashfield, *Rescuing Endangered Species* (1988), is an easy-to-understand book that covers success stories from salmon to the red wolf. It includes different programs and methods used to preserve and manage wildlife throughout the world. *Guardians of Wildlife* (1996), by Gary Chandler and Kevin Graham, covers similar ground by focusing on several different conservation groups.

SEE ALSO: Balance of nature; Captive breeding; Endangered species; Extinctions and species loss; Nature reserves; Poaching; Predator management; Wilderness areas; Wildlife refuges; Zoos.

Wildlife refuges

CATEGORY: Animals and endangered species

Wildlife refuges are regions of land or water set aside by governments or private organizations to protect and preserve one or more species of wildlife. The U.S. National Wildlife Refuge System has endured widespread congressional debate and public scrutiny involving environmental issues related to the societal, government, and commercial use of designated sanctuaries, culminating in the 1997 National Wildlife Refuge System Improvement Act and the transformation of America's refuges into multiple-use systems.

Prior to 1900, the federal government aggressively raised much-needed revenue and rewarded growing commerce by selling or giving away nearly 1 billion acres of land to states, homesteaders, veterans, railroads, and businesses. President Theodore Roosevelt initiated the protection of habitat for wildlife in 1903 when he set aside Pelican Island, a 3-acre ecosystem of barren sand and scrub in Florida's Indian River, as a federal reservation to protect birds from hunters supplying plumes to the fashion industry. Inspired while camping in California's Yosemite Valley with naturalist John Muir, Roosevelt established more than fifty wildlife refuges, five national parks, and eighteen national monuments, such as the Grand Canyon. He also increased the area of national forests by more than 150 million acres before leaving his second term in office. To preserve additional lands "for our children and their children's children forever, with their majestic beauty all unmarred," Roosevelt guaranteed land for future refuges by separating other federal public domain regions such as national forests and rangelands from the control of commercial interests. More than 90 percent of America's current refuge acreage resulted from Roosevelt's foresight, enabling the National Wildlife Refuge System to grow larger than the National Park System and entail nearly 4 percent of the surface area of the United States.

With vital assistance by private individuals and organizations such as the Nature Conservancy and the National Audubon Society, wildlife refuges have been established for waterfowl, big game, small resident game, and colonial nongame birds. Wildfowl refuges, easily the most plentiful, are geographically patterned to supply breeding, wintering, resting, and feeding areas along the four major North American migration flyways. The sportsmen who were essential in establishing many national refuges ensured that hunting would be permitted on most sanctuaries, with trapping allowed on many. Although the entire system logs nearly two million hunting visits annually, visitors who come for wildlife education and photography outnumber hunters and anglers by more than four to one.

The National Wildlife Refuge System, overseen by the Fish and Wildlife Service of the Department of the Interior, is the most comprehensive nature protection network in the world. The entire system includes nearly 90 million acres inside refuge boundaries, in addition to 3.3 million acres in areas specifically targeted for waterfowl or managed in cooperation with state agencies to protect important habitat. As of 1999, the United States owned about 510 separate wildlife refuges, with more continually being added. Nearly all of them are open to the public. Endangered and threatened species are supported on about fifty-five refuges created specifically for that purpose, and many urban refuges have been established near large cities. There is at least one wildlife refuge in each of the fifty states and several in overseas possessions from Puerto Rico to American Samoa in the South Pacific. The largest refuge is the 22-million-acre Alaska Yukon Delta National Wildlife Refuge, while the smallest is the 1-acre Mille Lacs National Wildlife Refuge in Minnesota.

Environmental Management Issues

Creating and maintaining a refuge to provide food, cover, and protection from human development for wildlife is considerably more difficult than simply sequestering an area and allowing nature to run its course. Continual management of resources is imperative to keep delicate ecosystems in balance; among the responsibilities are fixing broken floodgates and cleaning clogged ditches, battling for rapidly declining water supplies, seeding wildlife foods, plowing and burning areas that have overrun with un-

Cranes and crows share space at Bosque del Apache Wildlife Refuge in New Mexico. Such refuges provide essential stopovers for migrating birds and habitat for numerous plant and animal species. (AP/Wide World Photos)

wanted vegetation, and closing off areas from the public during sensitive periods for animals, such as mating and birthing seasons.

Although refuges often include areas just as spectacular as those within the National Park System, the National Wildlife Refuge System as a whole has not been well utilized by the public. This gave Congress the freedom to lease these public lands for commercial purposes such as grazing, farming, oil drilling, mining, logging of timber, military maneuvers, and motorized recreation. However, as public use of refuges increased during the 1980's and 1990's, Congress added eighty new refuges to the system, creating such a backlog of environmental preservation issues that Congress has contemplated selling some areas to pay for maintenance. With minimal budgets, refuge managers are charged with making certain that all activities are compatible with wildlife while still allowing potentially destructive activities such as off-road driving and motorcycling, power boating, and commercial fishing. Biodiversity has also become an important goal of managers, who care for about 700 species of birds, 220 species of mammals, 250 species of reptiles and amphibians, and 200 species of fish, in addition to innumerable species of plants.

Refuge areas owned by private corporations often sacrificed key habitat for short-term economic gain with little regard for long-term environmental and social consequences. However, business executives have realized that environmental issues are of genuine concern to most Americans. In response to increased environmental awareness and pressure by consumers, employees, and stockholders, many large businesses are now implementing stewardship strategies that protect natural resources, enhance wildlife habitat, and provide for public enjoyment of their underdeveloped land.

Key Legislative Actions

Following Roosevelt's initial work, new additions to the refuge system came slowly until the Dust Bowl years of the 1930's, when migratory bird populations became depleted. Congress then passed the 1934 Duck Stamp Act, which added a conservation fee onto waterfowl licenses purchased by hunters, enabling the Fish and Wildlife Department to acquire wetlands along major flyways. Additional monies to purchase refuges came from the Land and Water Conservation Fund set up in the 1960's to increase public space for outdoor recreation, which generated considerable revenues from offshore drilling leases. Passage of the 1980 Alaska National Interest Lands Conservation Act (ANILCA) enabled the refuge and park systems to double in size. Although 96 percent of all refuge units are outside Alaska, Alaska contains about 83 percent of the acreage.

The 1973 Endangered Species Act spurred managers of refuges to make more concessions for certain species of flora and fauna. Refuges at the close of the twentieth century harbored about 170 threatened or endangered species. Surveys in the late 1980's by the Fish and Wildlife Service and the General Accounting Office revealed that more than 60 percent of refuges permitted activities known to be harmful to wildlife. The most harmful practices, such as military activities and drilling, are not under the control of the Fish and Wildlife Service. The high-profile activist group Defenders of Wildlife organized a citizen's commission in 1992, which confirmed that the National Wildlife Refuge System was "falling far short of meeting the urgent habitat needs of the nation's wildlife" and was suffering from "chronic fiscal starvation and administrative neglect."

The National Wildlife Refuge System Improvement Act, signed into law by President Bill Clinton on October 9, 1997, dramatically shifted the priorities of the refuge system from its original sole purpose of protecting wildlife to the formation of a multiple-use system. The legislation redefined the mission statement regarding the conservation of habitat for fish, wildlife, and plants; designated priority public uses such as hunting, fishing, wildlife observation and photography, and environmental education and interpretation; and required that the environmental health of the refuge system be maintained. This monumental bill gave hunting, fishing, commercial trapping, and recreation equal status with the conservation of plants, birds, and animals. It also limited new or secon-

Milestones in Wildlife Protection

Year	Event
1870	The first state wildlife refuge is established in California.
1898	Kruger National Park is established in South Africa for the preservation of big game.
1900	The Lacey Act regulates the interstate commerce of birds and mammals; the act is supplemented by a similar act for black bass in 1926.
1903	The first federal bird sanctuary is established by President Theodore Roosevelt at Florida's Pelican Island.
1908	Theodore Roosevelt calls a conference of state governors and related officials to inventory natural resources in the United States.
1916	The National Park Service is established and forbids hunting within its jurisdiction.
1929	The Migratory Bird Conservation Act provides for a system of refuges along major flyways.
1934	The Duck Stamp Act requires hunters of migratory fowl to purchase duck stamps with their waterfowl licenses; proceeds are used to establish wildlife refuges.
1937	Taxes on arms and ammunition are used for wildlife preservation.
1940	The National Wildlife Refuge System is established by a consolidation of the Bureau of Biological Survey and the Bureau of Fisheries; its mission includes biological research and administration as well as enforcement of federal legislation.
1966	The National Wildlife Refuge System Administration Act mandates that all refuge uses be compatible with the primary purpose for which the refuge was established.
1970	The National Environmental Policy Act, as well as other legislation designed to combat pollution, is passed.
1973	The Endangered Species Act, which updates prior acts in 1966 and 1969, requires refuge managers to protect certain species of flora and fauna.
1980	The Alaska National Interest Lands Conservation Act doubles the amount of land in the U.S. refuge and park systems.
1987	A federal court rules that the U.S. Fish and Wildlife Service is responsible for policing the existing ban on spring hunting by native groups in Alaska.
1992	Research commissioned by Defenders of Wildlife finds that the National Wildlife Refuge System is grossly inadequate.
1997	The National Wildlife Refuge System Improvement Act establishes a revamped multiple-use mission statement for refuge habitat conservation.

dary refuge use to activities that are compatible with wildlife protection and made legislative changes more difficult for future congressional cycles. This principle of multiple use, however, will continue to allow and possibly increase mining, drilling, grazing, logging, and motorized recreation, in addition to increased military training, including bombing and tank and troop exercises. Upon signing the bill, Clinton stated that he "hoped and trusted that the process by which this bill was enacted will serve as a model for future congressional action on other environmental issues," with the future of the National Wildlife Refuge System to be shaped by the future of other bills such as the Clean Water Act, Wetlands Protection Act, and Endangered Species Act.

Daniel G. Graetzer

SUGGESTED READINGS: Among the books that examine wildlife refuges and the various methods used to maintain environmental balance on them are Dorothy H. Patent's *Places of Refuge: Our National Wildlife Refuge System* (1992), Laura Riley and William Riley's *Guide to the National Wildlife Refuges* (1993), and the National Geographic Society's *Animal Kingdoms: Wildlife Sanctuaries of the World* (1995). Articles in popular magazines that relate the current status of certain wildlife refuges with respect to environmental concerns include Douglas H. Chadwick and Joel Sartore, "Sanctuary: U.S. National Wildlife Refuges," *National Geographic* (October, 1996); Douglas H. Chadwick, "Blue Refuges: U.S. National Marine Sanctuaries," *National Geographic* (March, 1998); Susan Zakin, "The Misguided War over Refuges: Certain Groups Are Taking Advantage of Well-Intentioned Hunters and Fishermen," *Sports Afield* (March, 1995); and Diane Duffy and Carolyn S. Lavine, "Wildlife Sanctuaries on Corporate Lands," *Conservationalist* (June, 1994). For information on all refuges, contact the National Wildlife Federation, Department SA, 8925 Leesburg Pike, Vienna, Virginia, 22184-0001, 703-790-4000, http:///www.nwf.org.

SEE ALSO: Biodiversity; Ecosystems; Endangered species and animal protection policy; Forest and range policy; Land-use policy; Roosevelt, Theodore; Wildlife management.

Wilmut, Ian

BORN: July 7, 1944; Hampton Lucey, England
CATEGORY: Biotechnology and genetic engineering

Ian Wilmut, one of the foremost authorities on biotechnology and genetic engineering, conducted a landmark cloning experiment in 1996, which produced Dolly the sheep, the first mammal clone ever produced from adult cells.

Ian Wilmut earned his Ph.D. in animal genetic engineering from Darwin College at University of Cambridge in 1971. One of his first projects, conducted in 1973, involved the birth of the first calf ever reproduced from a frozen embryo. In 1974 Wilmut took a job with the Animal Breeding Research Station, later known as the Roslin Institute, a nonprofit organization affiliated with the University of Edinburgh in Scotland.

In 1990 Wilmut hired cell cycle biologist Keith Campbell to assist him with cloning studies at the Roslin Institute. Their first success came in 1995 with the birth of Megan and Morag, two Welsh mountain sheep cloned from differentiated embryo cells. The achievement came as the result of a new technique pioneered by Wilmut and Campbell that involved starving embryo cells before transferring their nuclei to fertilized egg cells. The technique synchronized the cell cycles of both cells and led Wilmut and Campbell to believe that any type of cell could be used to produce a clone.

In the landmark cloning experiment that produced the Finn-Dorset lamb Dolly, Wilmut removed udder cells from a six-year-old adult ewe and isolated the cells to starve them of nutrients, thereby arresting their growth and division. Next, he extracted an egg cell from another ewe and removed its nucleus, which contains the genetic material. Following the joining of the egg cell with one of the udder cells, Wilmut implanted the resultant embryo into a surrogate mother. Although Dolly was born on July 5, 1996, the success of the experiment was kept secret until February, 1997, in order for Wilmut and his scientific team to fully tabulate the results, confirm Dolly's survival, and secure a patent for their

Ian Wilmut (right) produced Dolly the sheep, the first clone of an adult mammal. (AP/Wide World Photos)

technique. Wilmut's breakthrough ended decades of skepticism among scientists who believed that cloning would never be possible for any animal.

On July 25, 1997, Wilmut and Campbell shocked the world again when they announced the birth of Polly, a lamb with a human gene in every cell of its body. Using a method similar to the one for cloning Dolly, Polly was cloned from a fetal skin cell that had a human gene. Wilmut and his research team had opened up a new frontier of science.

The announcements of the births Dolly and Polly stirred the scientific community and the public, kicking off a large-scale debate on the ethics and direction of cloning research. In particular, people feared the possibility of producing human clones. Wilmut, however, stated that he saw no reason for the pursuit of the first cloning of a human but pointed instead to the new possibilities for treating human disease. Cloned animals can act as the manufacturing plants for valuable human proteins, which are costly and difficult to produce in large amounts elsewhere. In addition, agriculture can benefit by cloning the best animals, such as the best milk-producing cows and the best wool-producing sheep.

Alvin K. Benson

See also: Biotechnology and genetic engineering; Cloning; Dolly the sheep; Genetically engineered organisms.

Wind energy

CATEGORY: Energy

Renewable and nonpolluting wind energy is sustainable and environmentally benign. Costs, however, are high, and economically viable energy falls far short of world needs.

Wind was one of the earliest energy sources exploited by humankind. Sailing ships and boats are illustrated in ancient Egyptian tomb paintings. Babylonian windmills pumped irrigation water as early as 1700 B.C.E. Windmills were abundant throughout Europe by the twelfth century; they kept the sea out of Holland, pumped mines, and milled grain. Sails drove ocean commerce.

In the late eighteenth century, steam engines began replacing windmills and sails. Electricity was first generated by wind turbines in 1890, and at the end of the nineteenth century, about 2,500 windmills generated approximately 30 megawatts of electricity in Denmark. Beginning around 1925, small (0.2 to 3 kilowatt), horizontal-axis, wind-powered generators were widely used to pump water and generate electricity. An estimated 6.5 million windmills were sold in rural America from the late nineteenth through the mid-twentieth centuries; in the 1930's their annual output was about 10 billion kilowatt hours. However, the Rural Electrification Administration, authorized in 1936, quickly constructed powerlines, relegating windmills and wind generators to remote, underdeveloped areas.

Winds flow from regions of high atmospheric pressure to those of low pressure. Cold, dense air forms areas of high air pressure, while warm air results in low pressure. Three belts of steady, strong winds characterize both the Northern and Southern Hemispheres: the polar easterlies, north of sixty degrees latitude; the prevailing westerlies, between 35 and 45 degrees latitude; and the trade wind belts, between 5 and 25 degrees latitude. These planetary wind belts are bounded by the relatively calm doldrums on the equator; the horse latitudes, around 30 degrees latitude; the subpolar lows, around sixty degrees latitude; and the polar highs. In addition, land is generally characterized by higher atmospheric pressure in the winter and lower pressure in the summer, contrasting with relatively uniform oceanic pressures. This may cause permanent or seasonal strong offshore or onshore wind. Thus wind energy development is more practicable in planetary wind belts and on favorable coasts. Topography further modifies wind distribution. Strong winds may concentrate in mountain passes, as in Hawaii and California. Flat plains open to planetary wind flow, as in Denmark and the American Midwest, are also favorable sites for wind generation.

Energy recoverable from wind is a function of three primary factors: wind velocity, volume of air collected, and air density. Velocity is the most important of these since the energy derivable from any given air mass increases as the cubic power of its velocity. Thus, high wind velocities are desirable. The volume of air exploited is a function of the size of the blades or vanes in the wind machine. Energy output increases as the square of blade size. Thus, larger blades or rotors are favored up to the maximum size that can operate and survive in the winds passing through the site. Air density is essentially constant under normal conditions and is an insignificant factor in wind machine design and location.

Windmills and wind generators may be mounted horizontally or vertically. Horizontally oriented rotors must face the wind to operate efficiently. Thus, they must be capable of rotating as the wind changes direction or operation will be interrupted. A machine with a vertical axis operates equally well with wind coming from any direction. Rotors of this design, however, are less efficient than horizontal rotors facing the wind. Twentieth century wind generator designs placed horizontally mounted, integrated rotors and generators atop towers or pylons.

Serious attempts at generating power for utility transmission lines began between 1935 and 1955. Many large, experimental generators were built, mostly in Europe. One of the largest wind turbines in the world was installed at Grandpa's Knob, near Rutland, Vermont. It was driven by a 53-meter (175-foot) propeller and generated 1.25 to 1.5 megawatts of power. In 1941 this generator became the first to feed power into a utility grid in the United States. It operated for only four and one-half years before a blade disintegrated, after which it was abandoned. Conditions during World War II, along with high costs, ended experimentation with large windmills in the United States, although research continued in Europe. In 1972 the Federal Wind Energy Program was initiated in the United States, which triggered development of wind energy machines under governmental subsidy. Several large horizontal-axis windmills and a few vertical turbines were built and tested. However, the most reliable and economical design proved to be smaller, three-bladed turbines clustered on wind farms. In the 1970's through the 1990's, wind farms proliferated in California. Development also expanded in Denmark and spread to most other European countries.

Wind generators do not emit carbon dioxide, sulfur dioxide, or nitrous oxides as do plants burning coal, oil, or natural gas. Thus there is no air pollution and no contribution to greenhouse gases. Wind generators produce no solid waste. Radioactivity is not involved, so there is no danger of catastrophic nuclear accidents. However, many people consider wind generators—especially wind farms, which may dominate extensive scenic landscapes—to be an offensive blight to the eyes. Many birds that try to fly through them are killed by the blades, and bird migration routes may be disrupted. Noise and subsonic vibrations generated by operating wind turbines can also be objectionable. Rotating metal rotors may generate enough electronic interference to affect television reception, and flying debris from disintegrating generators may be hazardous.

Ralph L. Langenheim

SUGGESTED READINGS: Paul Gipe, *Wind Energy Comes of Age* (1995), is a partisan summary of the development and future of wind energy. *Wind Energy in America: A History* (1996), by Robert W. Righter, portrays wind energy as sustainable and posing minimal environmental cost. *Wind Energy and the Environment* (1989), edited by D. T. Swift-Hook, addresses specific environmental problems from an engineering point of view.

SEE ALSO: Alternative energy sources; Energy policy; Solar energy.

Windmills in Altamont Pass in California. Wind generators provide a clean source of energy, but many consider them to be an eyesore, and they may interfere with the migratory routes of birds. (Deborah Cowder)

Windscale radiation release

DATE: October 10, 1957
CATEGORY: Nuclear power and radiation

On October 10, 1957, the Windscale nuclear reactor, located on the west coast of England, overheated. The resulting fire in the reactor core gave rise to one of the world's first serious nuclear power plant accidents. The release of significant amounts of radioactive material into the atmosphere caused short-term contamination of several hundred square miles of the surrounding countryside.

The first indication that radioactive material was escaping from the Windscale reactor came on the evening of October 10, when a nearby weather station detected an increase in background radiation. Health physicists considered the release of iodine 131 to be the most serious hazard. Iodine 131 falls to the ground, where it may be consumed by cows eating grass and concentrated in their milk. If humans drink this contaminated milk, iodine 131 concentrates in the human thyroid gland, where its radioactive decay can cause cancer. The government monitored milk from the region for evidence of iodine-131 contamination. Two days after the first release of radioactive material, milk samples from farms near the Windscale plant showed evidence of contamination. Initially, the government impounded milk supplies from a 3-kilometer (2-mile) circle around the plant. However, as iodine-131 contamination was detected over a wider region, milk produced over an area of about 500 square kilometers (200 square miles) was impounded. The contaminated milk was dumped into the sea, and milk for the people living near the Windscale reactor was trucked in from outside the contaminated region. Since iodine 131 decays rapidly, the ban on consumption of milk from the affected area lasted only a few weeks.

The fire also released a significant amount of polonium 210 into the atmosphere. Inhaling polonium 210 is a concern because it decays by emitting alpha particles, which are dangerous to the lungs. However, in the areas where the concentration of polonium 210 was highest, the additional exposure to radioactive decay was approximately equivalent to the average annual background rate of radioactive decay in the British Isles.

Filters in the stacks of the Windscale nuclear plant trapped a large fraction of the radioactive material released from the reactor. Thus, the design of the Windscale plant minimized the public health hazard of the event. The quick action by the British government to collect and destroy contaminated milk from the affected region also reduced the health effects of the release. Since the consumption of nonradioactive iodine reduces the amount of radioactive iodine absorbed by the thyroid, the distribution of iodine pills to the population of the region near the reactor would have further reduced the hazard.

The Windscale event released about 0.001 times the amount of iodine 131 into the atmosphere that was released by the fire at the Chernobyl nuclear reactor in 1986. A study conducted in 1997, forty years after the Windscale release, concluded that individuals who received the most serious exposure to the Windscale radiation should experience an increase in their likelihood of developing a fatal cancer by 6 in 1 million per year, compared to the normal fatal cancer risk of 1 in 300 per year. The conclusion was that the health effects of the Windscale release were minimal.

George J. Flynn

SEE ALSO: Nuclear accidents; Nuclear power; Radioactive pollution and fallout.

Wise-use movement

CATEGORY: Philosophy and ethics

The wise-use movement is an amorphous collection of antienvironmentalist groups promoting the economic exploitation of natural resources, particularly in the American West. Wise-use advocates typically have interests in logging, ranching, fishing, mining, motor sports, or real estate development. Members of such groups criticize environmentalists and environmental policies as excessive and elitist.

Wise-use activists maintain that true conservationists should seek a reasonable balance between human and natural values and that human values should come first. The diverse activists challenge environmental "extremists" and restrictive resource policies through obstruction, lobbying, advocacy in the media, litigation, and (rarely) violence. Founded in the West, where federal lands are extensive, wise-use concepts and organizations now have a national scope.

The antienvironmentalist movement developed as environmental policies of the 1960's and 1970's threatened historical values and patterns of natural resource development. Until the early 1970's, public land management promoted local resource development industries, particularly ranching, logging, and mining. Laws such as the Endangered Species Act of 1973 and various wilderness bills restricted access to federal land resources. Noneconomic and nonlocal values, such as wilderness preservation, advanced at the cost of traditional economic resource use. Environmental policies also affected private lands. The new environmental policies hurt the livelihood of workers and communities throughout the West, as well as the profits of large and small businesses.

The wise-use movement has strong roots in the Sagebrush Rebellion of the late 1970's, which occurred when western rural interests sought to counter national environmental politics and policies by shifting federal resources to state and local governments. However, Ronald Reagan's election as president of the United States in 1980 and the proexploitation tilt of appointees such as Secretary of the Interior James Watt undercut pressures for such change. The contemporary wise-use movement emerged in the late 1980's when conservative westerners felt that their rural interests had again been forgotten. The antienvironmental movement started as a creature of trade groups and large economic interests. Groups such as People for the West, for example, existed solely through the support of mining interests. Still, the movement attracted publicity, developed its strategy, and cultivated its ideological links with conservative think tanks and politicians.

Environmentalists initially underestimated the strength of the wise-use movement, perhaps because of its bombastic antienvironmentalist rhetoric and weak popular base. In the early 1990's, however, wise-use groups scored significant legal and political victories. Their powerful symbolic actions and organizational abilities attracted more popular support in the West. Broad public campaigns, such as that of the Yellow Ribbon Coalition to promote old-growth logging in the Pacific Northwest, demonstrated the new force and appeal of wise-use activism. Coalition activists attacked preservationist ethics by labeling the endangered species conflict as a matter of "people versus owls." Strong traditional interest groups such as the American Farm Bureau Federation came to favor wise-use ideology and strategies. By 1999 there were more than 250 groups associated with the wise-use label, and sympathizers claimed that the number was six times as many. The movement also enhanced its influence in Congress, especially after Republicans took control in 1994. However, continuing public concern over the environment restrained legislative action on wise-use proposals. Although willing to interfere in the implementation of laws such as the Endangered Species Act, Congress did not overturn any major environmental legislation.

In the early 1990's wise-use advocates broadened their national appeal by moving beyond public lands and traditional western issues to ally themselves with advocates in the property rights movement. In simple terms, the property rights movement argues that the Fifth Amendment requires that the government compensate for any loss of income arising from the federal regulation of property use. Such an interpretation could effectively preclude most significant environmental regulations, since the amount of compensation could be overwhelming. The United States Supreme Court has occasionally favored this view, such as in *Lucas v. South Carolina Coastal Council* (1992), but has not recognized all environmental regulatory costs as "takings" requiring compensation.

The wise-use movement appears to be a natural countermobilization against the environmentalist successes of the 1970's. The key question is

Advocates of wise use believe that public land should be made available for cattle grazing, as well as logging and mining. Opponents argue that such practices degrade areas that should be preserved in their natural state. (Ben Klaffke)

whether the movement can sustain its momentum. Continued public support for environmental regulations has thus far prevented even significant reform of the problematical Endangered Species Act and keeps the most radical wise-use concepts on the periphery of the debate. The public may never accept the wise-use perspective on the appropriate use of natural resources. Furthermore, wise-use proposals appeal primarily to rural interests and extractive industries, while the country continues to urbanize and shift toward a service economy.

On the other hand, the movement has had powerful allies in Congress, such as Republican senator Don Young, and the property rights issue generated significant public sympathy and business interest. The movement may benefit from the Supreme Court's changing view of property rights and its increased deference to government decisions. The fear of violating property rights may cripple environmental regulators, while groups seeking to stop the sale of federal natural resources may be less welcome in court.

Wise-use advocates have damaged the image of environmentalists without successfully selling their view of the appropriate balance between human and natural values. The movement continues to complicate environmental policy making without redefining the public purpose. The variety of wise-use activists ensures the movement's survival. Even the radical wing is diverse. For example, some Nevada wise users argue that the federal government lacks legal jurisdiction over counties and build illegal roads on federal lands. Others, like Ron Arnold of the Center for the Defense of Free Enterprise, seek to "destroy the environmental movement once and for all." Such extremists draw media attention and public fire. They also clear the ground for moderate groups to make seemingly reasonable inroads on environmental regulations. The movement may shed the "wise-use" label if it becomes identified with extreme acts, such as the bombings of Bureau of Land Management (BLM) offices in Nevada in the 1990's. Whatever the names and strategies, organized and sophisticated resistance to environmentalism has become an important part of the political landscape.

Mark Henkels

SUGGESTED READINGS: *The Wise Use Agenda* (1989), edited by Alan M. Gottlieb, summarizes the goals and perspectives of the wise-use movement as formulated at the 1988 Wise Use Strategy Conference. Most books on wise-use activism have an environmentalist view of the movement as a problem. *A Wolf in the Garden* (1996), edited by Philip D. Brick and R. McGreggor Cawley, provides diverse and thoughtful analyses of the movement's sources and impacts and considers what environmentalists can learn from wise-use advocates. Jacqueline Vaughn Switzer's *Green Backlash* (1997) comprehensively describes the dimensions of the counterenvironmentalist movement and analyzes both its strategies and policy goals. David Helvarg's unabashed attack on the wise-use movement, *The War Against the Greens* (1994), connects the wise-use movement to a broad range of arch-conservative activism, including the militia movement. For a much more complex view of the challenges facing contemporary environmentalism, including the battle for popular support, see Mark Dowie's *Losing Ground* (1995).

SEE ALSO: Antienvironmentalism; Grazing; Land-use policy; Privatization movements; Range management; Sagebrush Rebellion.

Wolf reintroduction

CATEGORY: Animals and endangered species

The reintroduction of wolves into former habitat areas to preserve the species and regain ecological balance has led to conflicts between environmentalists and ranchers.

Wolves have shared top-predator status with humans since prehistoric times. Wolves' wariness and the pack structure of their societies enabled them to coexist alongside human hunting and pastoral societies. However, large-scale agriculture and the rise of cities gradually but drastically shrank the wolves' forest ranges. The resultant raids on livestock led to the perception of wolves as vicious predators that should be exterminated. This process was well underway in Europe by medieval times. Wolves were extinct in the British Isles by 1776. In southern and eastern Europe, only a few survived in the most remote mountain ranges.

While Native Americans hunted individual wolves and displayed wolf trophies, their cultures usually respected wolves as worthy competitors and fellow creatures from whom humans might even learn spiritual lessons. As Europeans arrived in North America and spread across the continent, they brought the same dynamic and "vicious wolf" images that had operated in Europe. Settlers almost succeeded in eradicating wolves from the United States. By the 1960's, gray wolves, which had once ranged across most of North America, existed only in Alaska, Canada, and northern Minnesota.

The environmental movement of the 1960's prompted studies showing wolves as intelligent, social animals whose absence alters ecological balances. For example, without a natural predator, wild ungulate (hoofed) populations of elk or deer grow to unsustainable numbers and starve. When the Endangered Species Act was passed in 1973, wolves were among the best-known species listed. Further studies of wolf behavior and conservationist campaigns publicized the plight of the wolves. In 1991 the U.S. Congress held hearings on the possible reintroduction of wolves into wild areas in the nation's West. After gaining public and congressional support, the project moved forward. In 1995 wolves from Canada were released in two areas: Yellowstone National Park and central Idaho.

By most measures, these reintroductions succeeded. Both the Idaho and the Yellowstone wolves established multiple breeding pairs and packs and set in motion other intricate ecological effects. One example is a migration of elk to higher, timbered elevations. This preserves riverside plants, which provides more habitat for songbirds. Rereleased wolves are fitted with radio collars so their travels and fate can be monitored. Despite several deaths caused by vehicles, deliberate shootings, or nature, the wolf population has thrived and multiplied. By 1998 no added reintroductions were believed necessary in these two regions. The wolf presence has also boosted tourism, although from an environmental perspective this may be a mixed blessing. Success has inspired reintroduction proposals for other places, including Olympic National Park in Washington State, New York State's Adirondack Park, and even remote areas of Scotland and Japan. In a separate project, captive-bred red wolves were released into the southern Appalachians.

Despite compensation for livestock losses, many ranchers oppose wolf reintroduction and have gained some legal ground through a technicality of the Endangered Species Act.

Emily Alward

See also: Endangered species; Endangered Species Act; Predator management; Restoration ecology.

Wolman, Abel

Born: June 10, 1892; Baltimore, Maryland
Died: February 22, 1989; Baltimore, Maryland
Category: Water and water pollution

Engineer Abel Wolman pioneered water resource management strategies during the twentieth century.

Abel Wolman graduated from Baltimore City College in 1909 and Johns Hopkins University in 1913. He completed a second bachelor's degree in engineering at Johns Hopkins two years later and was employed as an engineer with the Maryland State Department of Health.

Wolman influenced the global environment by using engineering methods to improve public health. He was specifically concerned with water supplies, wastewater treatment, and sewage disposal, addressing the dangers of such waterborne diseases as typhoid. In 1919 Wolman worked with chemist Linn Enslow to standardize techniques to chlorinate municipal drinking water supplies. Many people believed that chlorine was poisonous, and Wolman convinced officials to adopt the procedure by explaining the disinfectant benefits. His chlorination methods were considered one of the most influential water management improvements for public health, resulting in decreased death rates from waterborne diseases.

Promoting a regional approach to water supply and sewage disposal, Wolman helped consolidate the Baltimore area into one water supply region. Named chief engineer for the Maryland State Department of Health in 1922, he analyzed municipal water supply needs and evaluated how to recycle wastewater. In 1935 President Franklin D. Roosevelt appointed Wolman as chairman of the Water Resources Committee of the Natural Resources Planning Board, which managed the federal government's water resources projects.

During the 1950's Wolman predicted possible environmental problems from unsafe disposal of nuclear wastes when private companies were granted access to nuclear energy through the Atoms for Peace program. Wolman served on the Reactor Licensing Board of the Atomic Energy Commission and insisted that concrete containment structures be built for the first commercial nuclear power plants in the United States. During the 1960's Wolman stressed the dangers of nonorganic environmental hazards to public health, noting that new technologically produced chemicals and contaminants had been introduced to water sources. He warned that humans must be held accountable for how they alter the environment and must envision how to protect resources.

Wolman was editor in chief of the *Journal of the American Water Works Association* (1921-1937) and *Municipal Sanitation* (1929-1935) and associate editor of the *American Journal of Public Health* (1923-1927). Johns Hopkins University presented Wolman with an honorary doctorate in 1937 when he established a Department of Sanitary Engineering at the university, serving as department chairman until he retired in 1962. Wolman received numerous awards, including the Tyler Prize for Environmental Achievement. A prolific author, Wolman's major articles were collected in *Water, Health, and Society: Selected Papers*, edited by G. F. White, in 1969. Wolman offered his environmental engineering expertise worldwide as a consultant, and he emphasized the responsibility of engineers to protect environmental quality. An advocate for the poor, Wolman testified against landlords who did not provide clean water for tenants. The *Baltimore Evening Sun* promoted environmental engineering during the International Drinking Water Supply and Sanitation Decade in the 1980's by eulogizing Wolman: "[He] envisioned a world in which the most basic of necessities, water to drink, would be safe and plentiful to all peoples of the world."

Elizabeth D. Schafer

SEE ALSO: Chlorination; Drinking water; Environmental engineering; Sewage treatment and disposal; Water conservation; Water pollution; Water quality; Water treatment.

World Fertility Survey

DATE: 1972-1984
CATEGORY: Population issues

The World Fertility Survey (WFS) used a series of interviews with women to document population growth and family-planning measures in various regions throughout the world. It was the first time such information had been gathered from developing regions such as sub-Saharan Africa. Most countries used the information to cope with and plan for population changes.

During the 1940's demographer Frank Notestein and others developed what has become known as "transition theory" to describe how population change occurred following modernization and industrialization. He noted that populations were dramatically increasing around the world, especially in the poorer regions of the world, as mortality declined in the context of high fertility. Concerns about overpopulation were voiced, but the nature of the change was poorly understood. The Knowledge, Attitude, and Practice (KAP) surveys in the late 1960's measured who wanted children but were soon criticized as inadequate in terms of methodology and regional coverage.

Maurice Kendall, a statistician, partly in an effort to revitalize the International Statistical Institute (ISI), proposed the WFS in 1971 and led the planning in 1972. The surveys followed in 1973. The WFS was a collection of comparable, high-quality interviews of 341,300 women from sixty-one countries on maternity and marital histories, contraceptive knowledge, work histories, and husbands' backgrounds in order to document population growth and family-planning measures. Households rather than families were surveyed, and only women under the age of fifty who had been married at some point were eligible. The standard questions could be supplemented by "modules" on abortion and economic or community factors. Hundreds of technical reports and research articles using information gained during the surveys were published, particularly in the 1980's. Although funding stopped in 1984, the work has been continued through the Demographic and Health Surveys (DHS), a program based in the United States.

The WFS revealed that many more women would use family-planning services if available, how infant mortality rises when intervals between births are shorter, and that breast-feeding can slow the arrival of the next child. The WFS also showed that fertility had significantly declined in regions of Asia, Latin America, and the Middle East, although it remained relatively high when contrasted to the United States and Europe. No decline in fertility was evident for sub-Saharan Africa, and transition theory did not account for these regional differences. Seventeen of the forty-two countries that participated in the WFS included a "community module" with the goal of determining whether reproductive behavior was in response to increased access to family-planning services or a function of declining mortality thanks to better primary health care services. Although community effects were clearly reflected in infant and child mortality patterns, they were not strong for reproductive behaviors. In particular, it was not clear that access alone was sufficient to change reproductive behaviors. Part of the problem was that certain concepts, such as household, had very different meanings in different areas (for example, in settings where husbands and wives

The World Fertility Survey, which occurred between 1972 and 1984, consisted of interviews with women around the world for the purpose of gathering data on population growth and family planning. (AP/Wide World Photos)

had separate residences). Furthermore, details on the health services available were often inaccurate. The earliest versions of the less well-funded DHS had fewer questions on these issues, a deficiency that has made it difficult to identify the motivations and mechanisms that drive fertility change.

Joan C. Stevenson

SEE ALSO: Ehrlich, Paul; Population-control and one-child policies; Population-control movement; Population growth.

World Heritage Convention

DATE: 1972
CATEGORY: Preservation and wilderness issues

The World Heritage Convention is a worldwide agreement by nations to protect important cultural, historic, and natural sites.

The United States proposed the World Heritage Convention, also known as the International Convention Concerning the Protection of the World Cultural and National Heritage, in 1972 to commemorate the one-hundredth anniversary of the establishment of Yellowstone National Park and was the first nation to sign it. The convention essentially promotes the U.S. national park concept worldwide. By the convention's twenty-fifth anniversary in 1997, nearly 150 nations had ratified the agreement and placed more than five hundred sites on the World Heritage list. Nations signing the World Heritage Convention voluntarily nominate sites within their nations for inclusion on the World Heritage list. Nominations are reviewed by the World Heritage Committee, a twenty-one-nation body elected from among the signatory nations. The committee also places sites threatened by natural disaster or civil strife on the World Heritage in Danger list.

Nations signing the World Heritage Convention pledge to identify, maintain, and preserve important natural and cultural sites within their territory as part of the universal heritage of humanity; they also pledge to promote and publicize these sites for worldwide public enlightenment. In addition, member nations assist one another with studies, advice, training, and equipment necessary to resolve problems, restore damaged areas, and establish programs to protect, preserve, and publicize the sites. The World Heritage Committee offers technical advice and assistance through its World Heritage Fund. Individual nations offer direct nation-to-nation assistance.

To be listed, sites must possess outstanding, universally recognized cultural and natural features. Sites include both human-made constructions and natural areas. Selected natural sites include areas that represent major stages in the earth's evolutionary history, ongoing geological processes or biological evolution, or human interaction with the environment; that contain unique, rare, or superlative national phenomena; or that provide habitats for rare or endangered plants and animals.

Each signatory nation maintains sovereignty over its sites and is responsible for site maintenance and protection. Listed sites include those owned by the national government (such as national parks and national historic landmarks), state or tribal governments, local governments, or private groups or individuals, with the owners pledging to protect their properties in perpetuity.

International sites include the Great Wall of China, the Taj Mahal in India, Ecuador's Galápagos Islands, the Tower of London, and the massive Spanish fortifications at San Juan, Puerto Rico. North American sites include twenty in the United States, including Grand Canyon National Park in Arizona; Everglades National Park in Florida; Independence Hall, in Philadelphia, Pennsylvania; Cahokia Mounds State Historic Site in Illinois; Taos Pueblo in New Mexico; President Thomas Jefferson's Monticello in Virginia; and the Statue of Liberty in New York.

Gordon Neal Diem

SEE ALSO: National parks; Preservation; Wilderness areas.

Y

Yokkaichi, Japan, emissions

CATEGORY: Air and air pollution

The heavy pollution of the Yokkaichi area by petrochemical plants caused widespread health problems for residents and led to landmark court cases and legislation.

Yokkaichi, a port city on Japan's Ise Bay, developed as a major industrial and petrochemical center in the early twentieth century. The demands of World War II and Japan's postwar recovery led to further industrial expansion in the area, and an oil-refinery complex known as the Yokkaichi Kombinato was created in the 1950's. Although the complex was an economic success, the pollution it generated was soon linked to breathing difficulties and a variety of other health problems in area residents. Researchers investigating the complaints found a high correlation between airborne sulfur dioxide and the incidence of bronchial asthma in children and of bronchitis in older people. Nevertheless, in 1963, a second industrial complex was opened in the region, and a third was added in 1973. In one district of Yokkaichi, airborne sulfur dioxide levels were found to be 800 percent above normal.

In the early 1960's, nearly half of the area's young children, nearly a third of its elderly, and approximately a fifth of its young adults had developed respiratory abnormalities.

In 1967, a group of Yokkaichi residents filed a suit against the Shiohama Kombinato, which ran one of the petrochemical complexes. In 1972, the plaintiffs were awarded nearly three hundred thousand dollars in damages. The award marked the first time that a group of Japanese companies had been held liable for damages, setting a precedent that made other kombinatos vulnerable to such litigation. As a result of the case and ensuing controversy, in 1967 Japan's government enacted a basic antipollution law. Within the next several years, additional laws were adopted spelling out redress rights for victims from the Yokkaichi area and also for residents of polluted areas near Kawasaki and Osaka. Regulations requiring refineries to adhere to pollution-abatement policies were also strengthened.

As a result of such measures, by the mid-1970's, airborne sulfur-dioxide levels in the Yokkaichi region had decreased by more than 60 percent, and the rate of respiratory complaints among area residents had also declined sharply. By the 1990's, nearly a hundred thousand Japanese citizens had been declared eligible for compensation under the new laws.

Alexander Scott

SEE ALSO: Air pollution; Ashio, Japan, copper mine; Environmental illnesses; Minamata Bay mercury poisoning; particulate matter.

Yosemite

CATEGORY: Preservation and wilderness issues

Yosemite is a scenic valley of the Merced River located on the western slope of the Sierra Nevada in east-central California. Yosemite Valley is a part of Yosemite National Park, which covers 761,000 acres and features alpine wilderness, lakes, waterfalls, towering mountains, huge granite formations, and giant sequoia trees.

Caucasians first entered Yosemite Valley in 1851 during conflicts between California gold miners and local American Indians. In 1864, after lobbying by early conservationists, U.S. president

Abraham Lincoln signed the Yosemite Land Grant, which gave Yosemite Valley and an area a few miles to the south called Mariposa Big Tree Grove—39,200 acres of federal land in all—to the state of California as a reserve to be used for public enjoyment and recreation. The state-supervised reserve was the first area specifically set aside by the United States to be preserved for all future generations. Its inception planted the seed for the national park system, although Yellowstone, not Yosemite, was the first site officially designated a national park in 1872.

Despite the protection provided by the 1864 act, the floor of Yosemite Valley was used for commercial purposes, including plowing and orchard planting, timber cutting, and grazing. The unprotected, high mountain country surrounding the valley was also logged and grazed. American preservationist John Muir, while exploring the area in the late 1860's, became concerned about this disturbance. For the next two decades, he publicized his concerns and worked to preserve the high country.

Muir's efforts paid off in 1890 when some 932,000 acres of the surrounding high country gained federal protection through the establishment of Yosemite National Park. However, the valley itself and Mariposa Grove remained under California's jurisdiction. Muir and others then worked to get Yosemite Valley transferred from the state to the federal government in order to protect it and consolidate the public holdings into a single, unified national park. To help rally public support for Yosemite Valley and other land in the Sierra Nevada, Muir and others founded the Sierra Club in 1892.

In 1903 Muir persuaded President Theodore Roosevelt that the valley needed federal protection; in 1906 California ceded the area back to the federal government. Yosemite Valley and Mariposa Grove thus became a part of Yosemite National Park. There was a cost, however: The

Yosemite Valley became a California state park in 1864. In 1906 the area was ceded to the federal government and became part of Yosemite National Park. (Douglas Long)

overall size of the park was reduced, and private mining and timber holdings were excluded from restrictions. In 1913, despite the opposition of Muir and other conservationists, Congress approved a project to dam and flood Hetch Hetchy Valley in the northwest corner of Yosemite National Park as a reservoir for the city of San Francisco.

Since its inception as a state park in 1864, Yosemite has been a magnet for tourists. By the late twentieth century, it was attracting millions of visitors—and their automobiles—each year. These visits have strained the local environment, bringing traffic congestion, smog, and damage to the land.

Jane F. Hill

SEE ALSO: Grazing; Hetch Hetchy Dam; Muir, John; National parks; Nature reserves; Sierra Club.

Yucca Mountain, Nevada, repository

CATEGORY: Nuclear power and radiation

The selection of Yucca Mountain as a repository for the permanent disposal of highly radioactive waste has been controversial because of a number of environmental concerns, including potential volcanic disturbances, earthquakes, rapid groundwater movement, hydrothermal activity, and the necessity to transport nuclear waste across the country.

The 1982 Nuclear Waste Policy Act began the process of establishing a permanent underground storage repository for highly radioactive waste. In 1987 Congress passed the Nuclear Waste Policy Amendments Act, which singled out Yucca Mountain as the only site to be studied for the national repository. Yucca Mountain is located about 160 kilometers (100 miles) northwest of Las Vegas, Nevada. Plans indicated that the storage facility would be 305 meters (1,000 feet) underground when completed. The repository was not expected to be completed until at least 2010.

Environmentalists have expressed concern about Yucca Mountain's volcanic history. The mountain was formed millions of years ago by a series of explosive volcanic eruptions. While the explosive type of volcano is extinct, scientists are studying seven small, dormant volcanoes in the area. Two of these volcanoes may have been active within the last ten thousand years, which is a relatively short time in geologic terms. However, Department of Energy (DOE) scientists have stated that the possibility of an eruption is only one in seventy million per year.

The Yucca Mountain area also contains more than thirty known earthquake fault lines, some of which have been active since work began on the repository. In 1992 an earthquake measuring 5.2 on the Richter scale caused considerable damage to the aboveground Yucca Mountain project field operations center. DOE scientists have responded that the threat of earthquakes to the underground repository is not a significant problem since underground structures can more easily withstand the ground motion generated by earthquakes.

There is also considerable concern about the movement of groundwater, which could transport any radioactive material that may leak from the repository. Evidence indicates a possible flow rate in excess of what is considered desirable. The DOE has stated that there is no cause for alarm, and its scientists are conducting additional testing to measure the scope of the problem.

Scientists have also discovered renewed evidence of hydrothermal activity at Yucca Mountain. Perched water (geothermal water that has risen to the surface and drained back down) has been found in several boreholes made in the area. Similar to groundwater movement, perched water can act as a transport mechanism for any nuclear waste material that may leak from the repository. DOE scientists have minimized the potential danger of perched water to the integrity of the storage facility.

Finally, there has been considerable concern exhibited by environmentalists over the transportation of highly radioactive waste to the permanent national repository at Yucca Mountain. Opponents of a centralized national repository have expressed concern over the possibility of an

A nuclear waste disposal site at Yucca Mountain, Nevada, in 1992. The plan to create a permanent repository for high-level radioactive waste at Yucca Mountain has been criticized for a number of reasons. (AP/Wide World Photos)

accident during the transport of the nuclear waste, which could result in the release of radiation in a populated area. Proponents of the project counter that the likelihood of radioactive contamination is slight and point to the existing successful safety record for the transport of nuclear material.

Kenneth A. Rogers and Donna L. Rogers

See also: Nuclear and radioactive waste; Nuclear regulatory policy.

Z

Zapovednik system

CATEGORY: Preservation and wilderness issues

In the first decade of the twentieth century, Russian zoologist G. A. Kozhernik developed the idea of nature reserves called zapovedniki *dedicated to the protection of entire ecosystems.*

The establishment and administration of nature reserves has generally mirrored the development of biology, ecology, and related sciences. The systematic development of nature reserves with a scientific basis accompanied the biological developments of the nineteenth century. During that time, Russian biologists related English evolutionist Charles Darwin's evolutionary theories to the environment. From such early scientific studies came a recognition of the importance of natural areas and their preservation.

During the twentieth century the popularity and support of Russian nature reserves waxed and waned, often influenced as much or more by the political and economic climate as scientific advances. As in other countries, environmental concerns were seldom of prime importance in the Soviet Union, a giant state that struggled for its own survival for more than seven decades before disintegrating into separate republics in 1991.

In January of 1919, while the new Soviet government was struggling for its existence, agronomist Nikolai N. Pod'iapol'skii proposed the establishment of the regime's first *zapovednik*, or nature reserve, at Astrakhan. The first five years of the Soviet Union also saw the organization of the All-Russian Society for the Protection of Nature, a volunteer conservation organization that had the effect of enhancing environmental awareness among citizens.

In 1921 Soviet leader Vladimir Lenin signed legislation called On the Protection of Monuments of Nature, Gardens and Parks. This empowered the Ministry of Education to declare parcels of nature with special scientific, cultural, or historical value as *zapovedniki.* Between 1919 and 1932 a total of 128 *zapovedniki* with a total area of 12.6 million hectares were created. The reserves represented 0.56 percent of the total area of the country. They varied in size from parcels of fewer than 100 hectares to a few with more than 1,000,000 hectares each. Most were in western Russia, Ukraine, or the Caucasus, with a smaller number in Siberia and the far east on the Pacific coast.

The *zapovedniki* were administered by two separate government agencies with different philosophies and sets of goals. The Ministry of Education maintained relatively pristine reserves for their aesthetic properties and as sites for preservation and scientific research. During the 1920's these reserves were utilized by several important Russian ecologists who pioneered studies in such areas as productivity, trophic relationships, and predator-prey interactions in ecosystems. Unlike national parks in the United States, tourism was not an important consideration.

In contrast, nature reserves under the management of the Ministry of Agriculture were maintained primarily as centers of agricultural production and experimentation. Scientific management policies were sought to maximize yields of timber, fur, and other products of value to the Soviet economy. These goals compromised conservation efforts.

The emergence of Joseph Stalin as dictator of the Soviet Union in 1929 and the outbreak of World War II in 1939 had disastrous effects on conservation efforts in general and the Soviet *zapovedniki* in particular. Stalin disbanded the Society for the Protection of Nature and introduced a vigorous program of industrialization.

Lip service was paid to conservation while polluting industries were allowed to operate with few regulations.

During this same period, the conservation movement in the Soviet Union was subjected to a similar fate as genetics: Scientific principles were abandoned in favor of incorrect, unsupported ideas favored by Marxist theorists. As a result, many unspoiled nature reserves established during the 1920's were dismantled or converted into agricultural enterprises. By 1952 only forty reserves with a total of 1.5 million hectares were left, representing 12 percent of that which existed before the war.

Under Leonid Brezhnev, first secretary of the Communist Party during the 1960's, a new interest in conservation emerged. An improved economy and a reduction in the restrictions on individual freedom led more citizens to become tourists. An increased appreciation of natural areas resulted from visitation by international tourists. By 1981 nature reserves had become popular, and the number had grown to 129, surpassing the number that had existed before World War II.

After the breakup of the Soviet Union in 1991, new challenges emerged. Faced with a legacy of widespread pollution and near bankruptcy, the former states of the Soviet Union struggled to maintain themselves. Many conservationists feared that the environment in general, including the system of 147 (as of 1999) *zapovedniki*, would suffer. The biodiversity of the reserves was threatened by both a lack of funds and impoverished local people who destroyed the flora and fauna in order to survive. Most observers believed that the assistance of outside agencies would be required to prevent their deterioration.

Thomas E. Hemmerly

Suggested readings: For an overview of science and its relationship to Russian history and culture, see *The Cambridge Encyclopedia of Russia and the Former Soviet Union* (1994), edited by Archie Brown, Michael Kaser, and Gerald S. Smith. Among the books that focus more specifically on ecological and environmental topics in Russia are Ruben A. Mnatsakanian's *Environmental Legacy of the Former Soviet Union* (1992); Andrew J. Belloso's *Saving the Heart of Siberia: The Environmental Movement in Russia and Lake Baikal* (1993); and Douglas R. Weiner's *Models of Nature: Ecology, Conservation, and Cultural Revolution in Soviet Russia* (1985).

See also: Biodiversity; Biosphere reserves; Ecology; Ecosystems; Nature reserves; Soviet Plan for the Transformation of Nature; Wilderness areas.

Zebra mussels

Category: Animals and endangered species

The zebra mussel is an endemic European freshwater bivalve that occurs in North America as an exotic. Research efforts have focused on its life history patterns in North America, its potential influence on aquatic systems, and potential control methods.

Zebra mussels (*Dreissena polymorpha*) are native to southern Russia and are thought to have been introduced into Lake Saint Clair—which lies between Michigan and Ontario, Canada—in 1986 via discharged ballast water. Since its introduction, the zebra mussel has become widely dispersed, occurring in all of the Great Lakes by 1990; by 1994 it had appeared in or adjacent to nineteen states. This rapid dispersal is largely caused by their ability to attach to boats that navigate these waters, as well as the ability of all life stages to survive overland transport (for example, living on the hulls of boats that are transported between lakes). The prognosis for continued dispersal appears good, with the expectation of increasing colonization of inland lakes.

Zebra mussels typically live three to five years, with shell sizes that average 25 to 35 millimeters (1 to 1.4 inches). Females usually reproduce during their second year and can produce more than forty thousand eggs in one reproductive cycle. After fertilization, veliger larvae emerge within three to five days and are free-swimming for up to one month. Dispersal during this time

Zebra mussels cling to a pier that has been removed from Lake Erie. After the species was introduced into North American waters in 1986, it rapidly spread throughout the Great Lakes region, displacing native unionid mussels. (Jim West)

is primarily caused by water currents. Larvae then settle to the bottom where they crawl via a foot, looking for suitable substrate (preferred to be hard or rocky). They secrete proteinaceous byssal threads for attachment from a byssal gland, located in the foot.

From an ecological and environmental perspective, one of the most important concerns is the role of zebra mussel colonization in reducing abundance and species diversity of native unionid mussels. Because native unionid mussel beds provide the type of hard substrate that zebra mussels prefer, zebra mussels readily colonize such areas, negatively influencing feeding, growth, locomotion, respiration, and reproduction of native unionids.

In addition, zebra mussels have important environmental influences because of their role as biofoulers. They colonize pipes and can restrict water flow, negatively affecting water supply to hydroelectric facilities, nuclear power plants, and public water supply plants. Zebra mussels can attach to the hulls of boats, leading to increased drag, and their weight can sink navigational buoys. Because zebra mussel densities have been measured as high as 700,000 mussels per square meter, they clearly can cause serious problems.

Despite their ability to negatively affect aquatic systems, zebra mussels can have positive effects on water quality through their role as biofilters. An adult zebra mussel can filter the phytoplankton from as much as 1 liter of water per day and can significantly alter water quality. However, even this influence has potential negative consequences by reducing the amount of food available for zooplankton and eventually for recruiting fishes. This can also lead to a change in food webs from phytoplankton-dominated systems to macrophyte-dominated systems. Given their filtering ability, zebra mussels also tend to bioaccumulate substances, making it possible to increase the concentration of toxic substances that are passed up the food web.

Dennis R. DeVries

SEE ALSO: Biodiversity; Introduced and exotic species.

Zero Population Growth

DATE: founded 1968
CATEGORY: Population issues

Zero Population Growth is a nonprofit organization based in the United States that uses advocacy and education to slow population growth in order to achieve a sustainable balance among people, resources, and the environment.

Zero population growth is a demographer's term that refers to a balance between population increases (births) with decreases (deaths) to prevent any further rise in the number of people competing for limited natural resources. In 1968 Stanford University professor Paul Ehrlich, Connecticut lawyer Richard Bowers, and Yale professor Charles Remington established an organization called Zero Population Growth (ZPG) shortly after the Sierra Club's publication of Ehrlich's book *The Population Bomb* (1968). The book, which dealt with the impact of the growing U.S. population on the environment, generated interest that was naturally channeled to ZPG.

ZPG has focused its efforts on the United States, which proportionally uses more resources than any other nation in the world. The group popularized its goals through the use of slogans, songs, posters, bumper stickers, magazine advertisements, and television public service announcements. The initial message, "Stop at Two," urged couples to have only two children. Ehrlich's numerous appearances on *The Tonight Show* were particularly effective in promoting ZPG, which has grown to more than fifty thousand members. It generates and distributes educational materials that incorporate population and environmental issues into classrooms ranging from kindergarten to high school.

Since its beginning, ZPG has sought to make the connection between the environment and population. While recognizing that stabilizing the number of people is only part of the solution, the group has emphasized that reduced reproduction is a contribution to global well being that anyone can make. In 1968 advocating reproductive restraint was controversial, as large families were generally considered desirable. Furthermore, the means of reproductive control were not readily available. Until a U.S. Supreme Court decision in 1965, contraceptive use was illegal for married couples, and it continued to be illegal for unmarried couples until 1972. Sterilization was severely restricted, and abortion was not legalized until the 1973 *Roe v. Wade* decision. ZPG worked with other reproductive rights and civil rights organizations to change laws so that people could control their own reproductive health. By 1975 fertility in the United States had decreased from 3.4 to 1.8 children per woman. ZPG was among the catalysts for this demographic transition.

ZPG's initial message was directed toward the American white middle class but eventually began targeting minority and immigrant groups, as well as the rich and the poor. It also began focusing more on environmental impact, consumption, and quality of life issues rather than numbers. The diversity of its audience and goals makes slogans less effective.

ZPG may now be considered mainstream, although it still has detractors who call for no restraints on reproduction short of sexual abstinence or who profess that free markets and technological improvements will find the resources to support an ever-increasing population. However, there are those who believe that the message of Zero Population Growth may have been too timid and that world population has already exceeded the sustainable carrying capacity of the earth.

James L. Robinson

SEE ALSO: Ehrlich, Paul; Population-control and one-child policies; Population-control movement; Population growth; Sustainable development.

Zoos

Category: Animals and endangered species

Zoos were originally entertainment venues, though many have shifted their focus to education and active conservation of endangered and threatened species.

Nearly 600 million people worldwide visit a zoo each year—roughly 10 percent of the global population. Modern zoos, often called wildlife conservation parks or natural wildlife parks, have replaced cages of concrete and steel with simulated natural environments, and new animals are obtained via selective breeding instead of being captured from the wild.

Early zoos were the sole province of the wealthy; the first recorded zoo in history belonged a Chinese emperor in 1100 b.c.e. It was not until the nineteenth century that zoos were open to the public. The word "zoo" derives from the phrase "zoological park," and that was what the first zoos were designed to be: afternoon diversions along the same lines as the amusement park or the circus. Exotic beasts from newly charted regions were captured and displayed with little regard for their health or emotional well being. Mortality was high, and display animals were constantly replaced with animals captured from the wild, of which there seemed to be an inexhaustible supply.

The first zoo to use moats to separate animals from visitors was established in Germany by Carl Hagenbeck in 1907. These moats provided visitors with an unobstructed view and, depending on their placement, made it seem as if the animals were free. While the bars were gone, the habitat was still nothing like what the animals were accustomed to in the wild. Those animals that did not spend their days sleeping often displayed near-psychotic behavior patterns, such as pacing, head butting, and even self-mutilation.

Two things changed the way zoos functioned during the twentieth century. First, movies and television allowed potential visitors to see the animals in their natural habitats, and suddenly giraffes, lions, and zebras were no longer quite so exotic. Second, wild animals were becoming more scarce, and words such as "conservation" and "endangered" entered the collective vocabulary. Acquiring specimens from the wilderness became more costly, and zoos began to look at internal breeding programs to replenish their stock. However, they found that animals kept in unnatural and in some cases inhumane conditions would not breed.

New zoo enclosures were designed to encourage natural behavior in animals by replicating their natural environment as much as possible while still ensuring the safety of both the animals and the zoo visitors. Animals began receiving healthier diets and, when possible, were allowed to feed in much the same way they would in the wild—by digging, foraging, or grazing. Human contact with orphaned and injured animals was kept to an absolute minimum, and some zoos took the additional step of not naming their animals to discourage anthropomorphism. By 1995, 80 percent of the mammals on display in zoos were born in captivity.

Zoo managers continue to struggle to balance science, conservation biology, scarce resource allocation, and ethics. Among the choices that must be made are whether predators should be offered the chance to exercise natural hunting behaviors by being offered live prey, or whether zoos should maintain potentially deadly animals that are necessary for breeding programs but are dangerous and difficult to control, such as macaques, many of which harbor the deadly hepatitis B virus, or adult male elephants. Another dilemma is the question of what should become of "surplus" animals that are inbred, unable to reproduce, or are otherwise genetically inferior.

Municipal bureaucracies can also hamper zoo conservation efforts. Zoo managers must often combat local governments and public opinion when dealing with unpopular issues, such as surplus animals and resource allocation. In addition, budget cuts have forced zoo managers to turn to the private sector for financial assistance. Fund-raising activities range from the traditional "adopt an animal" programs to the extraordinary commercial venture of selling "exotic compost."

There are those who question whether zoos should exist at all—whether it is cruel or unusual to take animals from their natural habitat and place them on display. The People's Republic of China "rents" giant pandas that would better serve their species by remaining in the wild or in a captive-breeding program to replenish their numbers. Critics claim that the money devoted to zoos and captive-breeding programs would be better spent on preserving the animals' natural habitats. To combat these types of criticism, some zoos began to change their focus from "collecting" wildlife to "protecting" wildlife, also known as field conservation. In these new exhibits, zoo visitors view exhibits linked with protection and conservation programs in natural habitats, allowing visitors to connect what they are seeing in captivity to what's worth saving in the wilderness. Some zoos have taken the additional step of "adopting" wildlife refuges.

Despite the criticism, the fact remains that there are many animals species that simply could not survive without the existence of zoos and captive-breeding programs. Ironically, where historical zoos replenished their stock from the wilderness, some zoos are now replenishing the wilderness with captive-bred animals.

P. S. Ramsey

SUGGESTED READINGS: *Zoos* (1992), by Daniel Cohen and Susan Cohen, provides an excellent overview of zoos and their development. For a variety of perspectives on zoos and their place in the conservation movement, see *Ethics on the Ark* (1995), edited by Bryon G. Norton, Michael Hutchins, Elizabeth F. Stevens, and Terry L. Maple. *The Animal Estate: The English and Other Creatures in the Victorian Age* (1987), by Harriet Ritvo, offers a historical view of the first public zoos. To learn more about captive breeding and what must be done to preserve endangered species, see *Last Animals at the Zoo* (1992), by Colin Tudge.

SEE ALSO: Animal rights; Animal rights movement; Captive breeding; Convention on International Trade in Endangered Species; Endangered species.

Timeline

Milestones in Modern Environmentalism

1862 The Homestead Act opens the western United States to settlement and exploitation
1872 Yellowstone is named the first U.S. National Park
1902 The U.S. Congress passes the Reclamation Act, providing for the irrigation of the West
1903 Pelican Island is named the first U.S. National Wildlife Refuge
1903 Theodore Roosevelt and John Muir visit the Yosemite area
1905 Gifford Pinchot is named to lead the U.S. Forest Service
1905 The National Audubon Society is founded in New York City
1906 The International Association for the Prevention of Smoke is founded
1908 Theodore Roosevelt bans mining in the Grand Canyon
1908 Roosevelt convenes U.S. governors to discuss conservation
1910 The U.S. government establishes the Bureau of Mines
1911 The Triangle Shirtwaist factory fire prompts workplace safety regulation
1912 The U.S. Public Health Service is established
1913 The U.S. Congress passes the Migratory Bird Act
1913 The Los Angeles Aqueduct is completed
1913 Construction of the Hetch Hetchy Dam is approved by the U.S. Congress
1914 Martha, the last passenger pigeon, dies in a Cincinnati zoo
1916 New York City institutes the first comprehensive zoning ordinance
1916 The U.S. National Park Service is established
1917 Mount McKinley National Park becomes the first national park added to the new National Park Service
1918 The U.S. approves the Migratory Bird Treaty with Canada
1919 Vladimir Lenin approves the creation of the first Soviet nature preserve
1919 The National Parks and Conservation Association is founded
1920 The Mineral Act is adopted to regulate exploitation of U.S. public lands
1922 The Izaak Walton League is formed
1923 The Federal Power Commission rules against the construction of dams on California's Kings River
1924 The Soviet Union establishes a nature-protection society
1924 The Gila National Forest is preserved as a wilderness area
1924 The Oil Pollution Act establishes civil and criminal penalties for polluters
1925 The Geneva Protocol bans the use of poison gas and bacteriological weapons
1926 Vladimir Vernadsky publishes *The Biosphere*
1927 The Indiana Dunes State Park is created
1927 The U.S. Food and Drug Administration is established
1929 The government of Tanganyika proclaims the creation of the Serengeti National Park
1930 Dutch Elm disease begins devastating trees in the United States
1930 Construction of the Hawk's Nest Tunnel begins, leading to hundreds of deaths
1930 The Canadian government passes the National Parks Act
1930 The Du Pont corporation introduces Freon, the first commercial chlorofluorocarbon
1933 The Civilian Conservation Corps is established, putting millions of people to work on forestry and beautification projects

1933 The U.S. government creates the Tennessee Valley Authority
1933 Robert Marshall advocates preservation in *The People's Forests*
1933 Construction of the Grand Coulee Dam, the world's largest hydroelectric project, begins
1934 The Dust Bowl begins to devastate the central United States
1934 The Migratory Bird Hunting Stamp Act is passed to fund waterfowl conservation
1934 The U.S. Congress passes the Taylor Grazing Act
1934 Robert Marshall and Aldo Leopold form the Wilderness Society
1935 The *Trail Smelter* arbitration of a dispute between the United States and Canada affirms national responsibility for transboundary pollution
1935 The U.S. Soil Conservation Service is established to deal with the ravages of the Dust Bowl
1936 Boulder Dam is completed, providing power to the American Southwest
1936 Ansel Adams begins lobbying Congress to preserve the Kings Canyon
1936 Jay "Ding" Darling founds the National Wildlife Federation
1937 Passage of the Pittman-Robertson Act provides funding for state wildlife protection
1938 Englishman George Callendar announces a connection between industry and carbon dioxide emissions
1940 The U.S. Fish and Wildlife Service is formed
1941 Aerosol containers are introduced
1942 The first controlled fission reaction heralds the nuclear age
1943 The Sierra Club successfully opposes repeal of the Antiquities Act
1943 "Black Wednesday" in Los Angeles demonstrates the dangers of smog
1943 The Alaska Highway is completed
1944 Norman Borlaug begins to work on the creation of high-yield wheat
1945 2,4-D, the first modern herbicide, is introduced
1945 Grand rapids, Michigan, becomes the first U.S. city to fluoridate its water supply
1945 The United States ends World War II by using atomic bombs to annihilate Hiroshima and Nagasaki
1946 The International Whaling Commission is formed
1946 Harry S Truman creates the U.S. Bureau of Land Management
1946 The United States establishes the Atomic Energy Commission to regulate nuclear technology
1946 The first cloud seeding is attempted
1947 Chloramphenicol, an antibiotic with toxic side effects, is discovered
1948 Henry Fairfield Osborn publishes *Our Plundered Planet*, anticipating the age of environmentalism
1948 The Soviet Union undertakes the "Plan for the Transformation of Nature"
1948 The U.S. Congress approves the first Water Pollution Control Act
1948 The World Conservation Union is founded
1948 Donora, Pennsylvania, suffers a deadly temperature inversion
1949 Aldo Leopold publishes *A Sand County Almanac*
1949 The Soviet Union adopts measures to reduce air pollution
1950 The first effects of the mercury poisoning of Japan's Minamata Bay become apparent
1950 Meteorologists make the first computerized weather prediction
1951 The Nature Conservancy is founded in Washington, D.C.
1952 David Brower is named the executive director of the Sierra Club
1952 John D. Rockefeller III founds the Population Council
1952 A severe episode of smog in London contributes to thousands of deaths
1952 An explosion occurs in Canada's Chalk River Nuclear Reactor
1953 Keep America Beautiful is founded
1954 The Soviet Union completes its first nuclear power plant

1954 U.S. nuclear testing at Bikini Atoll contaminates nearby islands

1954 The Atomic Energy Act grants the federal government regulatory authority over the U.S. nuclear industry

1954 The Humane Society of the United States is established to protect animals

1955 Diquat herbicide is developed for weed control

1955 The U.S. Congress passes the Air Pollution Control Act

1955 A meltdown occurs in the first breeder reactor

1956 The U.S. Congress strengthens the Water Pollution Control Act

1956 The Echo Park Dam proposal is defeated by environmental activists

1956 The first British nuclear power plant is opened

1957 Norman Cousins founds SANE to oppose nuclear weapons

1957 Dioxin causes chloracne in West German chemical workers

1957 An explosion of nuclear waste takes place in the Ural Mountains

1957 The Price-Anderson Act limits the liability of U.S. nuclear power producers in the event of an accident

1957 The Windscale Reactor releases radiation into the English countryside

1957 The first U.S. commercial nuclear plant opens in Pennsylvania

1959 The St. Lawrence Seaway is completed through the Great Lakes

1960 Congress passes the Multiple Use-Sustained Yield Act

1960 The Hazardous Substances Labeling Act becomes law

1961 Procter & Gamble introduces disposable diapers

1961 Leading nations sign a treaty regulating the development of Antarctica

1961 The World Wildlife Fund is established

1962 Rachel Carson's best-selling *Silent Spring* warns of the dangers of pesticides and sparks the modern environmental movement

1962 The United States begins spraying Agent Orange in Vietnam

1963 Glen Canyon Dam is completed, flooding one of the American Southwest's most scenic areas

1963 Congressman Stewart Udall discusses conservation issues in *The Quiet Crisis*

1963 The United States, the Soviet Union, and Great Britain sign the Limited Test Ban Treaty, the first international limitation on nuclear weapons testing

1963 The First Clean Air Act is passed by Congress

1964 A pipeline begins to carry water to Israel's Negev Desert

1964 The Atomic Energy Commission begins investigating health concerns at Washington's Hanford Nuclear Reservation

1964 The successful breeding of a new rice strain begins the Green Revolution

1964 Congress passes the Wilderness Act

1964 China explodes its first nuclear bomb

1965 The Federal Water Quality Act is adopted

1965 Congress approves the first Clean Air Act Amendments

1965 The Vehicle Air Pollution Control Act is passed

1965 The Highway Beautification Act limits the use of billboards

1966 A judge's ruling in the *Scenic Hudson* case stops construction of the Storm King Power Plant in New York State

1966 The Animal Welfare Act regulates the use of animals in experiments

1966 The U.S. Congress passes the Endangered Species Preservation Act

1966 The Welsh village of Aberfan is buried by mining debris

1967 Syukuro Manabe and Richard Wetherald predict the greenhouse effect

1967 The *Torrey Canyon* spills oil off the English Coast

1967 Scientists and lawyers form the Environmental Defense Fund

1968 Canada establishes the Experimental Lakes Area to study eutrophication

1968 A protracted drought begins in the Sahel region of Africa

1968 Paul Ehrlich's *The Population Bomb* is published
1968 The discovery of huge oil reserves in Alaska sparks controversy
1968 *The Whole Earth Catalog* is first published
1969 The Sierra Club blocks the construction of additional dams on the Colorado River
1969 The Soviet Union declares Lake Baikal a protected zone
1969 An offshore oil well blows out near Santa Barbara, California
1969 The Union of Concerned Scientists is founded
1969 Former Sierra Club president David Brower forms Friends of the Earth
1969 A pesticide spill poisons the Rhine River
1969 The Cuyahoga River bursts into flames in Cleveland, Ohio
1969 Canada bans the hunting of baby seals
1969 Canadian activists form Greenpeace to oppose nuclear testing
1969 The Federal Coal Mine Health and Safety Act is passed
1970 Jacques Cousteau announces a large decline in ocean life
1970 The Natural Resources Defense Council is founded
1970 President Richard M. Nixon establishes the Council on Environmental Quality
1970 Nixon signs the National Environmental Policy Act
1970 The Nuclear Nonproliferation Treaty goes into effect
1970 Canada bans commercial fishing in Lake St. Claire and Lake Erie
1970 The first Earth Day demonstrates the mass appeal of environmentalism
1970 Construction of the Trans-Amazon Highway is begun
1970 Congress passes the Resource Recovery Act to encourage recycling
1970 The U.S. Environmental Protection Agency is created
1970 The Occupational Safety and Health Act mandates safe workplaces
1970 The second Clean Air Act Amendments are adopted
1970 Congress approves the Mining and Minerals Act
1971 Egypt's Aswan High Dam is dedicated
1971 The Save the Whales campaign is launched
1971 The release of fungicide-tainted grain poisons thousands of Iraqis
1971 The Lead Poisoning Prevention Act regulates the use of lead-based paints
1972 The Club of Rome issues *The Limits to Growth*
1972 Congress revises the Water Pollution Control Act
1972 Earthwatch is founded
1972 The first U.N. environmental conference is held in Stockholm
1972 The World Fertility Survey is conducted
1972 The Great Lakes Water Quality Agreement is signed by Canada and the United States
1972 The U.S. Supreme Court rules against environmentalists in *Sierra Club v. Morton*
1972 The United States launches the Earth Resources Technology Satellite
1972 Oregon enacts the first U.S. bottle bill
1972 The Noise Control Act is passed
1972 The United States bans the use of DDT
1973 The Cousteau Society is founded
1973 The Ecology Party is formed in Great Britain
1973 The Convention on International Trade in Endangered Species (CITES) is signed
1973 The Chipko movement protects India's forests
1973 OPEC institutes an oil embargo, precipitating the first oil crisis
1973 Congress passes the Endangered Species Act
1974 Automakers introduce the catalytic converter
1974 Françoise D'Eaubonne coins the term "Ecofeminism"
1974 F. Sherwood Rowland and Mario Molina warn that chlorofluorocarbons are destroying the ozone layer

1974 Karen Silkwood becomes a martyr for the antinuclear movement
1974 The Safe Drinking Water Act mandates contaminant standards for U.S. water
1975 Edward Abbey's *The Monkey Wrench Gang* advocates "ecotage"
1975 The U.N. Global Monitoring System is inaugurated
1975 The Worldwatch Institute is founded
1975 The Hazardous Materials Transportation Act is passed
1975 Hells Canyon is preserved as a national recreation area
1976 The "Sagebrush Rebellion" begins
1976 The Teton Dam collapses in Idaho
1976 An explosion in a factory in Seveso, Italy, releases dioxin
1976 The National Forest Management Act is approved
1976 The Environmental Protection Agency is empowered to regulate toxic chemicals
1976 The United States bans polychlorinated biphenyls (PCBs)
1976 Congress recognizes the Bureau of Land Management
1976 The *Argo Merchant* spills oil off the New England Coast
1977 The Trans-Alaska Pipeline opens
1977 The Clean Air Act is amended again
1977 Paul Watson founds the Sea Shepherd Conservation Society
1977 Congress creates the Federal Energy Regulatory Commission
1977 The last natural case of smallpox in the world occurs in Somalia
1978 The U.S. Consumer Product Safety Commission bans lead paints
1978 The Public Utilities Regulatory Act promotes renewable energy
1978 The *Amoco Cadiz* runs aground, spilling oil off the coast of France
1978 Chlorofluorocarbons are banned in the United States
1978 Sun Day is held to promote solar energy
1978 The U.S. Supreme Court protects the snail darter, ruling against the completion of Tellico Dam
1978 Love Canal in New York State is declared a disaster area
1979 The Three Mile Island accident prompts nuclear industry reforms
1979 Two oil tankers collide near Tobago
1980 People for the Ethical Treatment of Animals is founded
1980 Mexico controls a huge leak in an offshore oil well
1980 Dave Foreman founds Earth First!, providing a radical fringe for the environmental movement
1980 The U.S. Supreme Court grants a patent for a living organism in *Diamond v. Chakrabarty*
1980 *The Global 2000 Report* is issued
1980 China conducts an atmospheric nuclear test
1980 The Alaska Lands Act is passed
1980 Superfund is created to clean up hazardous wastes
1981 Activists oppose deployment of the MX missile
1981 Lois Gibbs founds the Citizen's Clearinghouse for Hazardous Waste
1981 Julian Simon's *The Ultimate Resource* argues in favor of population growth
1981 U.S. interior secretary James Watt expands energy leasing on the outer continental shelf
1981 The Group of Ten meets for the first time
1981 Israel destroys an Iraqi nuclear reactor
1981 The United States announces the production of neutron bombs
1981 Construction of the Siberian Gas Pipeline begins
1982 Solar One begins operation in the Mojave Desert
1982 The United Nations adopts the Law of the Sea Treaty
1982 The Nuclear Waste Policy Act calls for the creation of a nuclear repository in the United States
1982 Dioxin contamination forces the evacuation of Times Beach, Missouri
1983 California's Kesterson National Wildlife Refuge is poisoned by agricultural runoff

1983 West Germany's Green Party obtains parliamentary seats
1983 The U.S. government reveals mercury releases at Oak Ridge, Tennessee
1983 Millions of Europeans join in antinuclear protests
1984 The first "deep ecology" platform is drafted
1984 Thomas E. Lovejoy III proposes the idea of the debt-for-nature swap
1984 The United Nations holds a population conference
1984 The Bhopal disaster kills and injures thousands in India
1985 Dave Foreman's *Ecodefense* advocates "monkeywrenching"
1985 European nations open Superphénix, a fast-breeder nuclear reactor
1985 The Rainforest Action Network begins a boycott of Burger King restaurants
1985 New Zealand closes its ports to U.S. nuclear warships
1985 Researchers discover a hole in the ozone layer
1985 French agents sink the Greenpeace ship *Rainbow Warrior*
1985 Dian Fossey is murdered over her efforts to protect mountain gorillas
1986 The International Whaling Ban goes into effect
1986 The United States and Canada issue a joint report on acid rain
1986 The Soviet Chernobyl nuclear plant undergoes a meltdown, causing the largest and most deadly nuclear accident in history
1986 Congress passes the Community Right-to-Know Act
1986 A fire in a Swiss warehouse causes toxic chemicals to spill into the Rhine
1987 Congress prohibits marine plastics dumping
1987 Genetically altered bacteria are released into the environment in California
1987 The garbage barge *Mobro* cruises the U.S. Atlantic and Gulf coasts in search of a place to unload
1987 *Our Common Future*, the "Brundtland Report," is published
1987 Radioactive powder from discarded medical equipment injures hundreds of Brazilians
1987 Yucca Mountain, Nevada, is designated as the first U.S. radioactive waste repository
1988 Iraq uses toxic gas in the Gulf War with Iran
1988 A tank collapse releases fuel into the Monongahela River in Pennsylvania
1988 Medical waste washes up on U.S. Atlantic beaches
1988 The president of Brazil announces plans to protect the rain forest
1988 The Global ReLeaf Program is initiated
1988 The Alternative Motor Fuels Act is passed
1988 The United Nations creates a panel to study climate change
1989 The Environmental Defense Fund begins the McToxics campaign
1989 The *Exxon Valdez* oil spill devastates Alaskan wildlife
1989 The president of Kenya burns a fortune in ivory to demonstrate opposition to the ivory trade
1989 Pope John Paul II issues an environmental message
1990 Publication of "An Anti-Environmentalist Manifesto" signals a backlash
1990 The Montreal Protocol is signed to combat ozone-layer depletion
1990 Threats to the northern spotted owl prompt a debate over old-growth timber in the Pacific Northwest
1990 Congress approves the Oil Pollution Act
1990 Killer bees reach the United States
1990 The Pollution Prevention Act is adopted
1990 Congress approves additional Clean Air Act amendments
1991 The Environmental Protection Agency publicizes the dangers of secondhand smoke
1991 The Supreme Soviet declares the Aral Sea a disaster area
1992 The Yankee Rowe Nuclear Plant in Massachusetts is shut down
1992 The Earth Summit convenes in Rio de Janeiro

1992 The United Nations bans the use of drift nets
1993 Norway resumes whaling in defiance of an international ban
1993 Oregon's Trojan Nuclear Plant is retired
1993 The *Braer* runs aground, spilling oil near the Shetland Islands
1993 President Bill Clinton convenes the Forest Summit
1993 The crew of Biosphere 2 exits after two years
1994 The first genetically engineered food reaches supermarkets
1994 The bald eagle is removed from the Endangered Species List
1995 Congressional Republicans begin an unsuccessful assault on the Endangered Species Act
1995 France and China spur worldwide protests with underground nuclear tests
1995 Greenpeace protesters occupy the *Brent Spar*
1995 The Environment for Europe Conference addresses Eastern European pollution
1995 The United Nations releases the Global Biodiversity Assessment
1996 The *Sea Empress* spills oil off the coast of Wales
1996 "Mad cow" disease strikes humans in Britain
1997 U.N. General Assembly meets in New York City to commemorate the fifth anniversary of the Earth Summit
1997 The Superphénix fast-breeder reactor is closed in France
1997 Kyoto Accords signed in Kyoto, Japan
1998 Brazil agrees to set aside 10 percent of its rain forest for conservation
1998 India and Pakistan conduct tests of nuclear weapons
1998 Madrid Protocol to the Antarctic Treaty goes into effect

Directory of Environmental Organizations

ACID RAIN FOUNDATION
1410 Varsity Drive
Raleigh, NC 27606
(919) 828-9113

AFRICAN WILDLIFE FOUNDATION
1717 Massachusetts Avenue NW #602
Washington, D.C. 20036
(202) 265-8393

AMERICAN CETACEAN SOCIETY
P.O. Box 2639
San Pedro, CA 90731
(310) 548-6279

AMERICAN COUNCIL FOR AN ENERGY-EFFICIENT ECONOMY
Suite 801
1001 Connecticut Avenue NW
Washington, D.C. 20036
(202) 429-8873

AMERICAN OCEANS CAMPAIGN
Suite 102
725 Arizona Avenue
Santa Monica, CA 90401
(310) 576-6162

AMERICAN RIVERS
Suite 400
801 Pennsylvania Avenue SE
Washington, D.C. 20003
(202) 547-6900

AMERICAN WIND ENERGY ASSOCIATION
122 C Street NW
Washington, D.C. 20001
(202) 408-8988

ANTARCTIC PROJECT
424 C Street NE
Washington, D.C. 20002
(202) 544-0236

BAT CONSERVATION INTERNATIONAL
P.O. Box 162603
Austin, TX 78716
(512) 327-9721

CENTER FOR ENVIRONMENTAL INFORMATION, INC.
50 W. Main Street
Rochester, NY 14614
(716) 262-2870

CENTER FOR MARINE CONSERVATION
Suite 500
1725 DeSales Street NW
Washington, D.C. 20036
(202) 429-5609

CENTER FOR RESPONSIBLE TOURISM
P.O. Box 827
San Anselmo, CA 94979
(510) 843—5506

CITIZEN'S CLEARINGHOUSE FOR HAZARDOUS WASTE
P.O. Box 6806
Falls Church, VA 22040
(703) 237-2249

CLEAN WATER ACTION
Suite 300
1320 18th Street NW
Washington, D.C. 20036
(202) 457—1286

COALITION FOR ENVIRONMENTALLY RESPONSIBLE ECONOMIES (CERES)
711 Atlantic Avenue
Boston, MA 02111
(617) 451-0927

CONGRESSIONAL LEGISLATION HOTLINE
(202) 225-1772

CONSERVATION INTERNATIONAL
Suite 1000
1015 18th Street NW
Washington, D.C. 20036
(212) 429-5660

CONSULTATIVE GROUP ON INTERNATIONAL AGRICULTURAL RESEARCH (CGIRR)
1818 H Street NW
Washington, D.C. 20133
(202) 477-1234

COUNCIL FOR RESPONSIBLE GENETICS
5 Upland Road
Cambridge, MA 02140
(617) 868-0870

CULTURAL SURVIVAL
215 First Street
Cambridge, MA 02142
(617) 621-3818

DEFENDERS OF WILDLIFE
1101 14th Street NW #1400
Washington, D.C. 20005
(202) 659-9510

DUCKS UNLIMITED
One Waterfowl Way
Long Grove, IL 60047
(708) 438-4300

EARTH FIRST!
P.O. Box 5176
Missoula, MT 59806

EARTH KIDS ORGANIZATION
P.O. Box 3847
Salem, OR 97302

EARTH ISLAND INSTITUTE
Suite 28
300 Broadway
San Francisco, CA 94133
(415) 788-3666

EARTHWATCH
P.O. Box 403
Mt. Auburn Street
Watertown, MA 02272
(800) 776-0188

ECOTOURISM SOCIETY
P.O. Box 755
North Bennington, VT 05257-0755
(802) 447-2121

ENVIRONMENTAL DEFENSE FUND
257 Park Avenue South
New York, NY 10010
(212) 505-2100

FOOD AND WATER, INC.
(800) EAT-SAFE

FRESHWATER FOUNDATION
Spring Hill Center
725 County Road 6
Wayzata, MN 55391
(612) 449-0092

FRIENDS OF THE EARTH
1025 Vermont Avenue NW
Washington, D.C. 20005
(202) 783-7400

GLOBAL GREEN USA
P.O. Box 21451
Columbus, OH 43221-0451
(805) 565-3485

GLOBAL RESPONSE: ENVIRONMENTAL ACTION NETWORK
P.O. Box 7490
Boulder, CO 80306-7490
(303) 444-0306

GREENHOUSE CRISIS FOUNDATION
Suite 630
1130 17th Street NW
Washington, D.C. 20036
(202) 466-2823

Greenpeace
1436 U Street NW
Washington, D.C. 20009
(202) 462-1177

Human Ecology Action League
P.O. Box 49126
Atlanta, GA 30359-1126
(404) 248-1898

Human Society of the United States
2100 L Street NW
Washington, D.C. 20037
(202) 452-1100

Institute for Consumer Responsibility
6506 28th Avenue NE
Seattle, WA 98115
(206) 523-0421

Institute for Local Self-Reliance
2425 18th Street NW
Washington, D.C. 20009
(202) 232-4108

International Alliance for Sustainable Agriculture
1701 University Avenue SE
Minneapolis, MN 55414
(612) 331-1099

International Eco-Agriculture Technology Association, Inc.
P.O. Box 998
Welches, 0R 97067
(800) 798-5513

International Federation of Organic Agriculture Movements
c/o Berne Ward Geier
Okozentrum, Imsbach W-6695
Tholey-Theley, Germany
6853 5190

International Institute for Energy Conservation
Suite 940
750 First Street NE
Washington, D.C. 20002
(202) 842-3388

International Society for Animal Rights
421 South State Street
Clark Summit, PA 18411
(717) 586-2200

International Whaling Commission
The Red House
135 Station Road
Histon, Cambridge CB4 4NP England
(0223) 233971

Institute for Food and Development Policy
398 60th Street
Oakland, CA 94618
(510)) 654-4400

Izaak Walton League of America
Level B
1401 Wilson Boulevard
Arlington, VA 22209-2318
(703) 528-1818

Jobs and the Environment Campaign
1168 Commonwealth Avenue
Boston, MA 02134
(617) 232-5833

Kids for Saving Earth
Suite 130
620 Mendelssohn
Golden Valley, MN 55427
(612) 525-0002

Land Stewardship Project
14758 Ostlund Trail North
Marine on the St. Croix, MN 55047
(612) 433-2770

National Audubon Society
700 Broadway
New York, NY 10003
(212) 979-3000

NATIONAL COALITION AGAINST THE MISUSE OF PESTICIDES
Suite 200
701 E Street SE
Washington, D.C. 20003
(212) 543-5450

NATIONAL PARKS AND CONSERVATION ASSOCIATION
1776 Massachusetts Avenue #200
Washington, D.C. 20036
(202) 223-6722

NATIONAL RECYCLING COALITION
Suite 305
1101 30th Street NW
Washington, D.C. 20007
(202) 625-6406

NATURAL RENEWABLE ENERGY LABORATORY
1617 Cole Boulevard
Golden, CO 80401
(303) 275-4065

NATIONAL SEED STORAGE LABORATORY
U.S. Department of Agriculture
Fort Collins, Colorado

NATIONAL WILDLIFE FEDERATION
1400 Sixteenth Street NW
Washington, DC 20036
(202) 797-6800

NATURAL RESOURCES DEFENSE COUNCIL
40 West Twentieth Street
New York, NY 10011
(212) 727-2700

NATURE CONSERVANCY
1815 Lynn Street
Arlington, VA 22209
(703) 841-5300

NUCLEAR REGULATORY COMMISSION
Washington, D.C. 20555
(301) 492-7000

OCEANIC SOCIETY EXPEDITIONS
Fort Mason Center, Building E
San Francisco, CA 94123
(800) 326-7491

PEOPLE FOR THE ETHICAL TREATMENT OF ANIMALS (PETA)
P.O. Box 42516
Washington, D.C. 20015
(301) 770-7444

POPULATION COUNCIL
One Dag Hammarskjold Plaza
New York, NY 10017
(212) 339-0500

RAINFOREST ACTION NETWORK
Suite 700
450 Sansome
San Francisco, CA 91111
(415) 398-4404

RAINFOREST ALLIANCE
65 Bleeker Street
New York, NY 10012-2420
(212) 677-1900

ROCKY MOUNTAIN INSTITUTE
1739 Snowmass Creek Road
Snowmass, CO 81654
(303) 927-3851

SAVE THE REDWOODS LEAGUE
Room 605
114 Sansome Street
San Francisco, CA 94104
(415) 362-2352

SEED SAVERS EXCHANGE
3076 North Winn Road
Decorah, Iowa 52101
(319) 382-5990

SEEDS OF CHANGE
621 Old Santa Fe Trail No. 10
Santa Fe, NM 87501
(505) 983-8956

SIERRA CLUB
730 Polk Street
San Francisco, CA 94109
(415) 776-2211

SOIL AND WATER CONSERVATION SOCIETY
7515 N.E. Ankeny Road
Ankeny, IA 50021
(800) THE-SOIL

SOLAR ENERGY INDUSTRIES ASSOCIATION
Fourth Floor
122 C Street NW
Washington, D.C. 20001
(202) 408-0660

UNION OF CONCERNED SCIENTISTS
2 Brattle Street
Cambridge, Massachusetts 02238
(617) 547-5552

UNITED NATIONS ENVIRONMENT PROGRAMME
Room 803
2 U.N. Plaza
New York, NY 10017
(212) 963-8138

U.S. AGENCY FOR INTERNATIONAL DEVELOPMENT
2201 C Street NW
Washington, D.C. 20520
(202) 647-4000

U.S. BUREAU OF LAND MANAGEMENT
Room 5600
1849 C Street NW
Washington, D.C. 20240
(202) 208-5717

U.S. BUREAU OF MINES
810 Seventh Street NW
Washington, D.C. 20241
(202) 501-9619

U.S. DEPARTMENT OF AGRICULTURE
Fourteenth Street and Independence
Avenue SW
Washington, D.C. 20250
(202) 720-8732

U.S. DEPARTMENT OF ENERGY
Forrestal Building
1000 Independence Avenue
Washington, D.C. 20585
(202) 586-5000

U.S. DEPARTMENT OF HEALTH, FOOD AND DRUG ADMINISTRATION (FDA)
5600 Fishers Lane
Rockville, MD 20857
(301) 443-1544

U.S. DEPARTMENT OF THE INTERIOR
1849 C Street NW
Washington, D.C. 20240
(202) 208-3100

U.S. DEPARTMENT OF TRANSPORTATION
Nassif Building
400 Seventh Street SW
Washington, D.C. 20590
(202) 366-4000

U.S. ENVIRONMENTAL PROTECTION AGENCY
401 M Street SW
Washington, D.C. 20460
(202) 260-2080

U.S. FISH AND WILDLIFE SERVICE
1849 C Street NW
Washington, D.C. 20240
(202) 208-5634

U.S. FOREST SERVICE
P.O. Box 96090
Fourteenth Street and Independence
Avenue SW
Washington, D.C. 20250
(202) 205-1760

U.S. PUBLIC INTEREST RESEARCH GROUP
215 Pennsylvania Avenue
Washington, D.C. 20003
(202) 546-9707

U.S. SENATE DOCUMENT ROOM
(202) 225-3456

WHITE HOUSE OFFICE OF ENVIRONMENTAL POLICY
Room 360
Old Executive Office Building
1600 Pennsylvania Avenue
Washington, D.C. 20501
(202) 456-6224

WILDERNESS SOCIETY
900 Seventeenth Street NW
Washington, D.C. 20006
(202) 833-2300

WILDLIFE CONSERVATION SOCIETY
New York Zoological Society
185th Street and Southern Boulevard
Bronx, NY 10460
(718) 220-5100

WORLD BANK (INTERNATIONAL BANK FOR RECONSTRUCTION AND DEVELOPMENT)
1818 H Street NW
Washington, D.C. 20433
(202) 477-1234

WORLD COUNCIL OF INDIGENOUS PEOPLES
555 King Edward Avenue
Ottawa, Ontario, Canada KIN 6NS

WORLD HEALTH ORGANIZATION (WHO)
Avenue APPIA, CH-1211
Geneva 27, Switzerland
(22) 7913105

WORLD NEIGHBORS
4127 NW 122nd Street
Oklahoma City, OK 73120
(405) 752-9700

WORLD WILDLIFE FUND
1250 24th Street NW
Washington, D.C. 20037
(202) 293-4800

WORLDWATCH INSTITUTE
1776 Massachusetts Avenue NW
Washington, D.C. 20036
(202) 452-1999

WORLD RESOURCES INSTITUTE
Suite 700
1709 New York Avenue NW
Washington, D.C. 20006
(202) 638-6300

ZERO POPULATION GROWTH
Suite 320
1400 16th Street NW
Washington, D.C. 20036
(202) 332-2200

Directory of U.S. National Parks

ALASKA

Denali National Park and Preserve
P.O. Box 9
McKinley Park, AK 99755

Gates of the Arctic National Park and Preserve
P.O. Box 74680
Fairbanks, AK 99707

Glacier Bay National Park and Preserve
P.O. Box 140
Gustavus, AK 99826

Katmai National Park and Preserve
P.0. Box 7
King Salmon, AK 99613

Kenai Fjords National Park
P.O. Box 1727
Seward, AK 99664

Kobuk Valley National Park
P.0. Box 1029
Kotzebue, AK 99752

Lake Clark National Park and Preserve
701 C Street
P.O. Box 61
Anchorage, AK 99513

Wrangell-St. Elias National Park and Preserve
P.O. Box 29
Glennallen, AK 99588

ARIZONA

Grand Canyon National Park
P.O. Box 129
Grand Canyon, AZ 86023

Petrified Forest National Park
Petrified Forest National Park, AZ 86028

ARKANSAS

Hot Springs National Park
P.O. Box 1860
Hot Springs, AR 71902

CALIFORNIA

Channel Islands National Park
1901 Spinnaker Drive
Ventura, CA 93001

Kings Canyon National Park
Three Rivers, CA 93217

Lassen Volcanic National Park
P.O. Box 100
Mineral, CA 96063

Redwood National Park
1111 Second Street
Crescent City, CA 95531

Sequoia National Park
Three Rivers, CA 93271

Yosemite National Park
P.O. Box 577
Yosemite National Park, CA 95389

COLORADO

Mesa Verde National Park
Mesa Verde National Park, CO 81330

Rocky Mountain National Park
Estes Park, CO 80517

FLORIDA

Biscayne National Park
P.O. Box 1369
Homestead, FL 33090

Everglades National Park
P.O. Box 279
Homestead, FL 33030

HAWAII

Haleakala National Park
P.0. Box 369, Makawao
Maui, HI 96768

Hawaii Volcanoes National Park
Hawaii National Park, HI 96718

KENTUCKY

Mammoth Cave National Park
Mammoth Cave, KY 42259

MAINE

Acadia National Park
P.O. Box 177
Bar Harbor, ME 04609

MICHIGAN

Isle Royale National Park
87 North Ripley Street
Houghton, MI 49931

MINNESOTA

Voyageurs National Park
P.0. Box 50
International Falls, MN 56649

MONTANA

Glacier National Park
West Glacier, MT 59936

NEVADA

Great Basin National Park
Baker, NV 89311

NEW MEXICO

Carlsbad Caverns National Park
3225 National Parks Highway
Carlsbad, NM 88220

NORTH DAKOTA

Theodore Roosevelt National Park
Route 1, P.O. Box 168
Medora, ND 58645
(701) 842-2333

OREGON

Crater Lake National Park
P.O. Box 7
Crater Lake, OR 97604

SOUTH DAKOTA

Badlands National Park
P.O. Box 6
Interior, SD 57750

Wind Cave National Park
Hot Springs, SD 57747

TENNESSEE

Great Smoky Mountains National Park
Gatlinburg, TN 37738

TEXAS

Big Bend National Park
Big Bend National Park, TX 79834

Guadalupe Mountains National Park
HC6O, P.O. Box 400
Salt Flat, TX 79847

UTAH

Arches National Park
P.O. Box 907
Moab, UT 84532

Bryce Canyon National Park
Bryce Canyon, UT 84717

Canyonlands National Park
125 West 200 South
Moab, UT 84532

Capitol Reef National Park
Torrey, UT 84775

Cedar Breaks National Park
P.O. Box 749
Cedar City, UT 84720

Zion National Park
Springdale, UT 84767

VIRGINIA

Shenandoah National Park
Route 4, P.O. Box 348
Luray, VA 22835

WASHINGTON

Mount Rainier National Park
Tahoma Woods, Star Route
Ashford, WA 98304

North Cascades National Park
2105 Highway 20
Sedro Woolley, WA 98284

Olympic National Park
600 East Park Avenue
Port Angeles, WA 98362

WYOMING

Grand Teton National Park
P.O. Drawer 170
Moose, WY 83012

Yellowstone National Park
P.O. Box 168
Yellowstone National Park, WY 82190

NATIONAL PARK SERVICE REGIONAL OFFICES

Alaska Region
National Park Service
2525 Gambell Street
Anchorage, AK 99503
(907) 257-2690

Mid-Atlantic Region
National Park Service
143 South Third Street
Philadelphia, PA 19106
(215) 597-7055

Midwest Region
National Park Service
1709 Jackson Street
Omaha, NE 68102
(402) 221-3431

National Capital Region
National Park Service
1100 Ohio Drive SW
Washington, D.C. 20242
(202) 619-7222

North Atlantic Region
National Park Service
15 State Street
Boston, MA 02109
(617) 223-5008

Pacific Northwest Region
National Park Service
Suite 212
83 South King Street
Seattle, WA 98104
(425) 553-5565

Rocky Mountain Region
National Park Service
P.O. Box 25287
Denver, CO 80225
(303) 969-2500

Southeast Region
National Park Service
Richard B. Russell Building
75 Spring Street SW
Atlanta, GA 30303
(404) 331-5185

Southwest Region
National Park Service
P.O. Box 728
Sante Fe, NM 87504
(505) 988-6100

Western Region
National Park Service
450 Golden Gate Avenue
Box 36063
San Francisco, CA 94102
(415) 556-0560

Glossary

Abatement: Reducing the degree or intensity of, or eliminating, pollution.

Absorption: The uptake of water, other fluids, or dissolved chemicals by a cell or an organism (as tree roots absorb dissolved nutrients in soil).

Acclimatization: The physiological and behavioral adjustments of an organism to changes in its environment.

Acid deposition: A complex chemical and atmospheric phenomenon that occurs when emissions of sulfur and nitrogen compounds and other substances are transformed by chemical processes in the atmosphere, often far from the original sources, and then deposited on Earth in either wet or dry form. The wet forms, popularly called "acid rain," can fall to earth as rain, snow, or fog. The dry forms are acidic gases or particulates.

Acid mine drainage: Drainage of water from areas that have been mined for coal or other mineral ores.

Activator: A chemical added to a pesticide to increase its activity.

Adaptation: Changes in an organism's physiological structure or function or habits that allow it to survive in new surroundings.

Adsorption: Removal of a pollutant from air or water by collecting the pollutant on the surface of a solid material.

Adulterants: Chemical impurities or substances that by law do not belong in a food, or pesticide.

Aeration: A process which promotes biological degradation of organic matter in water.

Aerobic: Life or processes that require, or are not destroyed by, the presence of oxygen.

Aerosol: A finely divided material suspended in air or other gaseous environment.

Afterburner: In incinerator technology, a burner located so that the combustion gases are made to pass through its flame in order to remove smoke and odors.

Agent Orange: A toxic herbicide and defoliant used in the Vietnam conflict, containing 2,4,5-trichlorophen-oxyacetic acid (2,4,5-T) and 2-4 dichlorophenoxyacetic acid (2,4-D) with trace amounts of dioxin.

Air permeability: Permeability of soil with respect to air.

Air pollutant: Any substance in air that could, in high enough concentration, harm humans, animals, vegetation, or material. Pollutants may include almost any natural or artificial composition of airborne matter capable of being airborne. They may be in the form of solid particles, liquid droplets, gases, or combinations thereof.

Airborne particulates: Total suspended particulate matter found in the atmosphere as solid particles or liquid droplets.

Alachlor: A herbicide, marketed under the trade name Lasso, used mainly to control weeds in corn and soybean fields.

Alar: Trade name for daminozide, a pesticide that makes apples redder, firmer, and less likely to drop off trees before growers are ready to pick them.

Algae: Simple rootless plants that grow in sunlit waters in proportion to the amount of available nutrients. They can affect water quality adversely by lowering the dissolved oxygen in the water.

Algal blooms: Sudden spurts of algal growth, which can affect water quality adversely and indicate potentially hazardous changes in local water chemistry.

Algicide: A substance or chemical used specifically to kill or control algae.

Allergen: A substance that causes an allergic reaction in individuals sensitive to it.

Ambient air: Any unconfined portion of the atmosphere: open air, surrounding air.

Anaerobic: A life or process that occurs in, or is not destroyed by, the absence of oxygen.

Animal dander: Tiny scales of animal skin, a common indoor air pollutant.

Anisotropy: In hydrology, the conditions under which one or more hydraulic properties of an aquifer vary from a reference point.

Aqueous solubility: The maximum concentration of a chemical that will dissolve in pure water at a reference temperature.

Aqueous: Something made up of water.

Aquifer: An underground geological formation, or group of formations, containing water.

Area source: Any source of air pollution that is released over a relatively small area but which cannot be classified as a point source. Such sources may include vehicles and other small engines, small businesses and household activities, or biogenic sources such as a forest that releases hydrocarbons.

Arsenicals: Pesticides containing arsenic.

Asbestos: A mineral fiber that can pollute air or water and cause cancer or asbestosis when inhaled.

Asbestos abatement: Procedures to control fiber release from asbestos-containing materials in a building or to remove them entirely, including removal, encapsulation, repair, enclosure, encasement, and operations and maintenance programs.

Asbestosis: A disease associated with inhalation of asbestos fibers. The disease makes breathing progressively more difficult and can be fatal.

Ash: The mineral content of a product remaining after complete combustion.

Assay: A test for a specific chemical, microbe, or effect.

Attenuation: The process by which a compound is reduced in concentration over time through absorption, adsorption, degradation, dilution, or transformation.

Background level: The concentration of a substance in an environmental media (air, water, or soil) that occurs naturally or is not the result of human activities. Also, in exposure assessment, the concentration of a substance in a defined control area during a fixed period of time before, during, or after a data-gathering operation.

Backyard composting: Diversion of organic food waste and yard trimmings from the municipal waste stream by composting them in one's yard through controlled decomposition of organic matter by bacteria and fungi into a humus-rich product.

Bacteria: Microscopic living organisms that can aid in pollution control by metabolizing organic matter in sewage, oil spills, or other pollutants. Bacteria in soil, water, or air can also cause human, animal, and plant health problems.

Bed load: Sediment particles resting on or near the channel bottom that are pushed or rolled along by the flow of water.

Beryllium: A metal hazardous to human health when inhaled as an airborne pollutant. It is discharged by machine shops, ceramic and propellant plants, and foundries.

Best available control measures (BACM): A term used to refer to the most effective measures (according to EPA guidance) for controlling small or dispersed particulates and other emissions from sources such as roadway dust, soot and ash from woodstoves and open burning of brush, timber, grasslands, or trash.

Best available control technology (BACT): The most stringent technology available for controlling emissions; major sources are required to use BACT, unless it can be demonstrated that it is not feasible for energy, environmental, or economic reasons.

Best demonstrated available technology (BDAT): As identified by the EPA, the most effective commercially available means of treating specific types of hazardous waste. BDATs may change with advances in treatment technologies.

Bioaccumulants: Substances that increase in concentration in living organisms as they take in contaminated air, water, or food because the substances are very slowly metabolized or excreted.

Bioassay: A test to determine the relative strength of a substance by comparing its effect on a test organism with that of a standard preparation.

Bioconcentration: The accumulation of a chemical in tissues of a fish or other organism to levels greater than in the surrounding medium.

Biodegradable: Capable of decomposing under natural conditions.

Biodiversity: The variety and variability among living organisms and the ecological complexes in which they occur.

Biological magnification: A process whereby certain substances such as pesticides or heavy metals move up the food chain, work their way into rivers or lakes, and are eaten by aquatic organisms such as fish, which in turn are eaten by large birds, animals, or humans. The substances become concentrated in tissues or internal organs as they move up the chain.

Biological stressors: Organisms accidently or intentionally dropped into habitats in which they do not evolve naturally.

Biological treatment: A treatment technology that uses bacteria to consume organic waste.

Biomass: All of the living material in a given area; often refers to vegetation.

Biome: The entire community of living organisms in a single major ecological area.

Bioremediation: The use of living organisms to clean up oil spills or remove other pollutants from soil, water, or wastewater; also, the use of organisms such as nonharmful insects to remove agricultural pests or counteract diseases of trees, plants, and garden soil.

Biosphere: The portion of Earth and its atmosphere that can support life.

Biostabilizer: A machine that converts solid waste into compost by grinding and aeration.

Biota: The animal and plant life of a given region.

Biotechnology: Techniques that use living organisms or parts of organisms to produce a variety of products (from medicines to industrial enzymes) to improve plants or animals or to develop microorganisms to remove toxics from bodies of water or act as pesticides.

Biotic community: A naturally occurring assemblage of plants and animals that live in the same environment and are mutually sustaining and interdependent.

Bottle bill: Proposed or enacted legislation which requires a returnable deposit on beer or soda containers and provides for retail store or other redemption.

British thermal unit (BTU): A unit of heat energy equal to the amount of heat required to raise the temperature of one pound of water by one degree Fahrenheit at sea level.

Buffer strips: Strips of grass or other erosion-resisting vegetation between or below cultivated strips or fields.

Buy-back center: Facility where individuals or groups bring reyclables in return for payment.

By-product: Material, other than the principal product, generated as a consequence of an industrial process or as a breakdown product in a living system.

Cadmium (Cd): A heavy metal that accumulates in the environment.

Cap: A layer of clay or other impermeable material installed over the top of a closed landfill to prevent entry of rainwater and minimize leachate.

Capillary action: Movement of water through very small spaces as a result of molecular forces called capillary forces.

Carbon monoxide (CO): A colorless, odorless, poisonous gas produced by incomplete fossil fuel combustion.

Carcinogen: Any substance that can cause or aggravate cancer.

Catalyst: A substance that changes the speed or yield of a chemical reaction without being consumed or chemically changed by the chemical reaction.

Catalytic converter: An air-pollution abatement device that removes pollutants from motor-vehicle exhaust, either by oxidizing them into carbon dioxide and water or by reducing them to nitrogen.

Central collection point: A location where a generator of regulated medical waste consolidates wastes originally generated at various locations; the term can also apply to community hazardous-waste collections and to industrial and other waste-management systems.

Chemical compound: A distinct and pure substance formed by the union or two or more elements in definite proportion by weight.

Chemical stressors: Chemicals released to the environment through industrial waste, auto emissions, pesticides, and other human activity that can cause illnesses and even death in plants and animals.

Chemical treatment: Any one of a variety of technologies that use chemicals or a variety of chemical processes to treat waste.

Child resistant packaging (CRP): Packaging that protects children or adults from injury or illness resulting from accidental contact with or ingestion of residential pesticides that meet or exceed specific toxicity levels.

Chlorination: The application of chlorine to drinking water, sewage, or industrial waste to disinfect or to oxidize undesirable compounds.

Chlorofluorocarbons (CFCs): A family of inert, nontoxic, and easily liquefied chemicals used in refrigeration, air conditioning, packaging, insulation, or as solvents and aerosol propellants. Because CFCs are not destroyed in the lower atmosphere, they drift into the upper atmosphere where their chlorine components destroy ozone.

Chlorosis: Discoloration of normally green plant parts caused by disease, lack of nutrients, or various air pollutants.

Clean fuels: Blends or substitutes for gasoline fuels, including compressed natural gas, methanol, ethanol, and liquified petroleum gas.

Clear cutting: Harvesting of all the trees in one area at one time, a practice that can encourage fast rainfall or snowmelt runoff, erosion, sedimentation of streams and lakes, and flooding and that destroys vital habitat.

Coal cleaning technology: A precombustion process by which coal is physically or chemically treated to remove some of its sulfur so as to reduce sulfur dioxide emissions.

Coal gasification: Conversion of coal to a gaseous product by one of several available technologies.

Coastal zone: Lands and waters adjacent to the coast that exert an influence on the uses of the sea and its ecology, or whose uses and ecology are affected by the sea.

Cogeneration: The consecutive generation of useful thermal and electric energy from the same fuel source.

Coliform index: A rating of the purity of water based on a count of fecal bacteria.

Coliform organism: Microorganisms found in the intestinal tracts of humans and animals. Their presence in water indicates fecal pollution and potentially adverse contamination by pathogens.

Colloids: Very small, finely divided solids that remain dispersed in a liquid for a long time because of their small size and electrical charge.

Combustion product: A substance produced during the burning or oxidation of a material.

Commercial waste: All solid waste emanating from business establishments such as stores, markets, office buildings, restaurants, shopping centers, and theaters.

Community: In ecology, an assemblage of populations of different species within a specified location in space and time.

Compaction: Reduction of the bulk of solid waste by rolling and tamping.

Comparative risk assessment: Process that generally uses the judgment of experts to predict effects and set priorities among a wide range of environmental problems.

Compost: The relatively stable humus material that is produced from a composting process in which bacteria in soil mixed with garbage and degradable trash break down the mixture into organic fertilizer.

Concentration: The relative amount of a substance mixed with another substance.

Conductivity: A measure of the ability of a solution to carry an electrical current.

Confined aquifer: An aquifer in which ground water is confined under pressure which is significantly greater than atmospheric pressure.

Contact pesticide: A chemical that kills pests when it touches them, instead of by ingestion.

Contaminant: Any physical, chemical, biological, or radiological substance or matter that has an adverse effect on air, water, or soil.

Contour plowing: Soil tilling method that follows the shape of the land to discourage erosion.

Contour strip farming: A kind of contour farming in which row crops are planted in strips, between alternating strips of close-growing, erosion-resistant forage crops.

Core: The uranium-containing heart of a nuclear reactor, where energy is released.

Corrosion: The dissolution and wearing away of metal caused by a chemical reaction such as between water and the pipes, chemicals touching a metal surface, or contact between two metals.

Corrosive: A chemical agent that reacts with the surface of a material causing it to deteriorate or wear away.

Cover crop: A crop that provides temporary protection for delicate seedlings or provides a cover canopy for seasonal soil protection and improvement between normal crop production periods.

Cradle-to-grave system: A procedure in which hazardous materials are identified and followed as they are produced, treated, transported, and disposed of by a series of permanent, linkable, descriptive documents.

Crop rotation: Planting a succession of different crops on the same land area, as opposed to planting the same crop time after time.

Cross contamination: The movement of underground contaminants from one level or area to another as the result of invasive subsurface activities.

Cultural eutrophication: Increasing rate at which water bodies "die" by pollution from human activities.

Cumulative exposure: The sum of exposures of an organism to a pollutant over a period of time.

Curbside collection: Method of collecting recyclable materials at homes, community districts or businesses.

Decay products: Degraded radioactive materials, often referred to as "daughters" or "progeny"; radon decay products of most concern from a public health standpoint are polonium-214 and polonium-218.

Dechlorination: Removal of chlorine from a substance.

Decomposition: The breakdown of matter by bacteria and fungi, changing the chemical makeup and physical appearance of materials.

Decontamination: Removal of harmful substances such as noxious chemicals, harmful bacteria or other organisms, or radioactive material from exposed individuals, rooms, and furnishings in buildings, or the exterior environment.

Defluoridation: The removal of excess fluoride in drinking water to prevent the staining of teeth.

Defoliant: An herbicide that removes leaves from trees and growing plants.

Degasification: A water treatment that removes dissolved gases from the water.

Degree-Day: A rough measure used to estimate the amount of heating required in a given area; is defined as the difference between the mean daily temperature and 65 degrees Fahrenheit.

Demineralization: A treatment process that removes dissolved minerals from water.

Denitrification: The biological reduction of nitrate to nitrogen gas by denitrifying bacteria in soil.

Density: A measure of how heavy a specific volume of a solid, liquid, or gas is in comparison to water. depending on the chemical.

Detection limit: The lowest concentration of a chemical that can reliably be distinguished from a zero concentration.

Destination facility: The facility to which regulated medical waste is shipped for treatment and destruction, incineration, and/or disposal.

Destratification: Vertical mixing within a lake or reservoir to totally or partially eliminate separate layers of temperature, plant, or animal life.

Desulfurization: Removal of sulfur from fossil fuels to reduce pollution.

Diazinon: An insecticide; in 1986, the EPA banned its use on open areas such as sod farms and golf courses because it posed a danger to migratory birds.

Diffusion: The movement of suspended or dissolved particles (or molecules) from a more concentrated to a less concentrated area.

Digestion: The biochemical decomposition of organic matter, resulting in partial gasification, liquefaction, and mineralization of pollutants.

Dike: A low wall that can act as a barrier to prevent a spill from spreading.

Dinocap: A fungicide used primarily by apple growers to control summer diseases.

Dinoseb: A herbicide that is also used as a fungicide and insecticide; banned by the EPA in 1986 because it posed the risk of birth defects and sterility.

Dioxin: Any of a family of compounds known chemically as dibenzo-p-dioxins. Tests on laboratory animals indicate that dioxin is one of the more toxic anthropogenic compounds.

Discharge: Flow of surface water in a stream or canal or the outflow of ground water from a flowing artesian well, ditch, or spring. Can also apply to discharge of liquid effluent from a facility or to chemical emissions into the air through designated venting mechanisms.

Disinfectant: A chemical or physical process that kills pathogenic organisms in water, air, or on surfaces.

Dispersant: A chemical agent used to break up concentrations of organic material such as spilled oil.

Disposables: Consumer products, other items, and packaging used once or a few times and discarded.

Dissolved solids: Disintegrated organic and inorganic material in water.

Downgradient: The direction that groundwater flows; similar to "downstream" for surface water.

Drainage basin: The area of land that drains water, sediment, and dissolved materials to a common outlet at some point along a stream channel.

Dump: A site used to dispose of solid waste without environmental controls.

Dystrophic lakes: Acidic, shallow bodies of water that contain much humus and/or other organic matter; contain many plants but few fish.

Ecological exposure: Exposure of a non-human organism to a stressor.

Ecology: The relationship of living things to one another and their environment, or the study of such relationships.

Economic poisons: Chemicals used to control pests and to defoliate cash crops such as cotton.

Ecosphere: The "bio-bubble" that contains life on Earth, in surface waters, and in the air.

Ecosystem: The interacting system of a biological community and its non-living environmental surroundings.

Ecotone: A habitat created by the juxtaposition of distinctly different habitats; an edge habitat; or an ecological zone or boundary where two or more ecosystems meet.

Ejector: A device used to disperse a chemical solution into water being treated.

Emission: Pollution discharged into the atmosphere from smokestacks, other vents, and surface areas of commercial or industrial facilities; from residential chimneys; and from motor vehicle, locomotive, or aircraft exhausts.

Emission standard: The maximum amount of air polluting discharge legally allowed from a single source, mobile or stationary.

Emissions trading: The creation of surplus emission reductions at certain stacks, vents, or similar emissions sources and the use of this surplus to meet or redefine pollution requirements applicable to other emissions sources.

Enclosure: Putting an airtight, impermeable, permanent barrier around asbestos-containing materials to prevent the release of asbestos fibers into the air.

End-of-the-pipe: Technologies such as scrubbers on smokestacks and catalytic convertors on automobile tailpipes that reduce emissions of pollutants after they have formed.

End user: Consumer of products for the purpose of recycling.

Endangered species: Animals, birds, fish, plants, or other living organisms threatened with extinction by anthropogenic or other natural changes in their environment. Requirements for declaring a species endangered are contained in the Endangered Species Act.

Enrichment: The addition of nutrients from sewage effluent or agricultural runoff to surface water; greatly increases the growth potential for algae and other aquatic plants.

Environment: The sum of all external conditions affecting the life, development, and survival of an organism.

Environmental impact statement (EIS): A document required of federal agencies by the National Environmental Policy Act for major projects or legislative proposals significantly affecting the environment. A tool for decision making, it describes the positive and negative effects of the undertaking and cites alternative actions.

Environmental sustainability: Long-term maintenance of ecosystem components and functions for future generations.

Erosion: The wearing away of land surface by wind or water, intensified by land-clearing practices related to farming, residential or industrial development, road building, or logging.

Eutrophication: The slow aging process during which a lake, estuary, or bay evolves into a bog or marsh and eventually disappears.

Evaporation ponds: Areas where sewage sludge is dumped and dried.

Evapotranspiration: The loss of water from the soil both by evaporation and by transpiration from the plants growing in the soil.

Exposure level: The amount (concentration) of a chemical at the absorptive surfaces of an organism.

Exposure: The amount of radiation or pollutant present in a given environment that represents a potential health threat to living organisms.

Facultative bacteria: Bacteria that can live under aerobic or anaerobic conditions.

Fecal coliform bacteria: Bacteria found in the intestinal tracts of mammals.Their presence in water or sludge is an indicator of pollution and possible contamination by pathogens.

Feedlot: A confined area for the controlled feeding of animals. Feedlots tend to concentrate large amounts of animal waste that cannot be absorbed by the soil and, hence, may be carried to nearby streams or lakes by rainfall runoff.

Fill: Human-made deposits of natural soils or rock products and waste materials.

Filtration: A treatment process, under the control of qualified operators, for removing solid (particulate) matter from water by means of porous media such as sand or a man-made filter; often used to remove particles that contain pathogens.

Floc: A clump of solids formed in sewage by biological or chemical action.

Flocculation: Process by which clumps of solids in water or sewage aggregate through biological or chemical action so they can be separated from water or sewage.

Floodplain: The flat or nearly flat land along a river or stream or in a tidal area that is covered by water during a flood.

Flume: A natural or human-made channel that diverts water.

Fluoridation: The addition of a chemical to increase the concentration of fluoride ions in drinking water to reduce the incidence of tooth decay.

Fluorides: Gaseous, solid, or dissolved compounds containing fluorine that result from industrial processes. Excessive amounts in food can lead to fluorosis.

Fly ash: Non-combustible residual particles expelled by flue gas.

Food chain: A sequence of organisms each of which uses the next-lower member of the sequence as a food source.

Food web: The feeding relationships by which energy and nutrients are transferred from one species to another.

Fossil fuel: Fuel derived from ancient organic remains; fossil fuels include peat, coal, crude oil, and natural gas.

Fresh water: Water that generally contains less than 1,000 milligrams per liter of dissolved solids.

Fuel efficiency: The proportion of energy released by fuel combustion that is converted into useful energy.

Fugitive emissions: Emissions not caught by a capture system.

Fume: Tiny particles trapped in vapor in a gas stream.

Fumigant: A pesticide vaporized to kill pests. Used in buildings and greenhouses.

Fungicide: Pesticides that are used to control, deter, or destroy fungi.

Fungistat: A chemical that keeps fungi from growing.

Garbage: Animal and vegetable waste resulting from the handling, storage, sale, preparation, cooking, and serving of foods.

Gasohol: Mixture of gasoline and ethanol derived from fermented agricultural products containing at least nine percent ethanol. Gasification: Conversion of solid material such as coal into a gas for use as a fuel.

Genetic engineering: A process of inserting new genetic information into existing cells in order to modify a specific organism for the purpose of changing one of its characteristics.

Geographic information system (GIS): A computer system designed for storing, manipulating, analyzing, and displaying data in a geographic context.

Grasscycling: Source reduction activities in which grass clippings are left on the lawn after mowing.

Greenhouse effect: The warming of the earth's atmosphere attributed to a buildup of carbon dioxide or other gases; some scientists think that this build-up allows the sun's rays to heat the earth, while making the atmosphere opaque to infrared radiation, thereby preventing a counterbalancing loss of heat.

Greenhouse gas: A gas, such as carbon dioxide or methane, which contributes to potential climate change.

Ground cover: Plants grown to keep soil from eroding.

Ground water: The supply of fresh water found beneath the earth's surface, usually in aquifers, which supply wells and springs.

Habitat: The place where a population (human, animal, plant, or microorganism) lives and its surroundings, both living and nonliving.

Halogen: A type of incandescent lamp with higher energy-efficiency that standard ones.

Hazardous substance: Any material that poses a threat to human health or the environment. Typical hazardous substances are toxic, corrosive, ignitable, explosive, or chemically reactive.

Hazardous waste: By-products of society that can pose a substantial or potential hazard to human health or the environment when improperly managed.

Heat island effect: A "dome" of elevated temperatures over an urban area caused by structural and pavement heat fluxes, and pollutant emissions.

Herbicide: A chemical pesticide designed to control or destroy plants, weeds, or grasses.

Herbivore: An animal that feeds on plants.

Heterotrophic organisms: Species that are dependent on organic matter for food.

Household hazardous waste: Hazardous products used and disposed of by residential as opposed to industrial consumers. Includes paints, stains, varnishes, solvents, pesticides, and other materials or products containing volatile chemicals that can catch fire, react, or explode, or that are corrosive or toxic.

Household waste: Solid waste, composed of garbage and rubbish, which normally originates in a private home or apartment house.

Hydrocarbons: Chemical compounds that consist entirely of carbon and hydrogen.

Hydrogeological cycle: The natural process of recycling water from the atmosphere down to and through the earth and back to the atmosphere again.

Hydrologic cycle: Movement or exchange of water between the atmosphere and earth.

Hydrology: The science dealing with the properties, distribution, and circulation of water.

Hydrolysis: The decomposition of organic compounds by interaction with water.

Ignitable: Capable of burning or causing a fire.

Imminent threat: A high probability that exposure is occurring.

Immiscibility: The inability of two or more substances or liquids to dissolve readily into one another, such as soil and water.

Impoundment: A body of water or sludge confined by a dam, dike, floodgate, or other barrier.

Incineration: A treatment technology involving destruction of waste by controlled burning at high temperatures.

Incinerator: A furnace for burning waste under controlled conditions.

Incompatible waste: A waste unsuitable for mixing with another waste or material because it may react to form a hazard.

Indicator: In biology, any biological entity or processes, or community whose characteristics show the presence of specific environmental conditions.

Indirect discharge: Introduction of pollutants from a nondomestic source into a publicly owned waste-treatment system.

Indirect source: Any facility or building, property, road, or parking area that attracts motor-vehicle traffic and, indirectly, causes pollution.

Indoor air: The breathable air inside a habitable structure or conveyance.

Indoor air pollution: Chemical, physical, or biological contaminants in indoor air.

Industrial sludge: Semiliquid residue or slurry remaining from treatment of industrial water and wastewater.

Industrial source reduction: Practices that reduce the amount of any hazardous substance, pollutant, or contaminant entering any waste stream or otherwise released into the environment.

Industrial waste: Unwanted materials from an industrial operation; may be liquid, sludge, solid, or hazardous waste.

Inert ingredient: Pesticide components such as solvents, carriers, dispersant, and surfactants that are not active against target pests.

Infectious agent: Any organism, such as a pathogenic virus, parasite, or bacterium, that is capable of invading body tissues, multiplying, and causing disease.

Infectious waste: Hazardous waste capable of causing infections in humans, including: contaminated animal waste; human blood and blood products; isolation waste; pathological waste; and discarded sharps (needles, scalpels or broken medical instruments).

Influent: Water, wastewater, or other liquid flowing into a reservoir, basin, or treatment plant.

Inhalable particles: All dust capable of entering the human respiratory tract.

Inorganic chemicals: Chemical substances of mineral origin, not of basically carbon structure.

Insecticide: A pesticide compound specifically used to kill or prevent the growth of insects.

Institutional Waste: Waste generated at institutions such as schools, libraries, hospitals, and prisons.

Integrated pest management (IPM): A mixture of chemical and other, nonpesticide, methods to control pests.

Inversion: A layer of warm air that prevents the rise of cooling air and traps pollutants beneath it; can cause an air pollution episode.

Ion: An electrically charged atom or group of atoms.

Irradiated food: Food subjected to brief radioactivity, usually gamma rays, to kill insects, bacteria, and mold, and to permit storage without refrigeration.

Irradiation: Exposure to radiation of wavelengths shorter than those of visible light (gamma, x-ray, or ultraviolet) for medical purposes, to sterilize milk or other foodstuffs, or to induce polymerization of monomers or vulcanization of rubber.

Irrigation: Applying water or wastewater to land areas to supply the water and nutrient needs of plants.

Irritant: A substance that can cause irritation of the skin, eyes, or respiratory system.

Isotope: A variation of an element that has the same atomic number of protons but a different weight because of the number of neutrons. Various isotopes of the same element may have different radioactive behaviors; some are highly unstable.

Isotropy: The condition in which the hydraulic or other properties of an aquifer are the same in all directions.

Jar test: A laboratory procedure that simulates a water treatment plant's coagulation/flocculation units with differing chemical doses, mix speeds, and settling times to estimate the minimum or ideal coagulant dose required to achieve certain water quality goals.

Joint and several liability: Under CERCLA, this legal concept relates to the liability for Superfund site cleanup and other costs on the part of more than one potentially responsible party.

Karst: A geologic formation of irregular limestone deposits with sinks, underground streams, and caverns.

Kinetic energy: Energy possessed by a moving object or water body.

Kinetic rate coefficient: A number that describes the rate at which a water constituent such as a biochemical oxygen demand or dissolved oxygen rises or falls, or at which an air pollutant reacts.

Latency: Time from the first exposure of a chemical until the appearance of a toxic effect.

LC 50: Lethal concentration; median-level concentration, a standard measure of toxicity; tells how much of a substance is needed to kill half of a group of experimental organisms in a given time.

LD 50: Lethal dose; the dose of a toxicant or microbe that will kill 50 percent of the test organisms within a designated period.

Leachate: Water that collects contaminants as it trickles through wastes, pesticides or fertilizers.

Leaching: The process by which soluble constituents are dissolved and filtered through the soil by a percolating fluid.

Lead (Pb): A heavy metal that is hazardous to health if breathed or swallowed. Its use in gasoline, paints, and plumbing compounds has been sharply restricted or eliminated by federal laws and regulations.

Legionella: A genus of bacteria, some species of which have caused a type of pneumonia called Legionnaires' Disease.

Limnology: The study of the physical, chemical, hydrological, and biological aspects of fresh water bodies.

Lindane: A pesticide that causes adverse health effects in domestic water supplies and is toxic to freshwater fish and aquatic life.

Liquefaction: Changing a solid into a liquid.

Lithology: Mineralogy, grain size, texture, and other physical properties of granular soil, sediment, or rock.

Mandatory recycling: Programs which by laws require consumers to separate trash so that some or all recyclable materials are recovered for recycling rather than going to landfills.

Manufacturing use product: Any product intended (labeled) for formulation or repackaging into other pesticide products.

Marine sanitation device: Any equipment or process installed on board a vessel to receive, retain, treat, or discharge sewage.

Marsh: A type of wetland that does not accumulate appreciable peat deposits and is dominated by herbaceous vegetation. Marshes may be either fresh or saltwater, tidal or non-tidal.

Maximum available control technology (MACT): The emission standard for sources of air pollution requiring the maximum reduction of hazardous emissions, taking cost and feasibility into account.

Maximum contaminant level: The maximum permissible level of a contaminant in water delivered to any user of a public system.

Mechanical separation: Using mechanical means to separate waste into various components.

Medical waste: Solid waste generated in the diagnosis, treatment, or immunization of human beings or animals, in research pertaining thereto, or in the production or testing of biologicals, excluding hazardous waste and household waste.

Methane: A colorless, nonpoisonous, flammable gas created by anaerobic decomposition of organic compounds; a major component of natural gas used in the home.

Methanol: An alcohol that can be used as an alternative fuel or as a gasoline additive.

Microbial growth: The amplification or multiplication of microorganisms such as bacteria, algae, diatoms, plankton, and fungi.

Microbial pesticide: A microorganism that is used to kill a pest but is of minimum toxicity to humans.

Microenvironments: Well-defined surroundings such as the home, office, or kitchen that can be treated as uniform in terms of stressor concentration.

Mining waste: Residues resulting from the extraction of raw materials from the earth.

Mitigation: Measures taken to reduce adverse impacts on the environment.

Mobile source: Any non-stationary source of air pollution such as cars, trucks, motorcycles, buses, airplanes, and locomotives.

Modified source: The enlargement of a major stationary pollutant source is often referred to as modification, implying that more emissions will occur.

Molecule: The smallest division of a compound that still retains or exhibits all the properties of the substance.

Monitoring: Periodic or continuous surveillance or testing to determine the level of compliance with statutory requirements or pollutant levels in various media or in humans, plants, and animals.

Morbidity: Rate of disease incidence.

Mortality: Death rate.

Muck soils: Earth made from decaying plant materials.

Mudballs: Round material that forms in filters and gradually increases in size when not removed by backwashing.

Mulch: A layer of material (wood chips, straw, leaves, etc.) placed around plants to hold moisture, prevent weed growth, and enrich or sterilize the soil.

Multiple use: Use of land for more than one purpose; e.g., grazing of livestock, watershed and wildlife protection, recreation, and timber production. Also applies to use of bodies of water for recreational purposes, fishing, and water supply.

Municipal discharge: Discharge of effluent from wastewater treatment plants that receive wastewater from households, commercial establishments, and industries in the coastal drainage basin.

Municipal sewage: Wastes (mostly liquid) originating from a community; may be composed of domestic wastewaters and/or industrial discharges.

Municipal solid waste: Common garbage or trash generated by industries, businesses, institutions, and homes.

Mutagen: An agent that causes a permanent genetic change in a cell other than that which occurs during normal growth.

Navigable waters: Traditionally, waters sufficiently deep and wide for navigation by all, or specified vessels; such waters in the United States come under federal jurisdiction and are protected by certain provisions of the Clean Water Act.

Necrosis: Death of plant or animal cells or tissues. In plants, necrosis can discolor stems or leaves or kill a plant entirely. Nematocide: A chemical agent which is destructive to nematodes.

Nephelometric: Method of measuring turbidity in a water sample by passing light through the sample and measuring the amount of the light that is deflected.

Netting: A concept in which all emissions sources in the same area that are owned or controlled by a single company are treated as one large source, thereby allowing flexibility in controlling individual sources in order to meet a single emissions standard.

Neutralization: Decreasing the acidity or alkalinity of a substance by adding alkaline or acidic materials, respectively.

Nitrate: A compound containing nitrogen that can exist in the atmosphere or as a dissolved gas in water and which can have harmful effects on humans and animals. Nitrates in water can cause severe illness in infants and domestic animals. A plant nutrient and inorganic fertilizer, nitrate is found in septic systems, animal feed lots, agricultural fertilizers, manure, industrial waste waters, sanitary landfills, and garbage dumps.

Nitric oxide (NO): A gas formed by combustion under high temperature and high pressure in an internal combustion engine; it is converted by sunlight and photochemical processes in ambient air to nitrogen oxide. NO is a precursor of ground-level ozone pollution, or smog.

Nitrogen oxide (NOx): The result of photochemical reactions of nitric oxide in ambient air; major component of photochemical smog.

Nitrogenous wastes: Animal or vegetable residues that contain significant amounts of nitrogen.

Nitrophenols: Synthetic organopesticides containing carbon, hydrogen, nitrogen, and oxygen.

Non-point sources: Diffuse pollution sources (without a single point of origin or not introduced into a receiving stream from a specific outlet).

Nonpotable: Water that is unsafe or unpalatable to drink because it contains pollutants, contaminants, minerals, or infective agents.

Nuclear winter: Prediction by some scientists that smoke and debris rising from massive fires of a nuclear war could block sunlight for weeks or months, cooling the earth's surface and producing climate changes that could, for example, negatively affect world agricultural and weather patterns.

Nuclide: An atom characterized by the number of protons, neutrons, and energy in the nucleus.

Nutrient: Any substance assimilated by living things that promotes growth. The term is generally applied to nitrogen and phosphorus in wastewater, but is also applied to other essential and trace elements.

Nutrient Pollution: Contamination of water resources by excessive inputs of nutrients. In surface waters, excess algal production is a major concern.

Odor threshold: The minimum odor of a water or air sample that can be detected after successive dilutions with odorless water.

Off-site facility: A hazardous waste treatment, storage or disposal area that is located away from the generating site.

Offsets: A concept whereby emissions from proposed new or modified stationary sources are balanced by reductions from existing sources to stabilize total emissions.

Offstream use: Water withdrawn from surface or groundwater sources for use at another place.

Oil fingerprinting: A method that identifies sources of oil and allows spills to be traced to their source.

Oil spill: An accidental or intentional discharge of oil which reaches bodies of water.

Oligotrophic lakes: Deep clear lakes with few nutrients, little organic matter and a high dissolved-oxygen level.

On-site facility: A hazardous-waste treatment, storage, or disposal area that is located on the generating site.

Onboard controls: Devices placed on vehicles to capture gasoline vapor during refueling and route it to the engine when a vehicle is starting so that it can be efficiently burned.

Onconogenicity: The capacity to induce cancer.

One-hit model: A mathematical model based on the biological theory that a single "hit" of some minimum critical amount of a carcinogen at a cellular target such as DNA can start an irreversible series events leading to a tumor.

Opacity: The amount of light obscured by particulate pollution in the air; clear window glass has zero opacity, a brick wall is 100 percent opaque. Opacity is an indicator of changes in performance of particulate control systems.

Open burning: Uncontrolled fires in an open dump.

Open dump: An uncovered site used for disposal of waste without environmental controls.

Oral toxicity: Ability of a pesticide to cause injury when ingested.

Organic: Referring to or derived from living organisms; in chemistry, any compound containing carbon.

Organic matter: Carbonaceous waste contained in plant or animal matter and originating from domestic or industrial sources.

Organism: Any form of animal or plant life.

Organophosphates: Pesticides that contain phosphorus; short-lived, but some can be toxic when first applied.

Organophyllic: A substance that easily combines with organic compounds.

Osmosis: The passage of a liquid from a weak solution to a more concentrated solution across a semipermeable membrane that allows passage of the solvent but not the dissolved solids.

Outfall: The place where effluent is discharged into receiving waters.

Overburden: Rock and soil cleared away before mining.

Oxidant: A collective term for some of the primary constituents of photochemical smog.

Oxidation: The chemical addition of oxygen to break down pollutants or organic waste.

Oxygenated fuel: Gasoline that has been blended with alcohols or ethers that contain oxygen in order to reduce carbon monoxide and other emissions.

Ozone hole: A thinning break in the stratospheric ozone layer.

Ozone layer: The protective layer in the atmosphere that absorbs some of the sun's ultraviolet rays, thereby reducing the amount of potentially harmful radiation that reaches the earth's surface.

Packed tower: A pollution control device that forces dirty air through a tower packed with crushed rock or wood chips while liquid is sprayed over the packing material.

Pandemic: A widespread epidemic throughout an area, nation, or the world.

Paraquat: A standard herbicide used to kill various types of crops, including marijuana. Causes lung damage if smoke from the crop is inhaled.

Particle count: Results of a microscopic examination of treated water with a special "particle counter" that classifies suspended particles by number and size.

Particulate loading: The mass of particulates per unit volume of air or water.

Particulates: Fine liquid or solid particles such as dust, smoke, mist, fumes, or smog, found in air or emissions; also, very small solids suspended in water.

Pathogens: Microorganisms that can cause disease in humans, animals, and plants.

Pathway: The physical course a chemical or pollutant takes from its source to the exposed organism.

Peak levels: Levels of airborne pollutant contaminants much higher than average or occurring for short periods of time in response to sudden releases.

Percent saturation: The amount of a substance that is dissolved in a solution compared to the amount that could be dissolved in it.

Periphyton: Microscopic underwater plants and animals that are firmly attached to solid surfaces such as rocks, logs, and pilings.

Permeability: The rate at which liquids pass through soil or other materials in a specified direction.

Permissible dose: The dose of a chemical that may be received by an individual without the expectation of a significantly harmful result.

Persistence: Refers to the length of time a compound stays in the environment once introduced.

Pest: An insect, rodent, nematode, fungus, weed or other form of terrestrial or aquatic plant or animal life that is injurious to health or the environment.

Pesticide: Substances or mixtures intended for preventing, destroying, repelling, or mitigating any pest. Also, any substance or mixture intended for use as a plant regulator, defoliant, or desiccant.

Petroleum: Crude oil or any fraction thereof that is liquid under normal conditions of temperature and pressure.

Petroleum derivatives: Chemicals formed when gasoline breaks down in contact with ground water.

pH: An expression of the intensity of the basic or acid condition of a liquid; may range from 0 to 14, where 0 is the most acid and 7 is neutral. Natural waters usually have a pH between 6.5 and 8.5.

Pharmacokinetics: The study of the way that drugs move through the body after they are swallowed or injected.

Phenols: Organic compounds that are by-products of petroleum refining, tanning, and textile, dye, and resin manufacturing. Low concentrations cause taste and odor problems in water; higher concentrations can kill aquatic life and humans.

Phosphates: Certain chemical compounds containing phosphorus.

Phosphorus: An essential chemical food element that can contribute to the eutrophication of lakes and other water bodies. Increased phosphorus levels result from discharge of phosphorus-containing materials into surface waters.

Photochemical oxidants: Air pollutants formed by the action of sunlight on oxides of nitrogen and hydrocarbons.

Photochemical smog: Air pollution caused by chemical reactions of various pollutants emitted from different sources.

Photosynthesis: The manufacture by plants of carbohydrates and oxygen from carbon dioxide mediated by chlorophyll in the presence of sunlight.

Phytoplankton: That portion of the plankton community comprised of tiny plants.

Phytotoxic: Harmful to plants.

Phytotreatment: The cultivation of specialized plants that absorb specific contaminants from the soil through their roots or foliage.

Plankton: Tiny plants and animals that live in water.

Plutonium: A radioactive metallic element chemically similar to uranium.

Point source: A stationary location or fixed facility from which pollutants are discharged, or any single identifiable source of pollution, such as a pipe, ditch, ship, ore pit, or factory smokestack.

Pollen: The fertilizing element of flowering plants; a background air pollutant.

Pollutant: Generally, any substance introduced into the environment that adversely affects the usefulness of a resource or the health of humans, animals, or ecosystems.

Pollution: Generally, the presence of a substance in the environment that because of its chemical composition or quantity prevents the functioning of natural processes and produces undesirable environmental and health effects.

Polonium: A radioactive element that occurs in pitchblende and other uranium-containing ores.

Polyelectrolytes: Synthetic chemicals that help solids to clump during sewage treatment.

Population: A group of interbreeding organisms occupying a particular space; the number of humans or other living creatures in a designated area.

Porosity: Degree to which soil, gravel, sediment, or rock is permeated with pores or cavities through which water or air can move.

Post-consumer recycling: Use of materials generated from residential and consumer waste for new or similar purposes.

Potable water: Water that is safe for drinking and cooking.

Precipitator: Pollution control device that collects particles from an air stream.

Precursor: In photochemistry, a compound antecedent to a pollutant.

Prior appropriation: A doctrine of water law that allocates the rights to use water on a first-come, first-served basis.

Process wastewater: Any water that comes into contact with any raw material, product, by-product, or waste.

Producers: Plants that perform photosynthesis and provide food to consumers.

Protoplast: A membrane-bound cell from which the outer wall has been partially or completely removed. The term often is applied to plant cells.

Purging: Removing stagnant air or water from sampling zone or equipment prior to sample collection.

Putrefaction: Biological decomposition of organic matter; associated with anaerobic conditions.

Pyrolysis: Decomposition of a chemical by extreme heat.

Quality assurance/quality control: A system of procedures, checks, audits, and corrective actions to ensure that all EPA research design and performance, environmental monitoring and sampling, and other technical and reporting activities are of the highest achievable quality.

Quench tank: A water-filled tank used to cool incinerator residues or hot materials during industrial processes.

Radiation standards: Regulations that set maximum exposure limits for protection of the public from radioactive materials.

Radiation: Transmission of energy through space or any medium. Also known as radiant energy.

Radioisotopes: Chemical variants of radioactive elements with potentially oncogenic, teratogenic, and mutagenic effects on the human body.

Radon: A colorless naturally occurring, radioactive, inert gas formed by radioactive decay of radium atoms in soil or rocks.

Raw water: Intake water prior to any treatment or use.

Recarbonization: Process in which carbon dioxide is bubbled into water being treated to lower the pH.

Receiving waters: A river, lake, ocean, stream, or other watercourse into which wastewater or treated effluent is discharged.

Receptor: An ecological entity exposed to a stressor.

Reclamation: In recycling, the restoration of materials found in the waste stream to a beneficial use that may be for purposes other than the original use.

Recycling: Minimizing waste generation by recovering and reprocessing usable products that might otherwise become waste.

Red tide: A proliferation of a marine plankton toxic and often fatal to fish, perhaps stimulated by the addition of nutrients. A tide can be red, green, or brown, depending on the coloration of the plankton.

Refueling emissions: Emissions released during vehicle refueling.

Release: Any spilling, leaking, pumping, pouring, emitting, emptying, discharging, injecting, escaping, leaching, dumping, or disposing into the environment of a hazardous or toxic chemical or extremely hazardous substance.

Remote sensing: The collection and interpretation of information about an object without physical contact with the object; e.g., satellite imaging, aerial photography, and open path measurements.

Removal action: Short-term immediate actions taken to address releases of hazardous substances that require expedited response.

Reservoir: Any natural or artificial holding area used to store, regulate, or control water.

Residential use: Pesticide application in and around houses, office buildings, apartment buildings, motels, and other living or working areas.

Residential waste: Waste generated in single and multifamily homes.

Residue: The dry solids remaining after the evaporation of a sample of water or sludge.

Resistance: For plants and animals, the ability to withstand poor environmental conditions or attacks by chemicals or disease. May be inborn or acquired.

Resource recovery: The process of obtaining matter or energy from materials formerly discarded.

Riparian rights: Entitlement of a landowner to certain uses of water on or bordering the property, including the right to prevent diversion or misuse of upstream waters.

Risk: A measure of the probability that damage to life, health, property, or the environment will occur as a result of a given hazard.

River basin: The land area drained by a river and its tributaries.

Rodenticide: A chemical or agent used to destroy rats or other rodent pests.

Rubbish: Solid waste, excluding food waste and ashes, from homes, institutions, and workplaces.

Run-off: That part of precipitation, snow melt, or irrigation water that runs off the land into streams or other surface water. It can carry pollutants from the air and land into receiving waters.

Safe water: Water that does not contain harmful bacteria, toxic materials, or chemicals, and is considered safe for drinking even if it may have taste, odor, color, and certain mineral problems.

Salinity: The percentage of salt in water.

Salt water intrusion: The invasion of fresh surface or ground water by salt water.

Salvage: The utilization of waste materials.

Sanitary sewers: Underground pipes that carry off only domestic or industrial waste, not storm water.

Sanitation: Control of physical factors in the human environment that could harm development, health, or survival.

Saturation: The condition of a liquid when it has taken into solution the maximum possible quantity of a given substance at a given temperature and pressure.

Scrap: Materials discarded from manufacturing operations that may be suitable for reprocessing.

Scrubber: An air pollution device that uses a spray of water or reactant or a dry process to trap pollutants in emissions.

Secondary effect: Action of a stressor on supporting components of the ecosystem, which in turn impact the ecological component of concern.

Secondary materials: Materials that have been manufactured and used at least once and are to be used again.

Sedimentation: Letting solids settle out of wastewater by gravity during treatment.

Sediments: Soil, sand, and minerals washed from land into water, usually after rain.

Seepage: Percolation of water through the soil from unlined canals, ditches, laterals, watercourses, or water storage facilities.

Selective pesticide: A chemical designed to affect only certain types of pests, leaving other plants and animals unharmed.

Septic system: An on-site system designed to treat and dispose of domestic sewage.

Septic tank: An underground storage tank for wastes from homes not connected to a sewer line.

Settleable solids: Material heavy enough to sink to the bottom of a wastewater treatment tank.

Settling tank: A holding area for wastewater, where heavier particles sink to the bottom for removal and disposal.

Sewage: The waste and wastewater produced by residential and commercial sources and discharged into sewers.

Sewer: A channel or conduit that carries wastewater and storm-water runoff from the source to a treatment plant or receiving stream.

Sewerage: The entire system of sewage collection, treatment, and disposal.

Sharps: Hypodermic needles, syringes (with or without the attached needle), Pasteur pipettes, scalpel blades, blood vials, needles with attached tubing, and culture dishes used in animal or human patient care or treatment, or in medical, research or industrial laboratories.

Silt: Sedimentary materials composed of fine or intermediate-sized mineral particles.

Silviculture: Management of forest land for timber.

Sink: Place in the environment where a compound or material collects.

Sinking: Controlling oil spills by using an agent to trap the oil and sink it to the bottom of the body of water where the agent and the oil are biodegraded.

Siting: The process of choosing a location for a facility.

Skimming: Using a machine to remove oil or scum from the surface of the water.

Sludge: A semisolid residue from any of a number of air or water treatment processes; can be a hazardous waste.

Slurry: A watery mixture of insoluble matter resulting from some pollution control techniques.

Smog: Air pollution typically associated with oxidants.

Smoke: Particles suspended in air after incomplete combustion.

Soft water: Any water that does not contain a significant amount of dissolved minerals such as salts of calcium or magnesium.

Solid waste: Nonliquid, nonsoluble materials ranging from municipal garbage to industrial wastes that contain complex and sometimes hazardous substances.

Soot: Carbon dust formed by incomplete combustion.

Source reduction: Reducing the amount of materials entering the waste stream from a specific source by redesigning products or patterns of production or consumption.

Source separation: Segregating various wastes at the point of generation (such as the separation of paper, metal and glass from other wastes to make recycling simpler and more efficient).

Spoil: Dirt or rock removed from its original location—destroying the composition of the soil in the process—as in strip-mining, dredging, or construction.

Sprawl: Unplanned development of open land.

Spray tower scrubber: A device that sprays alkaline water into a chamber where acid gases are present to aid in neutralizing the gas.

Stabilization: Conversion of the active organic matter in sludge into inert, harmless material.

Stack: A chimney, smokestack, or vertical pipe that discharges used air.

Stack effect: Air, as in a chimney, that moves upward because it is warmer than the ambient atmosphere.

Stagnation: Lack of motion in a mass of air or water that holds pollutants in place.

Stakeholder: Any organization, governmental entity, or individual that has a stake in or may be impacted by a given approach to environmental regulation, pollution prevention, energy conservation, and so on.

Standards: Norms that impose limits on the amount of pollutants or emissions produced.

Storm sewer: A system of pipes (separate from sanitary sewers) that carries water runoff from buildings and land surfaces.

Stratification: Separating into layers.

Stratigraphy: Study of the formation, composition, and sequence of sediments, whether consolidated or not.

Stressors: Physical, chemical, or biological entities that can induce adverse effects on ecosystems or human health.

Strip mining: A process that uses machines to scrape soil or rock away from mineral deposits just under the earth's surface.

Sulfur dioxide: A pungent, colorless, gas formed primarily by the combustion of fossil fuels; becomes a pollutant when present in large amounts.

Sump: A pit or tank that catches liquid runoff for drainage or disposal.

Surface runoff: Precipitation, snow melt, or irrigation water in excess of what can infiltrate the soil surface and be stored in small surface depressions; a major transporter of nonpoint-source pollutants in rivers, streams, and lakes.

Surface water: All water naturally open to the atmosphere (rivers, lakes, reservoirs, ponds, streams, impoundments, seas, estuaries).

Surfactant: A detergent compound that promotes lathering.

Suspended solids: Small particles of solid pollutants that float on the surface of, or are suspended in, sewage or other liquids.

Systemic pesticide: A chemical absorbed by an organism that interacts with the organism and makes the organism toxic to pests.

Tail water: The runoff of irrigation water from the lower end of an irrigated field.

Tailings: Residue of raw material or waste separated out during the processing of crops or mineral ores.

Teratogenesis: The introduction of nonhereditary birth defects in a developing fetus by exogenous factors such as physical or chemical agents acting in the womb to interfere with normal embryonic development.

Terracing: Dikes built along the contour of sloping farm land that hold runoff and sediment to reduce erosion.

Thermal pollution: Discharge of heated water from industrial processes that can kill or injure aquatic organisms.

Threshold level: Time-weighted average pollutant concentration values, exposure beyond which is likely to adversely affect human health.

Threshold: The lowest dose of a chemical at which a specified measurable effect is observed and below which it is not observed.

Toxic cloud: Airborne plume of gases, vapors, fumes, or aerosols containing toxic materials.

Toxic concentration: The concentration at which a substance produces a toxic effect.

Toxic dose: The dose level at which a substance produces a toxic effect.

Toxic pollutants: Materials that cause death, disease, or birth defects in organisms that ingest or absorb them. The quantities and exposures necessary to cause these effects can vary widely.

Toxic substance: A chemical or mixture that may present an unreasonable risk of injury to health or the environment.

Toxic waste: A waste that can produce injury if inhaled, swallowed, or absorbed through the skin.

Toxicant: A harmful substance or agent that may injure an exposed organism.

Toxicity assessment: Characterization of the toxicological properties and effects of a chemical, with special emphasis on establishment of dose-response characteristics.

Toxicity testing: Biological testing (usually with an invertebrate, fish, or small mammal) to determine the adverse effects of a compound or effluent.

Toxicity: The degree to which a substance or mixture of substances can harm humans or animals.

Toxicological profile: An examination, summary, and interpretation of a hazardous substance to determine levels of exposure and associated health effects.

Transboundary pollutants: Air pollution that travels from one jurisdiction to another, often crossing state or international boundaries. Also applies to water pollution.

Transpiration: The process by which water vapor is lost to the atmosphere from living plants. The term can also be applied to the quantity of water thus dissipated.

Trash: Material considered worthless or offensive that is thrown away. Generally defined as dry waste material, but in common usage it is a synonym for garbage, rubbish, or refuse.

Treated wastewater: Wastewater that has been subjected to one or more physical, chemical, and biological processes to reduce its potential of being a health hazard.

Troposhpere: The layer of the atmosphere closest to the earth's surface.

Tundra: A type of treeless ecosystem dominated by lichens, mosses, grasses, and woody plants.

Ultra clean coal (UCC): Coal that is washed, ground into fine particles, then chemically treated to remove sulfur, ash, silicone, and other substances; usually briquetted and coated with a sealant made from coal.

Unconfined aquifer: An aquifer containing water that is not under pressure; the water level in a well is the same as the water table outside the well.

Unsaturated zone: The area above the water table where soil pores are not fully saturated, although some water may be present.

Upper detection limit: The largest concentration that an instrument can reliably detect.

Urban runoff: Storm water from city streets and adjacent domestic or commercial properties that carries pollutants of various kinds into the sewer systems and receiving waters.

Use cluster: A set of competing chemicals, processes, or technologies that can substitute for one another in performing a particular function.

User fee: Fee collected from only those persons who use a particular service, as compared to one collected from the public in general.

Utility load: The total electricity demand for a utility district.

Vapor dispersion: The movement of vapor clouds in air caused by wind, thermal action, gravity spreading, and mixing.

Vapor plumes: Flue gases visible because they contain water droplets.

Vapor pressure: A measure of a substance's propensity to evaporate.

Variance: Government permission for a delay or exception in the application of a given law, ordinance, or regulation.

Vector: An organism, often an insect or rodent, that carries disease.

Vegetative controls: Non-point-source pollution control practices that involve vegetative cover to reduce erosion and minimize loss of pollutants.

Ventilation rate: The rate at which indoor air enters and leaves a building.

Venturi scrubbers: Air-pollution control devices that use water to remove particulate matter from emissions.

Vinyl chloride: A chemical compound, used in producing some plastics, that is believed to be oncogenic.

Virgin materials: Resources extracted from nature in their raw form, such as timber or metal ore.

Viscosity: The molecular friction within a fluid that produces flow resistance.

Volume reduction: Processing waste materials to decrease the amount of space they occupy, usually by compacting, shredding, incineration, or composting.

Vulnerability analysis: Assessment of elements in the community that are susceptible to damage if hazardous materials are released.

Vulnerable zone: An area over which the airborne concentration of a chemical accidentally released could reach the level of concern.

Waste: Unwanted materials left over from a manufacturing process; also, refuse from places of human or animal habitation.

Waste characterization: Identification of chemical and microbiological constituents of a waste material.

Waste minimization: Measures or techniques that reduce the amount of wastes generated during industrial production processes.

Waste reduction: Using source reduction, recycling, or composting to prevent or reduce waste generation.

Waste stream: The total flow of solid waste from homes, businesses, institutions, and manufacturing plants that is recycled, burned, or disposed of in landfills, or segments thereof such as the "residential waste stream" or the "recyclable waste stream."

Waste-treatment plant: A facility containing a series of tanks, screens, filters, and other processes by which pollutants are removed from water.

Wastewater: The spent or used water from a home, community, farm, or industry that contains dissolved or suspended matter.

Water pollution: The presence in water of enough harmful or objectionable material to damage the water's quality.

Water table: The level of groundwater.

Water well: An excavation where the intended use is for location, acquisition, development, or artificial recharge of ground water.

Watershed: The land area that drains into a stream; the watershed for a major river may encompass a number of smaller watersheds that ultimately combine at a common point.

Wetlands: An area that is saturated by surface or ground water with vegetation adapted for life under those soil conditions, as swamps, bogs, fens, marshes, and estuaries.

Wildlife refuge: An area designated for the protection of wild animals, within which hunting and fishing are either prohibited or strictly controlled.

Xenobiota: Any biotum displaced from its normal habitat; a chemical foreign to a biological system.

Yard waste: The part of solid waste composed of grass clippings, leaves, twigs, branches, and other garden refuse.

Zero air: Atmospheric air purified to contain less than 0.1 ppm total hydrocarbons.

Bibliography

Selected Recent Works of Interest on Environmental Issues

AIR AND AIR POLLUTION

Air Quality Management and Assessment Capabilities in Twenty Major Cities. London: MARC, 1996.

Allegrini, Ivo, and Franco De Santis., eds. *Urban Air Pollution: Monitoring and Control Strategies.* New York: Springer-Verlag, 1996.

Bardana, Emil J., Jr., and Anthony Montanaro, eds. *Indoor Air Pollution and Health.* New York: Marcel Dekker, 1997.

Benedick, Richard Elliot. *Ozone Diplomacy: New Directions in Safeguarding the Planet.* Cambridge, Mass.: Harvard University Press, 1998.

Brimblecombe, Peter. *Air Composition and Chemistry.* 2d ed. New York: Cambridge University Press, 1996.

Chilton, Kenneth W., and Christopher Boerner. *Smog in America: The High Cost of Hysteria.* St. Louis: Center for the Study of American Business, 1996.

Chilton, Kenneth W., and Stephen Huebner. *Has the Battle Against Urban Smog Become "Mission Impossible?"* St. Louis: Center for the Study of American Business, 1996.

Clearing the Air: An Updated Report on Emission Trends in Selected U.S. Cities. Washington, D.C.: AAA Government Relations, 1997.

Committee for a Study of Transportation and a Sustainable Environment. *Toward a Sustainable Future: Addressing the Long-Term Effects of Motor Vehicle Transportation on Climate and Ecology.* Washington, D.C.: Transportation Research Board, 1997.

Cummins, Jackie. *The Quest for Cleaner Air.* Denver: National Conference of State Legislatures, 1996.

Davison, G., and C. N. Hewitt, eds. *Air Pollution in the United Kingdom.* Cambridge, England: Royal Society of Chemistry, Information Services, 1997.

Diesel Fuel and Exhaust Emissions. Geneva: World Health Organization, 1996.

Elsom, Derek M. *Smog Alert: Managing Urban Air Quality.* London: Earthscan Publications, 1996.

Gammage, Richard B., and Barry A. Berven, eds. *Indoor Air and Human Health.* 2d ed. Boca Raton, Fla.: CRC Press, 1996.

Garrett, Mark, and Martin Wachs. *Transportation Planning on Trial: The Clean Air Act and Travel Forecasting.* Thousand Oaks, Calif.: Sage Publications, 1996.

Gryning, Sven-Erik, and Francis A. Schiermeier, eds. *Air Pollution Modeling and Its Application XI.* New York: Plenum Press, 1996.

Godish, Thad. *Air Quality.* 3d ed. Boca Raton, Fla.: CRC/Lewis, 1997.

Green, Kenneth. *Rethinking EPA's Proposed Ozone and Particulate Standards.* Los Angeles: Reason Foundation, 1997.

Howitt, Arnold M. *The New Politics of Clean Air and Transportation.* Washington, D.C.: U.S. Department of Transportation, Federal Highway Administration, 1997.

Huebner, Stephen, and Kenneth Chilton. *EPA's Case for New Ozone and Particulate Standards: Would Americans Get Their Money's Worth?* St. Louis: Center for the Study of American Business, 1997.

Klaassen, Ger. *Acid Rain and Environmental Degradation: The Economics of Emission Trading.* Brookfield, Vt.: Edward Elgar, 1996.

Krupa, Sagar V. *Air Pollution, People, and Plants: An Introduction.* St. Paul, Minn.: American Phytopathological Society, 1997.

Lee, Dwight R. *The Turf Fight for Indoor Air Quality Protection.* St. Louis: Center for the Study of

American Business, Washington University, 1996.

Liefferink, Duncan. *Environment and the Nation State: The Netherlands, the EU, and Acid Rain.* New York: Manchester University Press, 1996.

McCubbin, Donald R., and Mark A. Delucchi. *The Social Cost of the Health Effects of Motor-Vehicle Air Pollution.* Davis: Institute of Transportation Studies, University of California, 1996.

Martineau, Robert J., Jr., and David Novello, eds. *Clean Air Act Handbook.* Chicago: American Bar Association, 1997.

Mullen, Maureen A., et al. *The Emissions Impact of Eliminating National Speed Limits: One Year Later.* Washington, D.C.: Transportation Research Board, 1997.

An Office Building Occupant's Guide to Indoor Air Quality. Washington, D.C.: U.S. Environmental Protection Agency, Indoor Environments Division, Office of Air and Radiation, 1997.

Pilkington, Alan. *Emissions or Economics: The Status and Potential of Alternate Fuel Technology.* Davis: Institute of Transportation Studies, University of California, 1997.

Rajan, Sudhir Chella. *The Enigma of Automobility: Democratic Politics and Pollution Control.* Pittsburgh: University of Pittsburgh Press, 1996.

Scorer, R. S. *Dynamics of Meteorology and Climate.* New York: John Wiley, 1997.

Seinfeld, John H, and Spyros N. Pandis. *Atmospheric Chemistry and Physics: From Air Pollution to Climate Change.* New York: John Wiley, 1998.

Sher, Eran, ed. *Handbook of Air Pollution from Internal Combustion Engines: Pollutant Formation and Control.* Boston: Academic Press, 1998.

Shprentz, Deborah Sheiman. *Breath-Taking: Premature Mortality Due to Particulate Air Pollution in 239 American Cities.* New York: Natural Resources Defense Council, 1996.

Somerville, Richard. *The Forgiving Air: Understanding Environmental Change.* Berkeley: University of California Press, 1996.

Turco, Richard P. *Earth Under Siege: From Air Pollution to Global Change.* Foreword by Carl Sagan. New York: Oxford University Press, 1997

United States Environmental Protection Agency. *The Benefits and Costs of the Clean Air Act, 1970 to 1990.* Washington, D.C.: The Agency, 1997.

Wark, Kenneth, et al. *Air Pollution: Its Origin and Control.* 3d ed. Menlo Park, Calif.: Addison-Wesley, 1998.

Wiles, Richard, et al. *Smokestacks and Smoke Screens: Big Polluters, Big Profits, and the Fight for Cleaner Air.* Washington, D.C.: Environmental Working Group, 1997.

Wilson, Richard, and John D. Spengler, eds. *Particles in Our Air: Concentrations and Health Effects.* Cambridge, Mass.: Harvard School of Public Health, 1996.

Xepapadeas, Anastasios, ed. *Economic Policy for the Environment and Natural Resources: Techniques for the Management and Control of Pollution.* Brookfield, Vt.: Edward Elgar, 1996.

Yunus, Mohammad, and Muhammad Iqbal, eds. *Plant Response to Air Pollution.* New York: John Wiley, 1996.

ANIMALS AND ENDANGERED SPECIES

Beans, Bruce E. *Eagle's Plume: The Struggle to Preserve the Life and Haunts of America's Bald Eagle.* New York : Scribner, 1996.

Clark, Tim W. *Averting Extinction: Reconstructing Endangered Species Recovery.* New Haven, Conn.: Yale University Press, 1997.

Endangered Species Habitat Conservation Planning Handbook. Washington, D.C.: U.S. Fish and Wildlife Service, 1996.

Freese, Curtis H., ed. *Harvesting Wild Species: Implications for Biodiversity Conservation.* Baltimore: Johns Hopkins University Press, 1997.

Hyett, Barbara Helfgott. *The Tracks We Leave: Poems on Endangered Wildlife of North America.* Urbana : University of Illinois Press, 1996.

Ives, Richard. *Of Tigers and Men: Entering the Age of Extinction.* New York: Doubleday, 1996.

Miller, Brian, et al. *Prairie Night: Black-Footed Ferrets and the Recovery of Endangered Species.* Washington, D.C.: Smithsonian Institution Press, 1996.

Simmonds, Mark P., and Judith D. Hutchinson, eds. *The Conservation of Whales and Dolphins: Science and Practice.* New York: Wiley, 1996.

Taylor, Victoria J., and Nigel Dunstone, eds. *The Exploitation of Mammal Populations.* New York: Chapman & Hall, 1996.

CONSERVATION

August, Jack L., Jr. *Vision in the Desert: Carl Hayden and Hydropolitics in the American Southwest.* Introduction by Bruce Babbit. Fort Worth: Texas Christian University Press, 1999.

Baydack, Richard K., Henry Campa III, and Jonathan B. Haufler, eds. *Practical Approaches to the Conservation of Biological Diversity.* Washington, D.C.: Island Press, 1999.

Cohen, Nahoum. *Urban Conservation.* Cambridge, Mass.: MIT Press, 1999.

Kibel, Paul Stanton. *The Earth on Trial: Environmental Law on the International Stage.* New York: Routledge, 1999.

Michalson, Edgar L., Robert I. Papendick, and John E. Carlson, eds. *Conservation Farming in the United States: The Methods and Accomplishments of the STEEP Program.* Boca Raton, Fla.: CRC Press, 1999.

Olson, Richard K., and Thomas A. Lyson, eds. *Under the Blade: The Conversion of Agricultural Landscapes.* Boulder, Colo.: Westview Press, 1999.

Owens-Viani, Lisa, ed. *Sustainable Use of Water: California Success Stories.* Oakland, Calif.: Pacific Institute for Studies in Development, Environment, and Security, 1999.

Troeh, Frederick R., et al. *Soil and Water Conservation: Productivity and Environmental Protection.* 3d ed. Upper Saddle River, N.J.: Prentice Hall, 1999.

ECOLOGY AND ECOSYSTEMS

Athanasiou, Tom. *Divided Planet: The Ecology of Rich and Poor.* Boston: Little, Brown, 1996.

Baarschers, William H. *Eco-Facts and Eco-Fiction: Understanding the Environmental Debate.* New York: Routledge, 1996.

Baskin, Yvonne. *The Work of Nature: How the Diversity of Life Sustains Us.* Washington, D.C.: Island Press, 1997.

Beckerman, Wilfred. *Through Green-Colored Glasses: Environmentalism Reconsidered.* Washington, D.C.: Cato Institute, 1996.

Beder, Sharon. *Global Spin: The Corporate Assault on Environmentalism.* White River Junction, Vt.: Chelsea Green, 1998.

Brick, Philip D., and R. McGreggor Cawley, eds. *A Wolf in the Garden: The Land Rights Movement and the New Environmental Debate.* Lanham, Md.: Rowman & Littlefield, 1996.

Bryant, Raymond L., et al. *Third World Political Ecology.* New York: Routledge, 1997.

Chapman, Graham, et al. *Environmentalism and the Mass Media: The North-South Divide.* New York: Routledge, 1997.

Clinebell, Howard John. *Ecotherapy: Healing Ourselves, Healing the Earth.* New York: Haworth Press, 1996.

Collinson, Helen, ed. *Green Guerrillas: Environmental Conflicts and Initiatives in Latin America and the Caribbean: A Reader.* London: Latin American Bureau, 1996.

Conard, Rebecca. *Places of Quiet Beauty: Parks, Preserves, and Environmentalism.* Iowa City: University of Iowa Press, 1997.

Davis, Peter. *Museums and the Natural Environment: The Role of Natural History Museums in Biological Conservation.* New York: Leicester University Press, 1996.

De Steiguer, Joseph Edward. *Age of Environmentalism.* New York: McGraw-Hill, 1997.

Dion, Mark, and Alexis Rockman, eds. *Concrete Jungle.* New York: Juno Books, 1996.

Dunn, James R., and John E. Kinney. *Conservative Environmentalism: Reassessing the Means, Redefining the Ends.* Westport, Conn.: Quorum, 1996.

Earth Decade Reading List. 4th rev. and enl. ed. Washington, D.C.: The Library of Congress, 1996.

Ehrlich, Paul R., and Anne H. Ehrlich. *Betrayal of Science and Reason: How Anti-Environmental Rhetoric Threatens Our Future.* Washington, D.C.: Island Press, 1996.

Frome, Michael. *Chronicling the West: Thirty Years of Environmental Writing.* Seattle, Wash.: Mountaineers, 1996.

Garner, Robert. *Environmental Politics.* New York: Prentice Hall, 1996.

Gottlieb, Roger S., ed. *The Ecological Community: Environmental Challenges for Philosophy, Politics, and Morality.* New York: Routledge, 1997.

Gould, Kenneth Alan, et al. *Local Environmental Struggles: Citizen Activism in the Treadmill of Production.* New York: Cambridge University Press, 1996.

Graham, Kevin, and Gary Chandler. *Environmental Heroes: Success Stories of People at Work for the Earth.* Boulder, Colo.: Pruett, 1996.

Herndl, Carl G., and Stuart C. Brown, eds. *Green Culture: Environmental Rhetoric in Contemporary America.* Madison: University of Wisconsin Press, 1996.

Harper, Charles L. *Environment and Society: Human Perspectives on Environmental Issues.* Upper Saddle River, N.J.: Prentice Hall, 1996.

Helvarg, David. *The War Against the Greens: The "Wise-Use" Movement, the New Right, and Anti-Environmental Violence.* San Francisco: Sierra Club Books, 1997.

Hjelmar, Ulf. *The Political Practice of Environmental Organizations.* Brookfield, Vt.: Avebury, 1996.

Lafferty, William M., and James Meadowcroft, eds. *Democracy and the Environment: Problems and Prospects.* Brookfield, Vt.: Edward Elgar, 1996.

Mathews, Freya, ed. *Ecology and Democracy.* Portland, Ore.: Frank Cass, 1996.

Merchant, Carolyn. *Earthcare: Women and the Environment.* New York: Routledge, 1996.

Payne, Daniel G. *Voices in the Wilderness: American Nature Writing and Environmental Politics.* Hanover: University Press of New England, 1996.

Pepper, David. *Modern Environmentalism: An Introduction.* New York: Routledge, 1996.

Ridgeway, James, and Jeffrey St. Clair. *A Pocket Guide to Environmental Bad Guys.* New York: Thunder's Mouth Press, 1998.

Rowell, Andrew. *Green Backlash: Global Subversion of the Environmental Movement.* New York: Routledge, 1996.

Shabecoff, Philip. *A New Name for Peace: International Environmentalism, Sustainable Development, and Democracy.* Hanover: University Press of New England, 1996.

Smith, Joseph Wayne, et al. *Healing a Wounded World: Economics, Ecology, and Health for a Sustainable Life.* Westport, Conn.: Praeger, 1997.

Snyder, Gary. *Mountains and Rivers Without End.* Washington, D.C.: Counterpoint, 1996.

Tokar, Brian. *Earth for Sale: Reclaiming Ecology in the Age of Corporate Greenwash.* Boston,: South End Press, 1997.

Wells, Edward R., and Allan M. Schwartz. *Historical Dictionary of North American Environmentalism.* Lanham, Md.: Scarecrow Press, 1997.

Westra, Laura, and Thomas M. Robinson, eds. *The Greeks and the Environment.* Lanham, Md.: Rowman & Littlefield, 1997.

FORESTS AND PLANTS

Bruenig, E. F. *Conservation and Management of Tropical Rainforests: An Integrated Approach to Sustainability.* Wallingford, Oxon, England: CAB International, 1996.

Colfer, Carol J. Pierce. *Beyond Slash and Burn: Building on Indigenous Management of Borneo's Tropical Rain Forests.* Bronx, N.Y.: New York Botanical Garden, 1997.

Conrad, David Eugene. *The Land We Cared for: A History of the Forest Service's Eastern Region.* Milwaukee: USDA-Forest Service, Region 9, 1997.

Ferguson, I. S. *Sustainable Forest Management.* New York: Oxford University Press, 1996.

Gale, Fred P. *The Tropical Timber Trade Regime.* New York: St. Martin's Press, 1998.

Hill, I. *Incentives for Joint Forest Management in India: Analytical Methods and Case Studies.* Washington, D.C.: World Bank, 1998.

Johns, Andrew Grieser. *Timber Production and Biodiversity Conservation in Tropical Rain Forests.* New York: Cambridge University Press, 1997.

Kohm, Kathryn A., and Jerry F. Franklin, eds. *Creating a Forestry for the Twenty-first Century: The Science of Ecosystem Management.* Washington, D.C.: Island Press, 1997.

Landsberg, J. J. *Applications of Physiological Ecology to Forest Management.* San Diego: Academic Press, 1997.

McEvoy, Thomas J. *Legal Aspects of Owning and Managing Woodlands.* Washington, D.C.: Island Press, 1998.

Mater, Jean. *Reinventing the Forest Industry.* Wilsonville, Ore.: GreenTree Press, 1997.

Newell, Josh. *The Russian Far East: Forests, Biodiversity Hotspots, and Industrial Developments.* Tokyo, Japan: Friends of the Earth-Japan, 1996.

Peters, Charles M. *The Ecology and Management of Non-Timber Forest Resources.* Washington, D.C.: World Bank, 1996.

Rajala, Richard. *Clearcutting the Pacific Rain Forest: Production, Science, and Regulation.* Vancouver: UBC Press, 1998.

Reed, David D. *Resource Assessment in Forested Landscapes.* New York: John Wiley & Sons, 1997.

Robinson, Eva Cheung. *Greening at the Grassroots: Alternative Forestry Strategies in India.* Thousand Oaks, Calif.: Sage Publications, 1998.

Sponsel, Leslie E., et al., eds. *Tropical Deforestation: The Human Dimension.* New York: Columbia University Press, 1996.

State of the World's Forests, 1997. Rome: Food and Agriculture Organization of the United Nations, 1997.

Watkins, Charles, ed. *European Woods and Forests: Studies in Cultural History.* New York: Cab International, 1998.

ENVIRONMENTAL HEALTH

Al Khayat, Muhammad Haytham. *Environmental Health: An Islamic Perspective.* Alexandria, Egypt: World Health Organization, Regional Office for the Eastern Mediterranean, 1997.

Berkson, Jacob B. *A Canary's Tale: The Final Battle: Politics, Poisons, and Pollution Versus the Environment and the Public Health.* Washington, D.C.: Island Press, 1997.

Bertollini, Roberto, et al. *Environment and Health 1: Overview and Main European Issues.* Copenhagen: World Health Organization, Regional Office for Europe, 1996.

Bishop, Lari A., ed. *Environment, Health, and Safety: A Platform for Progress.* New York: The Conference Board, 1997.

Bridges, Olga. *Losing Hope: The Environment and Health in Russia.* Brookfield, Vt.: Ashgate, 1996.

Chatsworth, Jennifer, ed. *The Ecology of Health: Identifying Issues and Alternatives.* Thousand Oaks, Calif.: Sage Publications, 1996.

Clinton, Bill. *Reinventing the Regulation of Cancer Drugs: Accelerating Approval and Expanding Access.* Washington, D.C.: National Performance Review, 1996.

Colborn, Theo. *Our Stolen Future: Are We Threatening Our Fertility, Intelligence, and Survival?* New York: Dutton, 1996.

Cook, Allan R., ed. *Environmentally Induced Disorders Sourcebook: Basic Information About Diseases and Syndromes Linked to Exposure to Pollutants and Other Substances.* Detroit: Omnigraphics, 1997.

Crawford-Brown, Douglas J. *Theoretical and Mathematical Foundations of Human Health Risk Analysis: Biophysical Theory of Environmental Health Science.* Boston: Kluwer Academic, 1997.

De Serres, Frederick J., ed. *Ecotoxicity and Human Health: A Biological Approach to Environmental Remediation.* Boca Raton, Fla.: CRC Lewis, 1996.

Diesel Fuel and Exhaust Emissions. Geneva: World Health Organization, 1996.

Hill, Douglas, ed. *The Baked Apple?: Metropolitan New York in the Greenhouse.* New York: New York Academy of Sciences, 1996.

Koren, Herman. *Handbook of environmental Health and Safety: Principles and Practices.* 3d ed. Boca Raton, Fla.: Lewis Publishers, 1996.

__________. *Illustrated Dictionary of Environmental Health and Occupational Safety.* Boca Raton, Fla.: Lewis Publishers, 1996.

LaFountain, Courtney. *Health Risk Reporting: Getting the Story Straight.* St. Louis: Center for the Study of American Business, Washington University, 1996.

Moeller, D. W. *Environmental Health.* Rev. ed. Cambridge, Mass.: Harvard University Press, 1997.

Rapport, David, et al., eds. *Ecosystem Health.* Malden, Mass.: Blackwell Science, 1998.

Resource Guide on Children's Environmental Health. Emeryville, Calif.: Children's Environmental Health Network, 1997.

Sellers, Christopher C. *Hazards of the Job: From Industrial Disease to Environmental Health Science.* Chapel Hill: University of North Carolina Press, 1997.

Shahi, Gurinder S., et al. *International Perspectives on Environment, Development, and Health: To-*

ward a Sustainable World. New York: Springer, 1997

NUCLEAR ENERGY

Agarwal, Prashant. *India's Nuclear Development Plans and Policies: A Critical Analysis.* New Delhi, India: Northern Book Centre, 1996.

Bodansky, David. *Nuclear Energy: Principles, Practices, and Prospects.* Woodbury, N.Y.: American Institute of Physics, 1996. Chow, Brian G. *Civilian Nuclear Programs in India and Pakistan.* Santa Monica, Calif.: RAND, 1996.

Stanton, Neville, ed. *Human Factors in Nuclear Safety.* Bristol, Pa.: Taylor & Francis, 1996.

Ten Years After Chernobyl: What Do We Really Know? Austria: International Atomic Energy Agency, Division of Public Information, 1997.

United States. Congress. Commission on Security and Cooperation in Europe. *The Legacy of Chornobyl, 1986 to 1996 and Beyond: Hearing Before the Commission on Security and Cooperation in Europe, One Hundred Fourth Congress, Second Session, April 23, 1996.* Washington, D.C.: U.S. Government Printing Office, 1996.

RESOURCES AND RESOURCE USE

Anderson, Terry Lee. *Enviro-Capitalists: Doing Good While Doing Well.* Lanham, Md.: Rowman & Littlefield, 1997.

Auty, Richard M., and Katrina Brown, eds. *Approaches to Sustainable Development.* New York: Pinter, 1997.

Bergh, Jeroen C. J. M. van den. *Ecological Economics and Sustainable Development.* Brookfield, Vt.: Edward Elgar, 1996.

Bernard, Ted. *The Ecology of Hope: Communities Collaborate for Sustainability.* East Haven, Conn.: New Society Publishers, 1997. Brekke, Kjell Arne. *Economic Growth and the Environment: On the Measurement of Income and Welfare.* Brookfield, Vt.: E. Elgar, 1997.

Buell, John. *Sustainable Democracy: Individuality and the Politics of the Environment.* Thousand Oaks, Calif.: Sage Publications, 1996.

Dahl, Arthur L. *The Eco Principle: Ecology and Economics in Symbiosis.* Atlantic Highlands, N.J.: Zed Books, 1996.

Daly, Herman E. *Beyond Growth: The Economics of Sustainable Development.* Boston: Beacon Press, 1996.

Furze, Brian, et al. *Culture, Conservation, and Biodiversity.* New York: John Wiley, 1996.

Hackett, Steven C. *Environmental and Natural Resources Economics: Theory, Policy, and the Sustainable Society.* Armonk, N.Y.: M.E. Sharpe, 1998.

Jackson, Tim. *Material Concerns: Pollution, Profit, and Quality of Life.* New York: Routledge, 1996.

Lamb, Robert. *Promising the Earth.* New York: Routledge, 1996.

Lewis, Philip H. *Tomorrow by Design: A Regional Design Process for Sustainability.* New York: John Wiley, 1996.

Marien, Michael. *Environmental Issues and Sustainable Futures: A Critical Guide to Recent Books, Reports, and Periodicals.* Bethesda, Md.: World Future Society, 1996.

Mitchell, Bruce. *Resource and Environmental Management.* Harlow: Longman, 1997.

Noman, Omar. *Economic Development and Environmental Policy.* New York: Kegan Paul International, 1996.

Perman, Roger, et al. *Natural Resource and Environmental Economics.* New York: Longman, 1996.

Perrings, Charles. *Economics of Ecological Resources: Selected Essays.* Lyme, N.H.: Edward Elgar, 1997.

Shabecoff, Philip. *A New Name for Peace: International Environmentalism, Sustainable Development, and Democracy.* Hanover: University Press of New England, 1996.

WASTE AND WASTE MANAGEMENT

Andress, Carol. *Waste Not, Want Not: State and Federal Roles in Source Reduction and Recycling of Solid Waste.* Washington, D.C.: Northeast-Midwest Institute, 1989.

Cross, Frank L., et al. *Infectious Waste Management.* Lancaster, Pa.: Technomic Publishing, 1990.

Enander, Richard T. *Hazardous Waste Tracking and Cost Accounting Practice.* Boca Raton, Fla.: CRC Press, 1996.

Freeman, Harry. *Source Reduction as an Option for Municipal Waste Management.* Cincinnati, Ohio: Waste Minimization Branch, U.S. Environmental Protection Agency, Risk Reduction Engineering Laboratory, 1989.

Higgins, Thomas E., ed. *Pollution Prevention Handbook.* Boca Raton, Fla.: Lewis Publishers, 1995.

Hilz, Christoph. *The International Toxic Waste Trade.* New York: Van Nostrand Reinhold, 1992.

Jacobson, Timothy C. *Waste Management: An American Corporate Success Story.* Washington, D.C.: Gateway Business Books, 1993.

Jenkins, Robin R. *The Economics of Solid Waste Reduction: The Impact of User Fees.* Brookfield, Vt.: E. Elgar, 1993.

Moyers, Bill D. *Global Dumping Ground: The International Traffic in Hazardous Waste.* Washington: Seven Locks Press, 1990.

Nemerow, Nelson Leonard. *Zero Pollution for Industry: Waste Minimization Through Industrial Complexes.* New York: John Wiley, 1995.

Rebovich, Donald. *Dangerous Ground: The World of Hazardous Waste Crime.* New Brunswick, N.J.: Transaction Publishers, 1992.

Selke, Susan E. M. *Packaging and the Environment: Alternatives, Trends, and Solutions.* Rev. ed. Lancaster, Pa.: Technomic, 1994.

Theodore, Louis, et al. *Pollution Prevention.* New York: Van Nostrand Reinhold, 1992.

Walsh, Edward J., et al. *Don't Burn It Here: Grassroots Challenges to Trash Incinerators.* University Park: Pennsylvania State University Press, 1997.

WATER AND WATER POLLUTION

Abel, P. D. *Water Pollution Biology.* 2d ed. London: Taylor & Francis, 1996.

Abramovitz, Janet N., et al. *Imperiled Waters, Impoverished Future: The Decline of Freshwater Ecosystems.* Washington, D.C.: Worldwatch Institute, 1996.

Da Rosa, Carlos D., et al. *Golden Dreams, Poisoned Streams: How Reckless Mining Pollutes America's Waters, and How We Can Stop It.* Washington, D.C.: Mineral Policy Center, 1997.

Gallagher, Lynn M., et al. *Clean Water Handbook.* 2d ed. Rockville, Md.: Government Institutes, 1996.

Hunter, Susan, et al. *Enforcing the Law: The Case of the Clean Water Acts.* Armonk, N.Y.: M.E. Sharpe, 1996.

Larson, Steven J., et al. *Pesticides in Surface Water: Distribution, Trends, and Governing Factors.* Chelsea, Mich: Ann Arbor Press, 1997.

Mahan, John. *Lake Superior: Story and Spirit.* Gaylord, Mich.: Sweetwater Visions, 1998.

Mason, C. F. *Biology of Freshwater Pollution.* 3d ed. Harlow, Essex, England: Longman, 1996.

Mayer, Jim. *Layperson's Guide to Water Pollution.* Sacramento, Calif.: The Foundation, 1996.

Mueller, David K., et al. *Nutrients in the Nation's Waters: Too Much of a Good Thing?* Washington, D.C.: U.S. Government Printing Office, 1996.

Ongley, E. D. *Control of Water Pollution from Agriculture.* Rome: Food and Agriculture Organization of the United Nations, 1996.

Protection of Transboundary Waters: Guidance for Policy- and Decision-making. New York: United Nations, 1996.

Schnoor, Jerald L. *Environmental Modeling: Fate and Transport of Pollutants in Water, Air, and Soil.* New York: John Wiley, 1996.

Sindermann, Carl J. *Ocean Pollution: Effects on Living Resources and Humans.* Boca Raton, Fla.: CRC Press, 1996.

Webb, Bruce, ed. *Freshwater Contamination.* Wallingford, Oxfordshire, England: IAHS, 1997.

Encyclopedia of Environmental Issues

Index